Structural Engineering of Transmission Lines

There are reasonable places for transmission lines and there are useful but unreasonable places for transmission lines. Some are challenging and fun to engineer and some are yawners. Here's hoping that some of your days require tackling the unreasonable ones because they're fun!

Crossing Twin Peaks in British Columbia with the Kemano–Kitimat transmission line, 1950s.
(Courtesy of H. Brian White)

ice
Institution of Civil Engineers
publishing

Structural Engineering of Transmission Lines

Peter Catchpole PEng
with Buck Fife

Cover photograph by Erik Ruggeri: Crew lifting the middle phase to Catenary 2, Nebraska, 2008

Published by ICE Publishing, One Great George Street, Westminster, London SW1P 3AA

Full details of ICE Publishing sales representatives and distributors can be found at:
www.icevirtuallibrary.com/info/printbooksales

Other titles by ICE Publishing:

Structural Dynamics for Engineers, second edition
H.A. Buchholdt and S.E. Moossavi Nejad. ISBN 978-0-7277-4176-9
Structural Systems: Behaviour and Design
L. Stavridis. ISBN 978-0-7277-4105-9
Communication Structures
B.W. Smith. ISBN 978-0-7277-3400-6
Conceptual Structural Design
O. Popovic Larsen and A. Tyas. ISBN 978-0-7277-3235-4

www.icevirtuallibrary.com

A catalogue record for this book is available from the British Library

ISBN 978-0-7277-5988-7

ICE Publishing is a division of Thomas Telford Ltd, a wholly-owned subsidiary of the Institution of Civil Engineers (ICE).

Commissioning Editor: Jo Squires
Production Editor: Vikarn Chowdhary
Market Development Executive: Catherine de Gatacre

Typeset by Academic + Technical, Bristol
Index created by Pauline Davies
Printed and bound by CPI Group (UK) Ltd, Croydon CR0 4YY

This book is dedicated to the teachers and mentors that have influenced my work, most notably H. Brian White, and to the younger engineers who show a keen interest in the work and on whose intelligence, curiosity and pursuit of excellence relies the future of the engineering of transmission lines.

Brian (right) and I conferred at the initial stages of several projects. It was always useful

Contents

Foreword

Fellow students, I often wonder why it is that only a small subset of engineers are attracted to electrical transmission lines. What is even more interesting is the passion that this small group has for these seemingly simple facilities, while the rest of the engineering community spend their careers, some even becoming famous, on tall buildings, long-span bridges and flying machines of one sort or another.

Maybe most engineers, and the community at large, overlook transmission lines because they are as common to the landscape as highways, farm tractors and fast food restaurants, plus, and most importantly, seem to do their job without fail, in all types of weather and all without moving parts. Probably every type of major infrastructure has its engineering 'fanatics'; however, it seems as if the transmission line engineering fanatics are the smallest and closest knit group of engineering colleagues. As support for this statement, I suggest that there are only two and at most three degrees of separation between any two people doing real transmission line design anywhere around the world. For us students of transmission line engineering, it seldom takes more than one conversation to put ourselves on the trail of the state-of-the-industry technical solution to whatever problem we are facing.

Why are we so enamoured with transmission lines? On the surface they appear so simple, sets of wires, spanning the globe, held up by steel, wood or concrete structures in a wide variety of configurations. In fact, when my daughter was in grade school, I often remarked that the reason transmission line design engineers get so little respect is that transmission lines appear simple enough to design and construct that my daughter could do it ... and it would certainly carry electricity ... at least for a day. In all seriousness, for those of us who have seen the light, we realise that transmission lines can be simple and straightforward, but most are not. Most are complicated structural systems that are asked to do their job of powering the world economy over all types of terrain and in all types of weather.

I offer two thoughts on what attracts us to transmission line design. Firstly, the structural system is surprisingly complex. Although when viewing a transmission line all components seem static, in reality there is a never-ending balancing act in progress, all components delicately balanced so that the line appears motionless. Understanding this delicate balance and then applying this knowledge to design such that the system stays balanced and intact for more than 50 years is no simple feat of engineering.

Secondly, each span and/or each structure location presents its own unique engineering challenges. Although there is most often a common set of criteria for the entire line, on every project there is at least one structure or span (most often many more) that will challenge even the most talented of engineers. The exceptional engineers handle the easy sections of the project routinely, employing time-tested techniques and tools to quickly dispatch these straightforward sections such that talent, energy and enthusiasm can be focused on the generally small number of challenging (most often described as 'fun') parts of the project.

My final thought on why some of us are so enamoured with transmission lines is that I suppose the transmission line design industry is a little like being in a secret club. Our specialty is out there for all to see, yet only a few of us know the secrets that make these simple-looking systems perform so well. In this book, Peter describes some of the 'secrets' (many of them learned the 'hard way') that will enable our club to grow.

Most of this book is focused not on the benign sections of the transmission line but on the unique locations/spans where terrain, weather, restricted access, jurisdictional constraints and any number of other hazards conspire to challenge the design engineer to find elegant, yet technically sound and cost-effective, solutions. This book is not only a resource and a reminder of things important but should be used to further your education in the pursuit of personal improvement for the benefit of your project, your company and the transmission line design industry.

I believe that transmission lines are a thing of joy and beauty to behold forever. So, behold the beauty, even if only you can see it, embrace the joy of solving thousands of unique problems during each design, be better on each project, give back to an industry that has given so much to you, and finally be grateful to those who have gone before.

Enjoy and put into practice the experience, insight and enthusiasm that Peter and his co-authors (those named and those that provided influence) have brought to life. Good luck, be safe and be good!

Ronald J. Carrington PE
Student, transmission line design

Preface

To a few people, this book may be considered a long time coming. These people would be those that saw or possess a much lesser scope document that I first put together in 2000. I dared to call it a book because I did have the intention, even then, of making it into something broader in scope and presumably of actual value. I am relieved to have waited, because I keep learning useful things. But, the time has come, and the result is this.

The fundamental purpose of this book is to lay down my thoughts on the structural engineering of transmission lines. I believe that the book can provide to you a take on the subject that is not covered by other publications. There is a plethora of papers that are constantly being written for our industry by a very wide-ranging population of experts. Many of those papers are insightful and useful. Many are not. Far too many are regional in understanding and value. So, herein I offer a sideways view of things that I intend to be complementary and supplemental to all other things written. I have tried with this book to provide timeless truths and certain basic insights with limited commentary on issues of this day.

This book does not dwell on the elementary, in that we dive right into our industry's lingo without much concern for consistently providing the reader with definitions of all the terms that we present. I assume the reader has had reasonable exposure to the business and is armed with at least elementary knowledge of our industry's lingo. In other words, I am not trying to introduce the completely uninitiated to the business of engineering transmission lines. Rather, I am offering to take reasonably experienced engineers to a higher level of understanding of that which they already have some knowledge.

For the less initiated, I hope that, by reading the book, you sense the challenging nature of the work and the joy that committed engineers can feel upon accomplishing good things within the business. In other words, I hope that the book helps spark your interest in the business and that it becomes a valued reference for you for years to come.

Early in my own career, I had the pleasure to work with newer and younger engineers. I learned early on that, as much as I enjoyed doing this work, I got much greater satisfaction helping these young entrants to our industry improve their game – to be better next year than they are today. This book is designed to provide to me that satisfaction of being of real help to you.

The end result of this effort could have been a book that is purely technical in style such as any textbook, but I chose another style. This book includes injections in a first person style. Throughout, you are occasionally subjected to my experiences written as personal accounts. I hope this makes it just a bit more interesting to read than the subject naturally begets. It certainly made it easier to write.

During my third year in high school, I wrote all literary exercises assigned to me in the style of John Lennon's *In his Own Write* and *Spaniard in the Works*. It was exhilarating to write everything as fanciful nonsense riddled with words with at least triple entendre. Most of my friends would say that I still think and talk that way. My English teacher that year was a Brit on exchange from his homeland. He was a fish out of water in several ways, and that probably led him to appreciate my style. Maybe he lived down the street from John Lennon when he was a kid. I don't know. He kept saying how much he enjoyed the writing that I handed in, but counselled me that the style would probably not work well later in life in the real world. I warn you that I have not entirely vacated the style. I hope that you find this book very useful as a resource for your engineering career, but I also hope you enjoy reading it. I like unorthodox things!

I can count four men who have shaped my life and career to the extent that I would call them my mentors. In order of their appearance, they were my father, the aforementioned John Lennon, Frank Mackay and H. Brian White.

Only one of these four should beg a question, that being: *who is Frank Mackay?* Frank showed up in my life when I entered this business at the age of 28 at a small integrated utility. He was their VP of All Things That Matter with 42 years under his belt with the organisation. He was a dynamic, life-loving, short, fast-moving whirling dervish with a booming voice and a full head of bright white hair. I loved him!

I was in the business for only a few months and had been assigned a line design to develop beyond the point to which it had been taken by the new incoming company president. Two days after I started that job, the only other engineer with civil/structural engineering knowledge of lines or substations had a heart attack. He recovered sort of, but never really returned to work. I was alone! I needed guidance, and went to Frank the VP with an opinion regarding that line's design, 'I'm not sure this is going to work and I think we should do it another way.' His answer was, 'Well, if that's what you think then let's do it.' That comment to me, the brand new untested engineer, lay a foundation of confidence and a recognition of responsibility within me that I have carried with me ever since. In turn, I have tried to treat the young(er) kids that I meet on the job over the years with the same respect for the same reason.

I got into this business because the band broke up. I love that line but it's not true. The band sucked. My employer back then was offering a stint in the Algerian desert, and I chose to 'stay in town'. A friend suggested the local utility, and I was excited to go work on dams, fast-flowing rivers and the like. On day 2, I found myself standing ankle deep in mud on a thing called a right-of-way, looking at 115 kV lines above me and asking myself, 'Why are there always three wires?' I was that dumb.

That was 1977. I stood in that same place as a line engineer several months ago (2012). I worked at that utility for just under 4 years, but have been consulting to it from afar since 1988. It is my *alma mater*. Lesson one: don't burn bridges!

You can imagine that from a father and from John Lennon, I learned what to do, how to behave and what matters, but also what not to do and not to say. Not all valuable mentoring is a positive directive. After all, we do learn the best lessons from mistakes, and it is lovely when they are the mistakes of others. This is, in fact, the reason that the chapter on line failures is so important to the book and for the reader.

Of these four named individuals, Brian White is the name that you should pay attention to, given our subject at hand. Brian entered the transmission line engineering business in the early 1950s. He was going strong when I met him in 1985, and was still whacking at the piñata when 87 years old. His reputation in the industry is such that we need only refer to him by his first name.

Brian became a very good friend and mentor to a great many people in this business on many continents over the years. A number of my colleagues and I myself consider ourselves to be students of Brian. This book is deeply marinated in Brian's view of transmission line engineering, and it is primarily to Brian that I make the teacher and mentor reference in the Dedication.

This book actually began as a joint effort with Brian about 5 years ago. I wanted it in part to be a vehicle for getting Brian's wonderful teachings out to the new generation of transmission line engineers. Brian had a passion for this work like no one else. Sadly, the collaboration did not make it to the end of the effort, but you will find considerable amounts of Brian's work within. Sometimes, the words are his and other times they are my version of his thinking. Then, I hope that I have taken subjects to an even further level for you in some areas.

Brian passed away on 8 December 2012 at the age of 90. He was described by a colleague as a phenomenon of nature. His passion for this work and his dedication to the profession of engineering were the reasons that he spent a great deal of time teaching and shaping many young engineers in the business, including myself.

Brian's written and teaching efforts always came under the heading of 'understanding transmission line behaviour'. It would be a suitable title for this book, but the phrase belongs to Brian. Although transmission lines exist for the primary reason of transporting electricity, this book says very little of value about electricity. Instead, the focus is on the structural nature of these lines and on the business of conducting business under the heading of 'projects'.

> I have often said, 'I don't care if there is electricity in the conductors, my job is to design something that will not fall down.'

A very fundamental message of this book is that it takes much more than an engineer's efforts to create a transmission line but that the engineer's work is essential to a successful, well-behaved outcome. I have often said 'a person will never be good at something unless he or she enjoys doing it'. Brian spun the table and added, 'a person will never be good at anything without doing the hard work of learning the subject – after that, the satisfaction that comes from accomplishments is the source of the happiness'. Later, I described this flipped description to a colleague, and he suggested that the relationship between hard work, competence and satisfaction may be a circle, and describing that one comes first is futile. The presence of the three matters is something that you should ensure occurs in your career.

Good luck with that. Then, I'll suggest that if their presence is not happening for you, do something else, as in: *make it happen* for you.

Throughout the book, I will provide you with comments, observations, stories and even calculations that show you that perfection is not to be defined by the technically minded engineer but by the collaboration of all players. I will also pound the drum trying to tell you what I believe matters and what I believe does not matter. After all, we all want to be effective in what we do, so my objective is to point you in a direction away from what does not matter. I also believe that nothing will ever change for the better unless somebody thinks outside the box and charts new paths for others to follow. Certainly, the opportunities for improvement in our field of endeavour are plenty. Be the adventurer!

Over the years, I have gathered the quotes and phrases that impressed me and express my sentiments about what matters. The first and last two are long-standing mantras of Brian's.

> It is the mark of an educated mind to rest satisfied with the degree of precision that the nature of the subject admits, and not to seek exactness where only an approximation is possible.
>
> Aristotle

> That your only tool is a hammer does not mean that all of your problems can be treated successfully as if they are nails.
>
> Anon.

> Everything in nature and all events have a relationship with other things that involve opposing factors. These natural things and events are also in a continual state of change. The ideal state is when these opposing forces are in relative balance. Do not believe in absolutes or in the ideal, rather that everything is relative, flexible and changeable.
>
> A paraphrase of yin and yang

> It is an easy and fatal step to think that the accuracy of our arithmetic is equivalent to the accuracy of our knowledge about the problem at hand. We suffer from 'delusions of accuracy.' Once an enthusiast gets this disease, he and all who depend on his conclusions for

their welfare are damned. [alternative ending: ~~are damned~~. will continually find the wrong answers with great precision.]

M. J. Moroney, *Facts from Figures* (Penguin, 1951)

If at first the idea is not absurd, then there is no hope for it.

Albert Einstein

Look for the effect of these quotes on my thoughts as you read this book. Whether you find it acceptable to adopt any of these principles as guides for yourself is up to you. Regardless, it is my hope that you will find the contents of the book useful, worth reading and understandable.

Peter Catchpole PEng

Structural Engineering of Transmission Lines
ISBN 978-0-7277-5988-7

http://dx.doi.org/10.1680/setl.59887.001

ice
Institution of Civil Engineers
publishing

Chapter 1
Introduction

Before you get all excited, please understand that this book is about the civil and structural engineering of transmission lines, not the electrical engineering of same. Much of this book is focused on understanding a transmission line's structural behaviour after it is created, and the rest is on its creation: design and construction. I want you to *understand* what you are creating before you try to create it. There are many books and guide manuals that tell you what to do, but none tell you why you should do it. Here, we try to describe the important things that need understanding to engineer a *good* line – one that behaves as you would like it to.

Most of this book's content is technical, but the subject of this Introduction is otherwise. In the Introduction, we describe the value of being in a certain frame of mind, the value of being able to see things in a certain way and in a certain context. Chapter 2 deals with electrical issues. It's short – no pun intended but one acknowledged. That chapter was hard to write since I am a civil/structural engineer, so I called on a very bright young engineer whom I admire greatly: Buck Fife. Buck was hired into my employer's fold about 10 years ago as a drafter. He very quickly showed an intellect, a sense of curiosity and tons of energy for the work. So much so that our employer basically helped him to go to college to get an engineering degree. He chose electrical engineering. We limit the discussion to the essence of things that matter to a structural engineer.

To me, a transmission line's conductors and the other wires suspended between supports are the most important structural members of a very long and bizarre structure. In Chapter 3, we will study the mathematics that is not in plain sight for viewing in this computer age but that provides the basis for understanding wires suspended in spans on a planet with gravity. We will dwell on the behaviour of these wires because they are the most important structural members of this very large and bizarre structure. We move on to Chapter 4 to discuss in detail a transmission line as a structural system.

Chapter 5 on loads and strengths is important for two reasons. It is necessary to understand the nature of load sources, the definitions of strengths of materials, and how to combine loads and strengths by the proper application of factors of safety to either or both loads and strengths. We work in a convoluted environment on this point. Sometimes, $2 \times 4 \neq 4 \times 2$. We weigh in on the decades-old subject of deterministic-based design methods versus probability and reliability-based design methods. I have an opinion. Oh yeah!

Chapter 6 should be included in Chapter 4, but transmission line structures made almost entirely of cables were an opportunity that shaped Brian White's career and, by sheer luck, my own. Working with large cables was such a unique experience on each occasion, and understanding cables is of such value to a transmission line engineering career, that the subject is given its own chapter.

With the line understood electrically and structurally and with an understanding of loads and strengths, we move to the most instructive section of the book, Chapter 7. That is the chapter on line failures and the oh-so-instructive lessons that are to be gleaned from a proper study of them. I have never met a failure that did not instruct. This chapter provides stories meant to bring the contents of Chapter 3 to life.

Thereafter, we move to Chapter 8 on the application of this new-found knowledge to projects. After all of the aforementioned chapters make their subject matter seem so important in their own right, the projects chapter puts that information into another perspective. That being one in which the engineer views and performs their work in the company of so many other diverse players, each with their own agendas and concerns. We hope this chapter offers insights not often mentioned elsewhere.

> I recently wrote for a seminar that I was involved in that 'We are not going to show you the keys to the Magic Kingdom today.' There is no 'easy' button for executing transmission line engineering work. On the contrary, we are going to show you the intertwining of issues – the yin and yang of issues that make doing this work interesting. It is this complexity that makes it challenging, worth your while and fun.

We close with a chapter on sustainable development. This is a subject that is relatively new to us humans as we slowly come to terms with the fact that we are doing a poor job of leaving this finite planet in good shape for future generations. The sustainable development conversation within the energy business is rightly focused on generation and consumption. There is practically nothing useful said about the transportation of bulk power (transmission) component of the business. We weigh in.

A few definitions

. . . of value but not discussed directly within the book. Odd!

Engineering is the work concerned with putting scientific knowledge to practical uses. It can be called (is called, in some places) applied science. Scientists keep looking for the answers. Engineers solve problems in the absence of knowing all the answers.

Design is the development of an idea into a constructible and viable (practical) entity.

Analysis is the study of a developed idea or constructed entity to determine its functional status and/or capability – its usefulness. Analysis is an integral part of the design process when an entity is under development. Develop a concept, analyse it. Improve it, analyse it again, etc. Analysis is also a stand-alone exercise when studying an existing entity's functional status and/or capability. Typically, we design new things and analyse existing things. Both are included in the act of engineering.

Civil engineering is the engineering of physical works that are typically used in service to society and that have no or few moving, mechanically linked parts that are essential to its function. Examples: roads, buildings, bridges, harbours, open and closed channel hydraulic facilities, and utility facilities such as pipelines and – wait for it – transmission lines. Civil engineering is closely related to mechanical engineering, where the focus is on physical works that have many moving, mechanically linked parts that are essential to its function.

Structural engineering is a subset of civil engineering – a specialty if you like. It tends not to include hydraulics of open channels of fluid and pavement design, road layout, etc. Classically, structural engineering focuses on entities constructed of steel, concrete, wood and other, more exotic building materials.

Information: with the advent of Google and the like, it seems that we have all of the information in the world at our fingertips. The information may be correct or false, high quality or poor quality and complete or incomplete. But remember, it is just information.

Knowledge is information that you have studied to the point of being satisfied that it is good, correct and reasonably complete and that you understand it. When, by way of thought, study or testing, you have satisfied yourself that information is good, correct and reasonably complete and the information becomes well understood by you, it becomes knowledge – your knowledge. To drive the point home . . . there is no knowledge on the internet!

Units of measure

Canada went metric on 1 January 1978. I had turned 29 the day before. Actually, the country went metric – sort of. Some 30+ years later, a sheet of plywood is sold as 4 ft × 8 ft × 6 mm. I guess Canadians were not willing to tear down their 'imperial' homes and reshape them to allow a full conversion. What is the way out of that mess?

Anyway, I rolled up my dutiful sleeves that January day and undertook my first ever metric design chore: renovating the layout of the men's washroom at the office.

This book is written by a bi-dimensional engineer, living and working in the USA (imperial units have become known as US units), still performing work in Canada and other countries. Keeping track of units and conversions has been and remains a daily exercise. Here's a few useful conversions:

Length
1 in. = 25.4 mm (exactly)
1 ft = 0.3048 m (exactly)
1 mile = 1.609344 km (1.61 km, not exactly)

Weight or tension (force)
1 lbf = 4.44822 N (1 N ≈ 1 apple)
1 lb/ft = 1.488 kg/m
224.8 lb = 1 kN (≈ 1 lineman or 1000 apples)
1 dN (10 N) = 2.248 lb = 1.02 kg (1 dN ≈ 1 kgf)
1 ton (short ton) ≈ 1.1 tonnes (metric tons)

Area
1000 kcmil = 506.7074 mm^2 (or 1 mm^2 ≈ 2 kcmil)
1 acre = 0.404 ha

A comment on knowing

It is interesting that the Electric Power Research Institute (EPRI) 'red' book published in 1980 discusses the electric field at length but barely mentions the electromagnetic field (EMF). Shortly after the time of that book's publication, the EMF from electric lines became a major consideration for defining our line designs. Since 1997 or so, the EMF issue became generally regarded as 'not a concern' by the scientific community but the opponents of line projects hang onto the concern like pit bulls.

Are the issues of the electric field and magnetic field important? We, as engineers and a society in general, are fully capable of not knowing what is relevant. We are capable of knowingly and unknowingly wasting our time on unimportant issues and not recognising or facing very important issues. In fact, the psychiatric and social science communities have shown that we humans are fully capable of believing in things that are simply not true. That explains a lot!

When you consider that we claim to know a lot more about many things compared with what we knew 30 years ago – and we do, it should be easy to admit that we don't know much today compared with what we will know 30 years from now. Be aware that some design issues that you address exist for no better reason than being the result of someone's misguided guesses, desire for acclaim or victory, or for advancement of their own business, and be aware that some design issues that you should address are not even known to you.

This time in history is referred to as the Information Age. As just said, information becomes knowledge after you have done the work to prove to yourself that the information is true/accurate. Until you do that work, that which is information remains quite secondary to knowledge. Information runs the risk of being completely bogus until you turn it into knowledge. Be aware of the difference.

Understanding the distinction and turning information into knowledge is also one of the greatest sources of a sense of accomplishment in this work. It will make you worthwhile to others. This book will provide you with information. It is necessary that you do the work of turning this book's information into knowledge. To do that, you must be endlessly curious.

The value of curiosity

Curiosity is the fuel for learning. If you are not curious, you will see nothing but what is clearly on the surface, right before you. Compared with a totality, there is rarely much in that place. If you see so little of the totality as what is blatantly before you without digging deeper for more, you will not only be uninformed but probably misinformed. To be clueless is not much different from being wrong.

It is surprising and quite damaging to society that so few of us humans are curious. Take a look at the people that you work with. I would bet heavily that there is a solid relationship between each person's level of curiosity about things and their value to you and the organisation where you slave. Slav ... just poking fun at all of us!

> I work with a fellow who was educated in Costa Rica. That fact alone has much to do with making him 'unique'. He has, by our North American view, some weaknesses but he has a strength that is hard to recognise. He is endlessly curious. He tells me that the Costa Rican education system teaches that. A weakness in that education system and the man would seem to be that he does not know what to be curious about. He is curious about everything so seems to waste a lot of time on things that do not matter. The trick is to guide his curiosity to useful things without killing its valuable boundless nature. He is a very useful and very knowledgeable engineer because he is deeply curious.

We would suggest that you look at yourself to understand your relationship with curiosity. If you can improve your ability to be curious and to act on it, then I could say, 'it will serve you well'. We could say that but that would not be enough. I will say this ...

> Be *endlessly* curious or be left behind by those who are.

A comment on what drives and rewards you

Remember from above: 'you will be good at your work if you enjoy it' and 'If you do the hard work, you will be good at the work and then you will enjoy it' and that these three ingredients for success – hard work, competence and satisfaction – maybe should be related as if in a circle. Where you hop onto this circle of necessary ingredients for success depends on your personality. So, we leave you with this ... Do the hard work. Your willingness to do so may need to be fuelled by your curiosity for the subject. You *will* get good at the work. You will stick with it if you enjoy it.

Although this industry has been around for about a century, things keep changing and there really are opportunities for bright, inventive, curious and adventurous minds to find and conquer challenges. If conquering challenges of this sort will feed your soul, welcome to the party.

Your power and responsibilities as an engineer

Your responsibility as an engineer is to participate in the creation, construction, maintenance or operation of a facility in a manner that provides value to its owner and to society in general. Some engineering associations put as much or more focus on your duty to society than they do on your duty to yourself or your customer/client/employer. Keep that in mind because it will affect your choices.

Power and responsibility go hand in hand. They are inseparable if your actions are to be useful and productive. Power without responsibility is dangerous. Responsibility without power is destined to reap failure. If you are given responsibility, make sure it comes with the requisite power to affect any changes necessary to produce the outcome that you desire. If you delegate responsibility, be sure to delegate the power that must go with it. Otherwise, you have set the stage for failure.

Now, let's get daring. If you are given responsibility for something but are not sure if you have the power to guide the work towards a successful outcome, test it. Test it early, and if the power is not in place with you, make a change.

For the design and construction engineer

A question relevant to this book is 'What impact can you, as an engineer, have on the value to the owner and to society of a transmission line when exercising your role as design engineer, construction engineer, maintenance engineer or operating engineer?'

If you were to list all of the cost elements involved in the creation of a transmission line – from the cost of foundations to the cost of corporate overheads and legal service fees – you will find that you can affect the cost of some single elements radically – say by 50% – but your ability to affect the overall and total cost is usually less than 1–3%. Let's look at this more closely.

The capital cost of a project is 100%. The cost of owner overheads, legal support, permitting, easement acquisition and your services are in the neighbourhood of 40%. The rest of the cost is labour and materials to put the facility in place. These are generally split between 40%–60% to 60%–40%. Let's say 50%–50%: 30% labour and 30% materials, respectively, of the entire 100%. The domain generally considered under the designer's control is material choices. If you do an exercise to select the best conductor and structure combination for the project, you can easily best another engineer's poorer choices by about 10% on the project's total material costs. That would be 3% of the total project cost. Perhaps the choice reduces the expected construction cost by another 10% for a grand impact of about 6% on the project. However you try to run this exercise, you will find the outcome is similar but tempered to less than this impact by the yin and yang of things. In short, the design engineer will have great difficulty affecting – reducing or increasing – the cost of a project by more than a very few percentage points. Remember this as we come back to it in the chapter on projects.

The parts of the effort that the design engineer is not likely to control are the route, land costs, mitigation measures, the legal costs, the contractor selection, many material choices, and the contract terms and conditions. Choices made by others with any of these cost components can easily swing the total cost by much more than a few percentage points and more than erase the small impact that the design engineer can have on the project's cost when they focus on material and design feature choices with the aim of saving a few dollars here and there. In other words, the optimising choices that the design engineer makes regarding pole species, steel thickness limits, load case selection, bolt sizes and so on do not affect the cost of the project. To toil there is to distract you from what matters.

There is one cost element over which the engineer has primary control and that, if not managed well, can be very costly to the owner, the public and to you, the engineer. It is the cost of an electrical or structural failure of the line. A costly failure can adversely affect public safety, your reputation and your business's viability very much, and that is where your attention should be. The construction engineer should have the identical concern and ensure that the line is installed as specified or to a quality known by him/her to be more than sufficient.

It is not essential that you select a perfect conductor, tower type, foundation dimension, bolt size or installation method for the purpose of saving capital cost money. It is important that you choose these things reasonably well but then ... move on! Your primary purpose should be to minimise the possibility of a costly failure and that you do this in a cost-effective way. Being able to do so will distinguish you from other engineers in your field. Therein lies the real art of line engineering. It is the art of getting those to whom you report to recognise your engineering artistry, because it will be invisible for the most part. Good luck with that.

> I long ago worked with a chap who spent endless hours hiding in his little office with his feet up on the desk. His mantra was 'I will work hard when they pay me well.' I tried but failed to get him to understand that things work the other way around. He did not work there long.

Work hard, be very curious, understand and pay attention to what you can control. In the end, you will have accomplished something worthwhile. Due reward may follow. Otherwise ... not so much!

For the maintenance and operations engineer

Designing and constructing a facility can be compared to bringing a child into this world. A good diet and proper nurturing only goes so far because at some point the child/facility is handed over to the education system/community/friends/partner/maintenance engineer who will guide that child/facility through years of life/service. After that hand-over, the parent/design engineer can only watch, hope and wish.

Like a school teacher, doctor or best friend, it is the responsibility of the maintenance engineer to understand the facility sufficiently so that their actions facilitate a long and satisfactory service life. There is much to understand, most notably: materials, conductor behaviour with time, temperature and loadings, structural member loads, electrical limits, etc.

Being part of a team

Although the design engineer might spend most paid hours of their life in isolation in an office away from other players in the creation of a transmission line, few of their actions fail to affect those others. At the highest level, the players include the owner and all of its staff, the owner's legal team, environmental and permitting staff/consultants, the material suppliers, the contractors, the maintenance and operating staff/companies, landowners and the public as a whole.

At a near-daily level, the design engineer's work requires input from the electrical, civil, structural, mechanical, chemical and environmental disciplines and responds to the system designers'/planners' needs. Ignorance or the ignoring of these other invested players invites trouble. It is not likely that any one person has reasonable knowledge in more than one of these fields, so teamwork is essential.

Staying in touch and keeping your eyes wide open for useful input during the design process is the point. Is your organisation shaped to allow these interactions to occur efficiently or at all? Some are not. Does your organisation actively promote such interaction as a necessary ingredient for success? Many do not. When an organisation is not structured to allow or even promote interaction, frustration and failure can follow. At best, mediocrity will put its dour face on the work produced in such an environment.

Not your grandfather's transmission line

Finally, a new catch phrase has been showing up in engineering and marketing language with reference to our work. That being, '(Isn't it time to realise that) we should no longer be building our grandfather's transmission lines.' The phrase is meant to be a reminder that this industry that so easily and routinely fails to express innovation should get on with doing the right (new) thing.

> In the photo (Figure 1.1), you see, in the middle, my paternal grandfather, Leslie. He was a construction superintendent, and retired in 1957 from Ontario Hydro in Canada. This photo was taken a few years earlier. In the background is the steelwork for a substation that he built near Niagara Falls. So, you are looking at my grandfather's transmission work. That substation still stands and works just fine. The phrase 'to not build our grandfather's lines' got me thinking. What is it that we are doing that has us earn that remark?

The remark was from a marketer who was telling a group that tubular poles are the future and latticed towers are a relic of grandpa's era. Well, that is not a rational conclusion, so let's correct it. A recently retired marketer from a 'top of the line' pole manufacturer in the USA used to say that his company campaigned hard during the 1970s and 1980s to develop a pole market in the US transmission industry. Why? To sell their product, of course. It worked very well. By the time of this book's writing, and before that, there are at least a half dozen pole manufacturers in the USA and no latticed tower manufacturers with significant capacity.

In part, their timing for the marketing effort was excellent. Large project development was waning after the 1970s and the opposition to transmission lines was gaining intelligence in its methods for opposing lines. The aesthetics of poles suited people more than the aesthetics of latticed towers. The smaller projects, especially those that did not hide in

Figure 1.1 My grandfather between two young engineers, pre-1957

the vast wilderness between urban areas, preferred and sometimes required the smaller footprint of poles as well. By the turn of the century, poles were seen as nearly the only option in the USA, and fabricating facilities for latticed towers were all offshore and often far overseas, causing a host of delivery, quality and 'preferred' source concerns, sometimes real and sometimes not.

Over the last few years, the first ever tubular guyed-V structure for a 500 kV project has been in development. The reason: the cost of the labour to assemble the typical latticed steel guyed-V structure is higher and more than offsets the higher material cost of the tubular structure. Meanwhile, most of the world continues to construct with lattice steel supplied from India, China, Korea, Europe, South America and the Middle East. What is going on? Is the USA more advanced as the marketer implied? The answer to that lies in part in the fact that these large tubular structures are very structurally unmanageable and may turn out to be viewed as, 'we better not do that again'. What was wrong with my grandfather's transmission line, anyway?

The essence of change that really needs to take place, as the grandfather remark invites, is simply the implementation of technologies and innovations through the application of knowledgeable and creative minds. The right choice is related to the circumstances of the situation, and will vary with time and place and project characteristics. The reason to recoil at the marketer's suggestion that tubular poles are the structure of the future is that he was thinking selfishly and regionally. That's fine if that is your working environment but, to be a globally useful engineer, that is not a sufficient basis for decision-making.

We have organisations in this business that only think locally, that do not promote and sometimes do not even allow new technologies or innovations, and thus do not foster or even employ knowledgeable and creative minds. It is not about the use of one type of structure over another but why one type is used over another. There are reasons that will drive the choice in either direction on a case-by-case basis. The application of knowledgeable and creative minds to problems leads to the best solutions. Stagnation is the result otherwise.

I bet you that grandpa was knowledgeable and creative. After all, he constructed things that we often so far do not feel a need to improve or replace after having been in service for two to three times their intended service life. Well done, gramps! I suggest that you be knowledgeable and creative as well so your grandchildren can say as much about you.

Structural Engineering of Transmission Lines
ISBN 978-0-7277-5988-7

http://dx.doi.org/10.1680/setl.59887.011

Chapter 2
A transmission line in an electrical network

As a civil engineer, the analogies of water in a hose for electricity in a wire and foam on the beer work for me. That is until I mention this to an electrical engineer and I get that look saying: 'idiot'. Still, the analogies provide a suitable sense of ampacity, voltage and resistance. It even helps with surges, flashover and corona (pinhole leaks under high pressure!).

When it comes to vars (it is the foam on the beer), capacitance, induction, inductance, impedance, electric field, magnetic field, radio interference losses, TV interference losses, harmonics, hysteresis and the like, things get real dicey fast, and I need to call on another person and I need to rely heavily on their expertise. There is a reason that I am not an electrical engineer.

Now, I introduce one of my favourite young engineers – Buck Fife. Buck is a born and raised small-town Idaho boy hired out of the local college as a drafter in my employer's transmission line engineering department. He works hard, is deadly smart. So much so that POWER offered to help him get an engineering degree. Five years later, Buck got his electrical engineering degree. He got the degree in the context of being a colleague to a group of structural and civil engineering people like me. So, he understands us. I asked for his help with this chapter. I also got his input anywhere else in the book that he felt useful.

Unlike our electrical engineer brethren, civil/structural line design engineers think of a transmission line as an entity in isolation from other lines and substations in the electrical network into which it is likely located. To do so is to miss out on understanding some important characteristics inherent to a line and that are somewhat under the control of the design engineer and that are essential to the line's role in the network.

Before going further, let's clarify two words: line and circuit. We use these two words too sloppily and interchangeably. A circuit is the electrical path. In the AC transmission world, it almost always involves three phases. In the DC world, it involves one or two poles and a return (path). Phases or poles are the same thing to the structural engineer. There can be one, two or even more than three circuits supported on one set of structures. Single, double, triple and quad circuit lines are all fairly common entities. A line is a series of structures set between two points that supports the conductors and related wires for any number of circuits. In this text, we will try to speak of each properly.

The subject of electrical aspects of overhead transmission lines is immense. This chapter does not attempt to cover the subjects in great detail, intentionally. However, we expose the aspects directly linked to structural issues. These are issues that a structural engineer in the transmission line industry should understand.

Why should you care about electricity?

The main objective of a transmission line is to carry electrical power from a source to a load. Not only should the line be designed to avoid structural failure, it must also perform electrically. The required electrical performance does impact the structural design of a line, and the structural design choices will affect the electrical performance as well.

The line's voltage, normal and abnormal, will influence the required clearance from the energised conductors to other features and the insulation required to achieve the clearance. The electrical load (voltage and current) on the line will drive the conductor size, type, bundle design and temperature, affecting sags and tensions. A line's relationship to the system within which it resides will define its performance requirements.

Being part of a system

A transmission circuit is radial like a single branch of a tree or integral to the network like one thread within a cobweb. A radial circuit is a sole path for power to flow from a source to a load. Circuits integral to the network provide a path as an option to another path between two points. The laws of physics determine the path that the power will take within the network, and the design engineer can affect somewhat the characteristics that are the input for that physics.

A circuit carries a base load or an intermittently light–heavy load. A circuit is essential to the network's stability/viability or it is not. A circuit is in physical proximity to other circuits on the network or it is not. Its proximity to other circuits puts it and the network at risk or it does not. The purpose of a circuit changes with time or it does not. Whew!

All of these situations exist. Understanding the situation should affect the design engineer's choices, and some characteristics are affected by their choices. Compare, for example, the consequences of a structural failure on a 20 km, 230 kV AC line that serves an industrial facility to a failure on a 1000 km, 600 kV DC line that joins two regional networks. The consequences do not compare, and the design criteria for each should also not compare, as it is the design criteria and the implementation of the design choices that set the bar for the circuit's performance.

Planners and designers

Just as structural-minded line design engineers think of a transmission line with its embodied circuit(s) as an entity in isolation from the electrical network in which it is likely located, system and planning engineers do not. In fact, there is very little overlap in understanding and concern between these two very essential players in the development and installation of a transmission line. CIGRÉ (2010) describes in detail the relationship between line planners and designers.

Essentially, it notes that the planners identify a system or circuit concern and ask designers to analyse the ability of circuits on lines in the network to handle the concern and, if they cannot do so, then to design a solution. The solution can be renovations to existing circuits/lines or the installation of a new circuit/line(s). The planners' questions to the design engineers are largely electrical performance questions, but there are impacts to the physical design as electrical choices affect conductor temperatures and therefore sags, tensions and clearances. Now, you are in the purview of structural engineering.

The reference recommends a very close working relationship between designers and the planners if best solutions are to be found and implemented. So, we ask the questions: Is your organisation shaped to allow these interactions to occur efficiently or at all? Many are not. Does your organisation actively promote such interaction between planning, electrical and structural engineers as a necessary ingredient for success? Many do not.

Planners and designers acting in isolation will deny the owner/operator the best solutions. This is because neither the planner nor the designer understands the other's work, and both are needed to affect best solutions. If we can recommend anything to an organisation, it is to make changes to your organisation so that the answer to the two questions above is *yes*.

Normal and contingency loads

To the planner and system operator, normal means everything is functioning as intended. Contingency means a state in which something is not functioning (something is 'off line') and the remaining components of the network are doing temporary extra duty picking up the power delivery or security work of the dysfunctional component. When one component of the system is off, the system engineer calls this an 'N-1' condition. Two things off is 'N-2', etc.

I once studied the structural failure of a line of which the initial events included:

1. Shutting off a 500 kV circuit for fear of a tower's imminent collapse. This created an N-1 condition.
2. Nearby in electrical terms, a raven – as bad luck would have it – flew into a distribution substation and shorted out a critical piece of equipment. Was this establishing an N-2 condition?
3. Then, a tower of a nearby 230 kV line did collapse. Entire sections of the local population suffered a power outage. Was this an N-3 event?
4. Would the outage impact be different if the 500 kV line had not been shut off as a precaution?

I do not know the answer to these questions because it is not clear to me that the relationship between nearby distribution circuits and a 500 kV backbone circuit are that strong. But, it is a curiosity.

Figure 2.1 Conductor temperature versus time

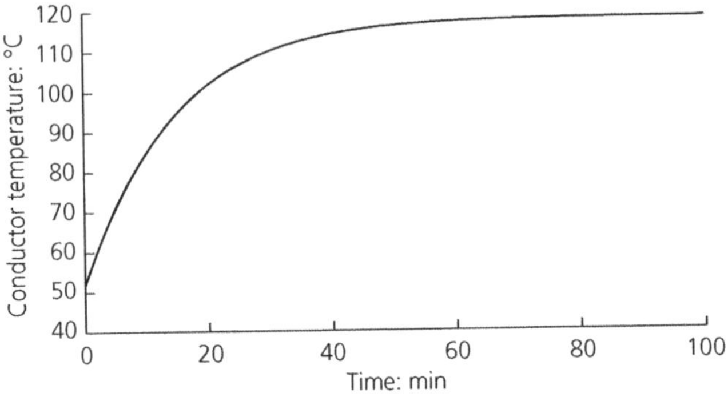

The design engineer's definition of normal and contingency is different. The design engineer is concerned only with a steady state or transient ampacity calculation that relates conductor temperature and sag to a moment's ampacity. The two definitions are not always related.

A contingency condition that does add amperes to a line can last from a few cycles – a fraction of a second – to several days or weeks, depending on the nature of the system and the ability of the owner to bring the network's operation back to the normal state. The design engineers' definition focuses on thermal ampacity, and requires that the contingency condition last more than about 30 min for the conductor to reach its steady state temperature due to an increase in current.

Figure 2.1 shows that it can take a conductor up to 30 min or so to approach a thermal steady state after the onset of a heat-developing higher amperes load. This figure assumes constant conditions with a sudden increase of current in the conductor. If the contingency event lasts less than 30 min or so, then the conductor temperature as determined by the transient calculation may be less than the final steady state temperature reached if the condition were to last longer. Therefore, it is necessary for a contingency event to last for more than about 30 min for the design engineer to label it as a steady state condition. When a contingency event will and always will last for only a very few minutes, the structural effects are considerably reduced. Watch for the difference in definition while communicating with planners/system engineers.

Should you design line clearances for the contingency condition or the normal condition? The contingency condition may cause an increase in sag over the normal condition. Since sags of conductors are assessed for no other reasons than public and worker safety and for operational integrity, it seems prudent to design for the larger, contingency, sag. The calculation of the contingency temperature requires a specific definition that is based on the network's material and operational characteristics – either a temperature target specified by the planners or it requires specific input parameters, namely a starting temperature and amperes load plus a higher load and a maximum duration for its occurrence that allow a calculation to be made. For example:

Scenario 1:

- conductor contingency ampacity: 1000 A
- pre-event ampacity: 75% of contingency capacity
- duration of contingency ampacity: 10 min
- selected ambient atmospheric conditions.

Scenario 2:

- conductor contingency ampacity: 1000 A
- pre-event ampacity: 50% of contingency capacity
- duration of contingency ampacity: 2 days
- selected ambient atmospheric conditions (hot summer, no wind).

Scenario 1 may produce a contingency condition temperature in the conductor that is less than the steady state calculation would produce for a 1000 A load. The second scenario may be a contingency situation to the planner, but the ampacity calculation of the contingency ampacity is a steady state calculation for 1000 A. The difference is governed by the pre-event power flow and the duration of the contingency event.

Ampacity

Ampacity is a statement of the amount (number) of amperes that can flow through a conductor under the constraints of the heat developed in the conductor by the sun and atmospheric conditions, its own resistance, its ability to dissipate that heat and under the limitation of a temperature constraint set by you. Stated otherwise, it is a calculation of the temperature that a conductor will reach based on the amperes flowing through it, its resistance to that flow and the variety of conditions existing at the time. It is a heat balance calculation. The heat created by current flow through electrical resistance, ambient air temperature and heat from the sun is balanced by the heat-dissipating work of wind and the wire's own heat-radiating (cooling) capability.

The amount of heat that the sun imparts to the conductor varies with the angle of the conductor to the sun, thus by its location and position on the Earth and by the time of day. It is cooled by the wind and its ability to radiate the heat away. Wind cooling varies with wind angle to the conductor and by wind speed. In detail, the variables used in the calculation are:

- conductor resistance
- conductor diameter and core diameter
- wind speed and angle to the conductor
- time of day and day of year
- latitude position on the Earth
- direction of the line (east–west or north–south)
- ambient air temperature
- maximum allowable conductor temperature
- current (amperes).

Figure 2.2 Conductor temperature versus wind speed and current

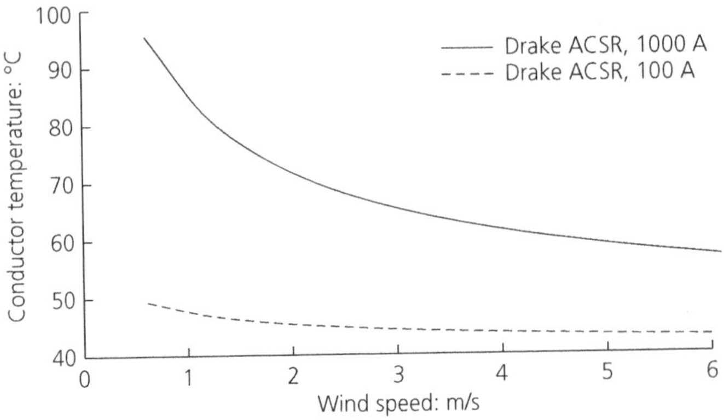

The result of an ampacity calculation is more sensitive to some of these variables than others. With high current flow, the heat balance formula becomes more sensitive to the cooling effects of wind. With low current flow, the wind has a smaller effect as there is less heat to cool. Figure 2.2 describes this relationship.

Since the calculation of temperature is actually for the temperature rise above the ambient air temperature, it is necessary to state the ambient air temperature as the starting point. The Institute of Electrical and Electronics Engineers (IEEE) standard 738 (IEEE, 2006) and *CIGRÉ Brochure 207* (CIGRÉ, 1992) define these calculations and it is important to note that the two calculations are not identical. It is equally important to note that the formulas for the various components of the heat balance calculation are approximations of reality.

Reality is of course more complex than these forms of calculations can ever represent. Despite the efforts of the IEEE and CIGRÉ, some people claim correctly that the temperature of a conductor is not what these standards assume because it is not a constant through the conductor from its core to its outer surface, nor along the length of a span and definitely not along the length of a line. The calculations' answer is of course an approximation or a generalisation. There are occasions when the application of the generalisation is suitable and times when it is not.

I describe our ability to be accurate with ampacity calculations when made as part of a line rating analysis in Catchpole (2002). Line rating analyses are made to determine the 'rating' for a line as used by the operator whose goal is a safe operation from the point of view of clearances. Other work has shown the same conclusion. That reference states that 'It is possible to predict the location of electrified parts of a line (the conductors) and the things in proximity to it that limit its maximum electrical capacity within about six inches [15 cm] – but no better.'

In other words, even with the best measuring and calculating tools and attention paid, you will not know the separation between in-span electrified parts and obstructions in their vicinity more accurately than this and most often with much less accuracy. The ampacity calculation itself is one of the reasons why. Since 6 in. (15 cm) of sag is typically caused by about 6–7°C (10°F) of temperature change, you can relate this accuracy to amperes and see that, with a large conductor, the ability to declare the ampacity of a line that complies with the criteria set is indeed crude. In light of this fact, it would be useful to remind yourself of your obligations as an engineer and of the nature of best practice and standard of care.

When a well-meaning engineer wants to argue with you that their calculation of ampacity is 943.7 A and yours is incorrect because you say it is 937.1 A, smile politely and say, 'OK'.

Impedance and line loadability

The ability of a circuit to transmit power, the circuit's 'loadability', is defined by three limits: thermal ampacity, voltage drop and steady state stability. The thermal limit is related to a conductor's ampacity, the ability to heat the conductor's materials without causing damage to them or without creating electrical clearance violations. The voltage drop and steady state stability limits are related to a line's impedance. As an AC circuit is lengthened, its impedance increases and reduces the amount of power that it can deliver. A system can tolerate only so much voltage drop.

Figure 2.3 illustrates the ability of a 115 kV circuit to transmit power as a function of its length. When the circuit is short, the voltage drop and stability constraints are trumped by the thermal ampacity of the conductor. The loadability is defined by the lower of the two constraints.

Increasing a line's conductor ampacity may or may not increase the line's loadability. For example, replacing the ACSR conductor (i.e. steel reinforced) with an ACSS conductor (i.e. steel supported) can permit a significant conductor ampacity increase. The ACSS conductor achieves increased ampacity by allowing operation at higher conductor temperatures. The effects on the impedance limits are negligible because

Figure 2.3 Loadability versus line length for an ACSR conductor

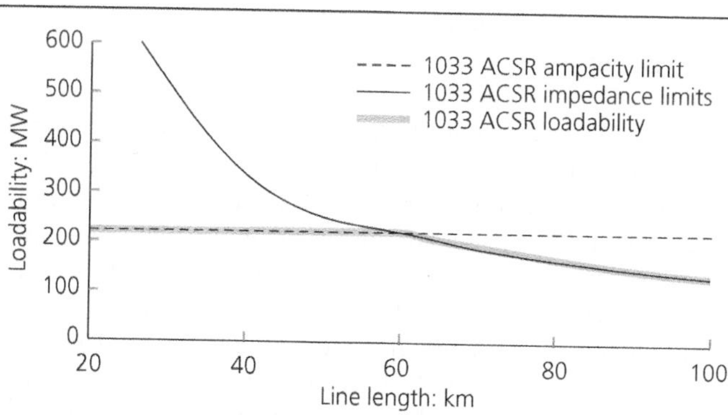

Figure 2.4 Loadability versus line length for ACSR and ACSS conductors

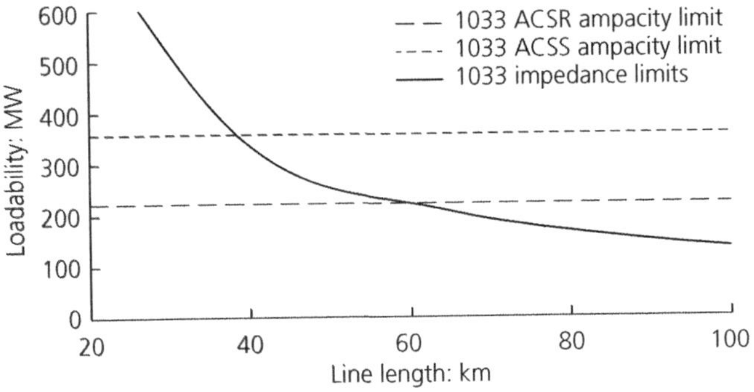

the internal geometry of the conductor has a minimal effect on transmission line inductance.

Figure 2.4 includes the added ACSS ampacity limit. Notice in the figure that replacing the ACSR conductor for the higher-ampacity ACSS conductor will only increase the line loadability for lines less than 60 km long. The full advantage of the ACSS ampacity is only achieved for lines less than 40 km long. Achieving the additional ampacity on longer lines may require reactive compensation at the line ends.

The impedance limits for both the ACSS and ACSR conductors are the same. This describes the minimal effect that internal conductor geometry has on transmission line inductance. The most significant parameters related to transmission line inductance are structure dimensioning (phase spacing), line length, bundle quantity and configuration, and, to a lesser extent, the conductor diameter.

If the voltage drop can be reduced per unit length by lowering the impedance, then the capacity over that distance is improved. This matters on long lines and particularly on very high-voltage lines where the basic, economical length of the line can be quite long. With bundled conductors in a circuit's phases, the impedance of the circuit is lowered by increasing the dimensions of the bundle and less so by reducing the phase spacing within the circuit.

Another way to think of a line's impedance is in terms of reactive power. A common analogy for reactive power is the low-density foam (head) on a beer. Real power is the drinkable liquid beer. The total (apparent) power is the total amount of beer, including liquid and foam. Think of a transmission line as an empty glass. For the same volume, the height of the glass will increase its tendency for more foam. Have you ever poured a beer into a very tall, narrow glass? This tendency for foam is similar to a line's characteristic impedance. More foam in the glass equates to less total beer in the glass to drink. Of course, this is a bad thing. As beer drinkers know, some foam is also good. There is an ideal amount of foam for each consumer, to provide both the right amount of foam and the right amount of real beer to drink. In an ideal world, the glass's propensity for foam would perfectly meet the consumer's desire.

In a transmission line's ideal world, the load on a line matches the line's reactive power characteristics. This ideal loading is called surge impedance loading (SIL). When a line is loaded by the SIL, the line's reactive power (foam) is neither excessive nor deficient for the consumer. In 1953, H. P. St Clair published a paper (Dunlop *et al.*, 1979) that generalised loadability curves across voltages by plotting loadability in terms of per unit of surge impedance loading (pu of SIL). For example, a value of 2.0 pu of SIL is the total beer contained in two glasses with the perfect amount of liquid beer and foam. Plotting line loadability in these terms facilitates a single curve for any voltage. In the analogy, larger voltages are simply larger-volume glasses, and longer lines are simply taller, narrower glasses with greater tendency for foam. For very tall glasses the less-dense foam will result in less total beer to consume. Similarly, for very long lines, the increased reactive power will result in less total power transmitted. This curve became very useful for system planners as a quick reference in understanding practical loadability for a line of a given voltage and length. Figure 2.5 provides a version of the 'St Clair curve'.

Unfortunately, the realities of line loadability are not as simple as presented. The provided discussion and loadability curves include the assumption that the line is radial. Note that a low-impedance circuit in electrical parallel to a high-impedance circuit will draw power flow from the higher-impedance circuit, shifting the balance of work that they share. Loadability of lines within a larger system will vary from this discussion, but it is conceptually accurate and very useful in understanding the various limits related to a line's loadability.

Lines that are configured to lower impedance by managing the bundle dimensions and phase separations are labelled high surge impedance loading (HSIL) lines. Is your network large enough to have long extra-high-voltage circuits in it to warrant such design features? Many are not. Most discussions on HSIL lines come out of Brazil, where the effects of such design features yield valued results.

Figure 2.5 The St Clair curve

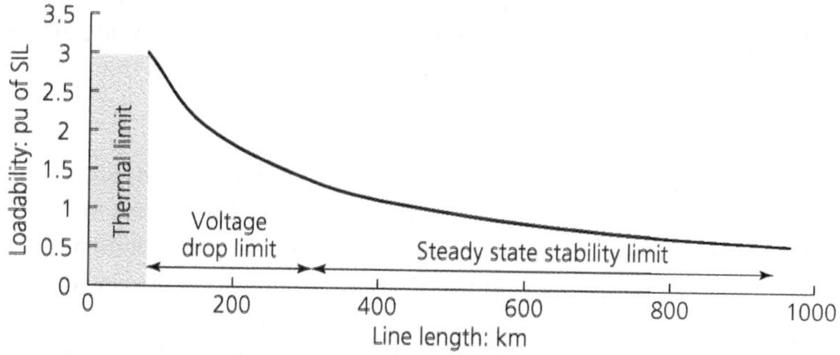

Reactive compensation can be used to change a given line's loadability. This approach changes a line's loadability limitations without modifying the line in any way; rather, its end points are modified to balance the reactive power flow.

In support of a client developing a non-traditional high-temperature conductor, a statement was made indicating that this conductor type increased the line's impedance. The statement was made by a utility that was replacing an existing conductor with this new conductor. Both conductors were similar in diameter. The towers were to be left as is, therefore the phase spacing is unchanged. Remembering that transmission line inductance is primarily a function of phase spacing, line length and bundle quantity/configuration. How could the line's impedance be increased by simply changing the conductor type?

The system operator had determined that the line's capacity must be increased to carry the loads required by the system. The line engineer applied a newly developed, non-traditional conductor type to the line. The new conductor type increased the ampacity by operating at higher temperatures, and could be installed on the existing structures without any modifications. The line engineer is happy, he has found a cost-effective solution to increasing the line's ampacity. The system operator is happy because the line engineer can provide the ampacity increase.

But the line's impedance has not changed, therefore its voltage drop and stability limits also remain unchanged. The line was previously limited by conductor ampacity, but the new, higher ampacity results in the line capacity being governed by the other limits. The line's impedance did not increase; rather, a new limit governs the line's loadability, making impedance an issue when it previously was not.

Power loss

Power loss on overhead transmission lines comes in three forms: resistive, corona and induced. For lines with corona issues, either caused by line configuration or location, corona losses can be significant. Induced losses are a result of induced currents on overhead shieldwires and power lost through resistive heating of the shieldwires. Large induced losses can be reduced by isolating the shieldwires from the grounding system. Typically, corona losses and induced losses are only considered at voltages of 345 kV or higher. When these power loss components are considered, they are of secondary concern. The primary component of power loss is resistive losses and they are not related to voltage.

Resistive loss is the heat (watts) that is lost into the air when current flows through something with resistance described by the function of power loss = current^2 × resistance ($P = i^2R$). For a three-phase transmission line, $P_{\text{loss}} = 3 \times i^2R$. Fundamentally, conductors have a very low resistance per unit length, but the sum of these very low resistances over the length of a typical transmission line can become large. In addition to length, conductor resistance is a function of the material cross-sectional area and resistivity,

Figure 2.6 Resistance versus conductor area and temperature

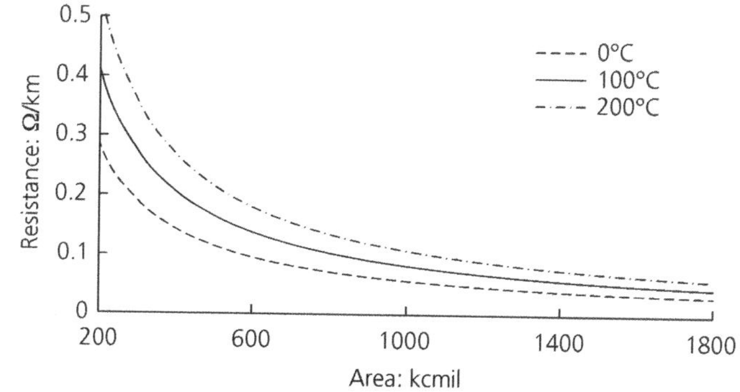

which varies with temperature. Figure 2.6 describes the resistance relationship with area and temperature for 1350 aluminium.

Two conductors with the same ampacity do not necessarily have the same resistance or the same power loss. For example, 1750 kcmil ACSR Chukar and 795 kcmil ACSS Drake can both carry about 1600 A under their respective steady state maximum operating temperatures of 100°C and 200°C. ACSS Drake has about half of the aluminium and approximately twice the resistance per metre at 75°C of ACSR Chukar. Additionally, ACSS Drake will operate at a higher temperature. At 1000 A, ACSS Drake will operate at approximately 100°C with a resistance of 0.09 Ω/km, while ACSR Chukar will operate at approximately 70°C with a resistance of 0.04 Ω/km. These two cases have been added to Figure 2.7. ACSS Drake will have approximately 2.25 times more power lost due to its resistivity than ACSR Chukar.

The exponential relationship of power loss to current is an obvious indication that the current selected for calculating power loss is an important parameter. Loss calculations should try to consider the current that will happen most often. Ultimately, the objective

Figure 2.7 Comparative resistance losses with conductor size

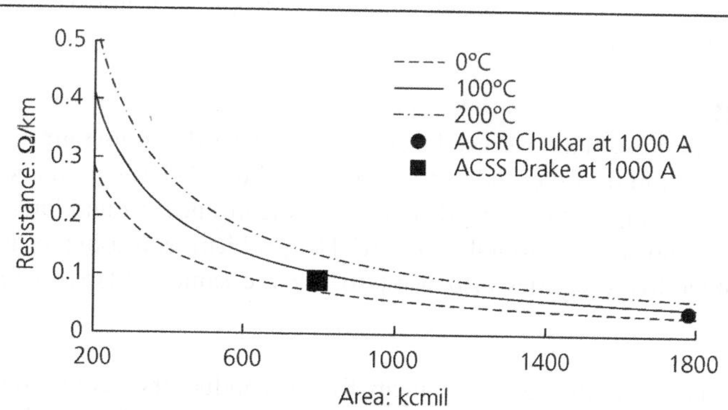

Figure 2.8 Comparative losses versus assumed currents

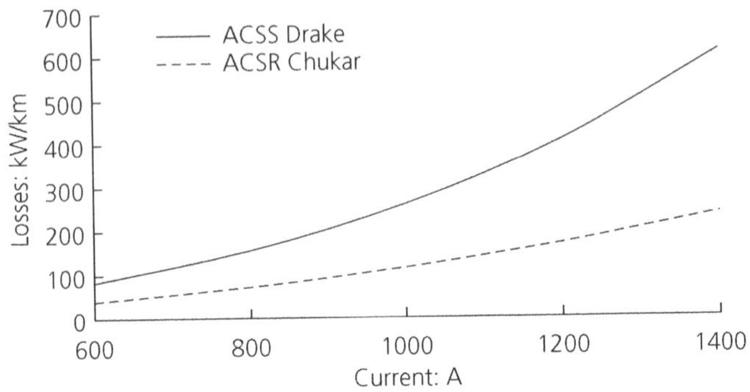

is to understand the financial value of the power lost over a certain period of time, typically 20–30 years. To understand the power lost over a period of time, you must predict the current flow that will occur on the line over the time frame that you care about. It's difficult to predict the future, and a sensitivity analysis should be considered when calculating power losses to understand the impacts of poor predictions. At the end of the day, you guessed despite your choice of formulas and parameters.

Figure 2.8 describes the power lost by ACSR Chukar and ACSS Drake as a function of the current assumed when making the calculation. The need to accurately predict the future is obviously important. Not only will this assumed current significantly impact the losses over the course of the line's life but it may also impact the conductor choice if that is the reason for the calculation.

Later, we will be posing an argument that says 'never mind the fine point of this calculation'. There are other things at stake.

> When I look back at the work of others in previous generations or my own work in the early decades, it is easy to understand the phrase 'Things were so much simpler then.' That does not make what was done back then wrong. Back then, Brian used to say, 'Put up all the aluminium you can, it's cheap.' That seems to be sage advice.

Clearances

The very basic premise of an overhead transmission line is that it uses air as the insulator between its wires and the other things that you care about. What's good about air is that it is dirt cheap – cheaper than using dirt! That's a dig at underground lines and *that* is a play on words. Did we have to tell you that? The problem with using air is that people and many other living entities also want to use the same air (space) for their own purposes.

In keeping a line's electrified parts – principally, the conductors – away from these other air users requires understanding two separation components. First, it is necessary to

Figure 2.9 Breakdown voltage versus strike distance

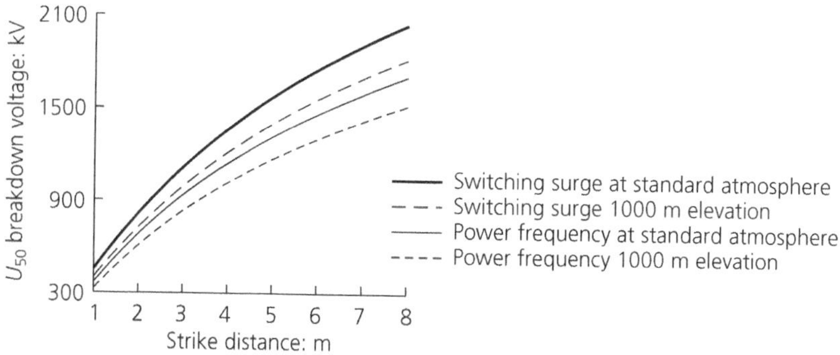

understand how far electrons at a high voltage can jump to ground from an electrified component of the line under various circumstances. That is physics. Second, it is necessary to understand how much room all of these other air users need or want under various circumstances when in the proximity of a line's electrified components. That is not physics. Add the two together to get the required minimum clearance under those circumstances.

The voltage that defines the required electrical component of clearance is not a hard-fast constant. Switching surges and lightning-caused overvoltage require clearances above and beyond the clearance that is required for the normal, power frequency, operating voltages. If the insulation value of the air space provided between the conductors and other features is insufficient, a flashover will occur.

This insulation value of air is dependent on the air's density, the type of item at either end of the air gap (rod, plane, wire, tower, etc.), the frequency and the polarity of the voltage across the air gap. The insulation value of air also varies non-linearly with gap distance. Figure 2.9 describes the relationship of air's insulation value in terms of U_{50} (i.e. the voltage that will result in flashover 50% of the time) and strike distance (gap distance) for different-frequency events and elevations above sea level.

A probabilistic approach is taken for the clearance between structure faces and energised conductors supported by insulator I strings that swing freely. This approach considers the combined probability for insulator swing (a function of weather conditions) and for voltage (a function of system operations) required to cause a flashover.

For example, the required clearance may be reduced under a very high wind event with the expectation that a high-voltage event will have a very low probability of occurring concurrently to the improbable insulator swing. Meanwhile, the required clearance between a conductor and a structure under a more common wind event is larger, to accommodate the increased probability of an overvoltage occurring concurrently to these common weather events. Figure 2.10 describes the common approach, which

Figure 2.10 Swing angle limits

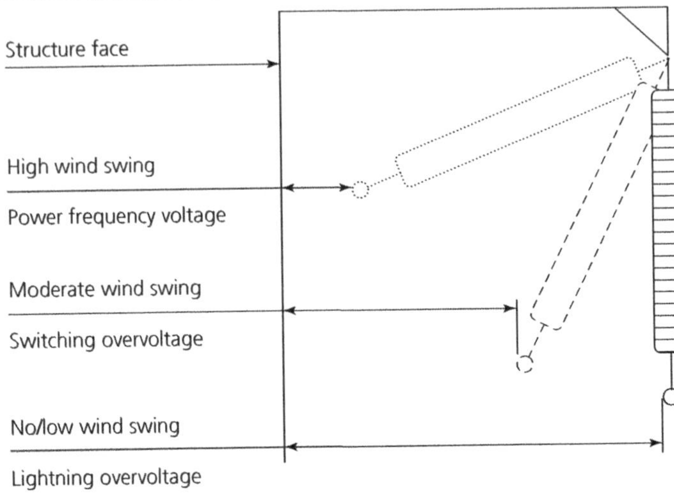

considers three possible voltage and wind combinations. Higher voltages will consider similar wind and voltage combinations.

Clearances to objects below the line, such as buildings, roads and signs, are inherently large given the range of possible types and uses of such features. Safety codes typically provide required minimum values to be achieved by design and construction. The clearances required by safety codes are commonly calculated assuming some conservative level of switching overvoltage. Survey, design and construction inaccuracies will impact the clearances achieved by the construction of a line. Understand the possible inaccuracies involved in this process and apply appropriate margins to the required minimum clearances.

Insulation

An insulator is the structural component that holds the energised conductors in their insulating air space while maintaining electrical isolation from the structure. Insulators come in a variety of material types, mechanical strengths and electrical strengths. The electrical strength of an insulator is related in part to the length of the insulator. Selecting the correct electrical strength of insulators for a given line requires the coordination of insulators, clearances and structure grounding systems.

An insulator flashover occurs when the voltage across the insulator exceeds the voltage withstand strength of the insulator. Insulation coordination is an effort to set a limit on the number of flashovers to define and achieve a desired performance. The measure of a line's flashover performance is commonly called the flashover rate (FOR), and is generally measured in flashovers per unit length of the line per year. There are four flashover modes, including:

- lightning overvoltage from shielding failure (SFFOR)
- lightning overvoltage from back-flashover (BFR)
- switching surge overvoltage flashover (SSFOR)
- power frequency flashover from contamination.

At voltages of 230 kV and below, for shielded transmission lines the BFR goals will define the required insulator lengths. For voltages above 230 kV, the SSFOR goal will determine the required insulator lengths. The local area's susceptibility to contamination will drive the requirement for an insulator's leakage distance. The quantity and position above the conductors of a line's shieldwires will be determined to achieve SFFOR goals. Air gap clearances (i.e. conductor-to-structure clearances) must be coordinated to insulator lengths for conditions (wind and temperature) that are expected to occur concurrently with lightning or switching events.

Airborne contaminates can build up on the surface of an insulator and can allow current flow along the surface of the insulator. The insulator sheds increase the length of the insulator surface significantly when compared with the insulator length itself. This surface length is known as the leakage or creepage distance. In most areas, the level of contamination will not drive insulation selection, but coastal or industrial areas with high levels of salt or industrial contamination may require insulators longer than normal or insulators with shed designs that increase the overall leakage distance. The IEC 60815 standard suggests a creepage distance between 27 and 54 mm/kV rms line to ground, dependent on the expected amount of contamination.

If shieldwires are not present or they don't properly shield the circuit's conductors, lightning can directly strike the conductors. The large amount of voltage injected into the conductors will result in an overvoltage on the conductors. If the overvoltage is sufficiently large, a flashover will occur. This type of event is described on the left of Figure 2.11.

Figure 2.11 Lightning strike failure modes

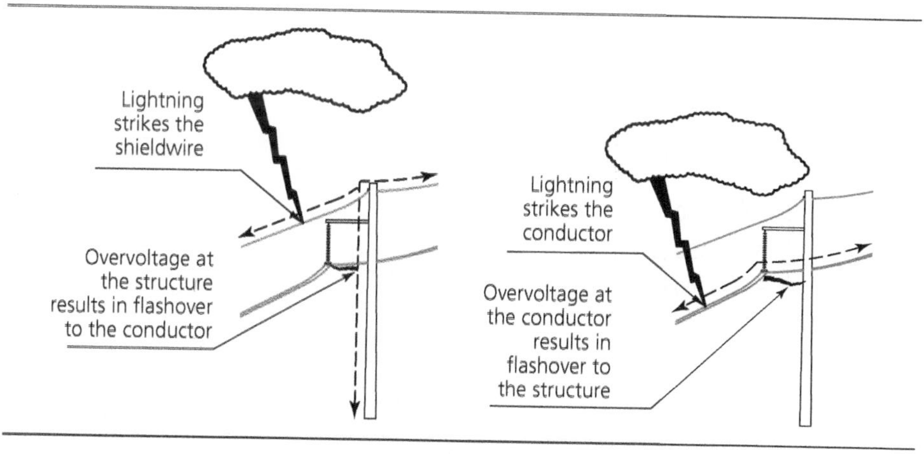

Proper shieldwire placement can result in very low SFFOR values. If this is achieved, back-flashover becomes the major component of lightning-related flashovers. Back-flashover is the result of lightning striking the shieldwire(s) and/or structure. The strike carries a current that must be dispersed through the structure and the attached grounding system. In fact, the dispersion wants to disperse itself to ground very near to where it struck the line. Lightning strikes are of a nature that they will not readily travel along the line's wires any distance to find an easy path to ground. They will insist in going to ground where the strike occurred.

Not that lightning can actually 'want' or 'insist' as it were!

The resistance of the structure and its grounding system may be fairly high and, if so, will impede this current flow, resulting in a large voltage at the structure. This overvoltage between the structure and conductors can be large enough to result in flashover to the conductors. This type of event is described on the right of Figure 2.11.

The significance of lightning in some regions is greater than others. Regional factors affecting the importance of lightning include the number of lightning strikes, the soil resistivity and nearby features that may shield the line. Amounts of lightning are weighed by the number of thunderstorm days per year (isokeraunic level) or by the number of cloud-to-ground strikes per square kilometre per year (ground flash density, GFD). The type and magnitude of lightning strikes also varies regionally. Some locations encounter numerous strikes of relatively low current; while other locations may experience strikes with extraordinary currents. Also, as air density decreases with increasing elevation, the withstand strength of air and insulators is reduced.

Line-specific factors that significantly affect lightning performance include insulation critical flashover values (CFOs), structure height, shieldwire location in relation to the conductors and structure footing resistance. Very tall structures attract more strikes than shorter structures. Figure 2.12 describes the number of strikes that will terminate on a line's shieldwires for a common configuration as a function of the line's structure height for varying values of GFD.

Figure 2.12 Flashover frequency versus height and GFD

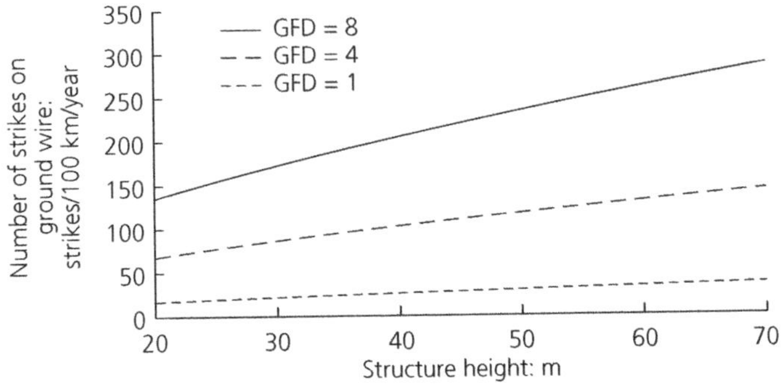

Figure 2.14 Conductor noise versus diameter and distance

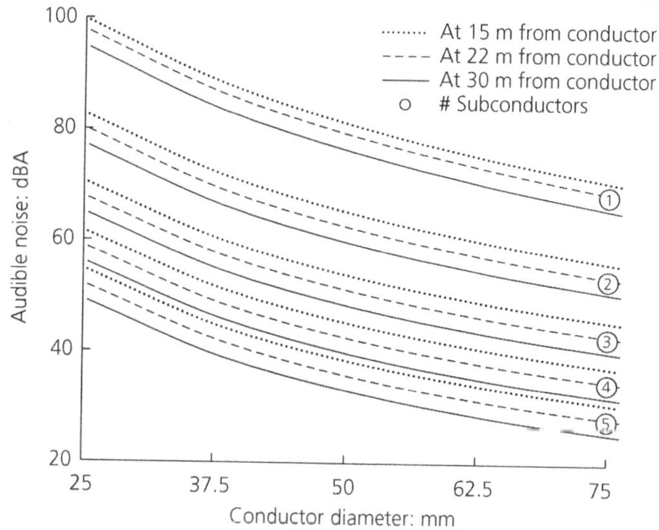

complaints. The noise level that results in public annoyance will vary with how often the noise is produced, the time of the day and time of year the noise is produced, and the background noise created by other noise creators. Typical audible noise limits are stated in terms of the 'equivalent day–night sound level' (L_{dn}) at the edge of the right-of-way (ROW). The L_{dn} is an average level with a penalty assigned to night-time noise when the public is more likely to be annoyed. Below L_{dn} levels of 50 dB, very few public complaints are received. Above 50–60 dB, public complaints become more common. L_{dn} levels above 60 dB can result in extensive complaints and possible legal actions.

During foul weather (rainy) conditions, corona and audible noise will increase. The landing of rain drops will create their own audible noise and may mask the increased audible noise from corona to some extent. The need to consider foul weather corona in a design will depend on the percentage of time the region receives rain.

Ultimately, audible noise is a function of the line's voltage and conductor configuration. Obviously, mitigation of a noisy transmission line is no easy feat as it's difficult to change the voltage and/or conductor configuration following construction. At high voltages, strict audible noise limits can govern conductor choices and configurations and/or ROW widths. Figure 2.14 describes the audible noise as a function of conductor diameter for varying subconductor quantities – single and up to five subconductors and ROW widths for a typical 500 kV line.

Electric and magnetic fields

Electric and magnetic fields (EF and MF, but collectively EMF) are invisible force fields, similar to gravitational force fields. Gravitational fields are created by an object of mass, electric fields by an electric charge and magnetic fields by the flow of electric current.

EMF commonly occurs in our natural environment and in our created environment. For transmission lines, the voltage creates electric fields and the current flow creates magnetic fields.

Despite their everyday occurrence in our environment, the high voltages and currents that accompany some transmission lines have brought forth public concern for adverse health effects caused by EMF produced by transmission lines. The significant studies on such adverse health effects have failed to conclusively indicate a direct causal relationship between the fields and health effects. In any case, the public concern exists, and transmission owners make an attempt to mitigate this concern.

Given that no direct link has been made between EMF and health effects, no distinct design threshold exists to manage them. Design limits can be mandated by federal, state or local agencies. These design limits attempt to limit EMF at or below those of existing transmission lines. Typically, they are limited by some magnitude at the ROW edge or in the ROW. Given that a line's voltage and current flow drive the magnitude of electric field and magnetic fields, respectively, there are few parameters that a line designer can change to reduce EMF. Typically, EMF limits will impact ROW widths and in some cases the required clearance above ground. The intent is to increase the distance from the public to the high-voltage conductors carrying currents.

HVDC lines

A vast majority of transmission lines are of the alternating-current (AC) type; however, in recent years there has been a shift in focus to direct-current (DC) applications at high voltages. In AC lines, the currents and voltages on the conductors are alternating above and below zero 50–60 times per second (i.e. 50–60 Hz). Alternatively, the currents and voltages on the conductor for DC lines are essentially constant.

Comparisons of AC and DC have been made since the inception of power transmission. The electrical power market originated as a DC market with the oversight of Thomas Edison. Soon after, George Westinghouse and Nikola Tesla developed and proposed transmission with the use of an AC system. At the time, the AC system had numerous advantages, including ease of voltage transformation, that allowed for very high voltages. These high voltages are essential when transmitting power over long distances. The product of voltage and current equates to power, therefore for the same power transfer a higher voltage results in lower current. Lower current means smaller I^2R losses and smaller voltage drop. These issues are important factors over long distances.

Eventually, the AC technology was proven superior to the DC alternative. Despite this historical result, the DC system has some advantages that are relevant today, given the very long transmission projects being considered. An HVDC line at a voltage of similar magnitude to an HVAC line will have the following advantages:

- larger power transfer with fewer wires and smaller structures
- smaller structures resulting in smaller ROWs and reduced land disturbances
- improved line loadability without the need for reactive compensation

Figure 2.15 HVDC break-even point on cost

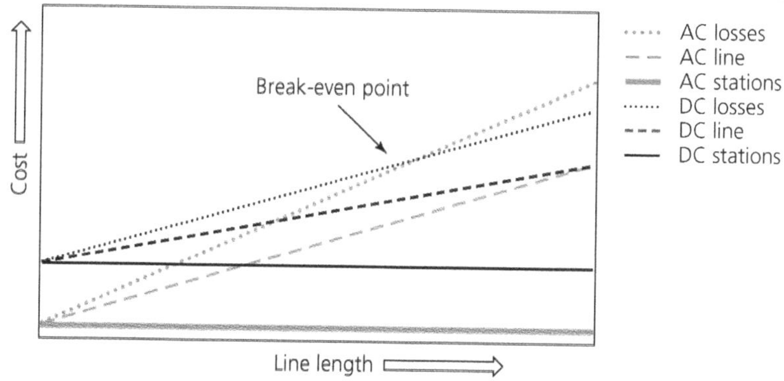

- conductor DC resistance is smaller than AC resistances, resulting in smaller power losses.

In the past, HVDC lines were used sparsely, given the difficulty to transform the voltage high enough for transmission over long distances and then transform the voltage low enough for consumption at the load. These difficulties in voltage transformation result in very expensive end terminals at each end of the HVDC line. The construction and operating cost savings found for HVDC transmission lines are countered by the expensive terminals.

The feasibility of HVDC transmission is dependent on the line's length. If the line is long enough for the per length cost savings to overcome the expensive terminal points, then the HVDC option is feasible. The length of line for which HVDC costs are equal to HVAC costs is often called the break-even point (Figure 2.15). Technology advancements are reducing the cost of these terminal points and reducing the break-even length.

The major differences between AC and DC are handled at the end points and in the system operations. The key differences concerning transmission line design are the number of poles/phases and insulation selection. The selection of DC insulation is often driven by leakage distance because the constant voltages of DC lines attract contamination differently than AC systems.

Features and controlling factors

Conductors, insulators, structural dimensions and clearances will be guided by electrical requirements. There is a give and take relationship between the electrical requirements and structural components of a transmission line. For example, a very large conductor may reduce power losses, but such a conductor may require very heavy structures with large material and construction costs. The best solution considers both viewpoints.

In some instances the relationship between those defining the electrical requirements and those designing the structural components does not allow for consideration of both viewpoints. Often one side does not understand nor consider the other. This is a great recipe for a very good batter and a very good frosting but a horrible-tasting cake.

We discussed the value of having the design engineers communicate and work with the planners, but we suggest also that there can be trouble within the design team. Within the design team there will be electrical engineers and structural engineers, and it is a rare thing to have these two camps understand each other's work well enough to make the cake a tasty one.

Table 2.1 Relationships between electrical requirements and line characteristics

Input parameter	Output effect							
	Input change	Electric field at ROW edge	Magnetic field at ROW edge	Audible noise at ROW edge	Lightning FOR	Conductor ampacity	Voltage drop and stability limits	Resistive losses
Line characteristics								
Voltage	↑	↑		↑	↓		↑	
Line length	↑						↓	↑
ROW width	↑	↓	↓	↓				
Line load	↑		↑					↑
No. of subconductors	↑			↓		↑	↑	
Structure characteristics								
Structure height	↑	↓	↓	↓	↑			
Phase height	↑	↓	↓	↓	↑			
Phase offset	↑	↑	↑					
Phase spacing	↑	↑	↑	↓			↓	
Footing resistance	↑				↑			
Shielding	↑				↓			
Insulator CFO	↑				↓			
Conductor characteristics								
Conductor diameter	↑			↓		↑	↑	↓
Conductor resistance	↑					↓		↑
Allowable temperature	↑					↑		
Bundle spacing	↑	↑		↑				
Environmental characteristics								
Wind velocity	↑					↑		
Wind angle to conductor	↑					↑		
Elevation	↑			↑	↑	↓		
GFD	↑				↑			

↑ Indicates an increase in the subject value, while ↓ indicates a decrease

The often dysfunctional relationship can be improved with some basic understanding from one side. Given the theme of this text, we promote an understanding of the electrical issues by the structural engineer. To aid in this understanding, Table 2.1 describes the give-and-take relationship of the various electrical requirements and structural components. This should arm you, the structural engineer, with the ability to qualitatively understand the impacts of the electrical issues that are affecting your structural design.

BIBLIOGRAPHY

Catchpole P (2002) Rating/uprating engineering of transmission line electrical capacity. *NWPPA 2002 Engineering and Operations Conference*, Reno, NV, USA.

CIGRÉ (1992) *Guidelines for the Evaluation of the Dielectric Strength of External Insulation. CIGRÉ Technical Brochure 72*. CIGRÉ, Paris, France.

CIGRÉ (2010) *Increasing Capacity of Overhead Transmission Lines – Needs and Solutions. CIGRÉ Technical Brochure 425*. CIGRÉ, Paris, France.

Dunlop RD, Gutman R and Marchenko PP (1979) Analytical development of loadability characteristics for EHV and UHV transmission lines. *IEEE Transactions on Power Apparatus and Systems* **98(2)**: 606–617.

Glover JD, Overbye TJ and Sarma MS (2012) *Power System Analysis and Design*, 5th edn. Cengage Learning, Stamford, CT, USA.

IEEE (Institute of Electrical and Electronics Engineers) (1997) IEEE 1243-1197. IEEE guide for improving the lightning performance of transmission lines. IEEE, New York, NY, USA.

IEEE (1999) IEEE 1313.2-1999. IEEE guide for the application of insulation coordination. IEEE, New York, NY, USA.

IEEE (2006) IEEE 738-2006. IEEE standard for calculating the current-temperature of bare overhead conductors. IEEE, New York, NY, USA.

Keast D (1980) Assessing the impact of audible noise from AC transmission lines: a proposed method. *IEEE Transactions on Power Apparatus and Systems* **99(3)**: 1021–1031.

Leman J (2009) Reactive compensation. *POWER Engineers Power Delivery Design Conference*, Hailey, ID, USA.

Phillips D (2007) Corona tutorial. *IEEE Joint Technical Committee Meeting*, Orlando, FL, USA.

Structural Engineering of Transmission Lines
ISBN 978-0-7277-5988-7

http://dx.doi.org/10.1680/setl.59887.035

Chapter 3
The nature of wires in spans

If we point you to two chapters that we consider the major point of this book, they are this one and Chapter 7 on lessons from failures. This chapter provides a host of formulas and insights that are the foundation for understanding a transmission line from a structural viewpoint. But, you will find that we don't even mention a line's support structures (poles and towers) in this chapter. That is because the most important structural component of a transmission line is all of the cables that span the support structures. All of these cables – conductors, overhead shield or ground wires and any communication cables – define and control the behaviour of the line as a structural entity. This will be described and displayed later. Chapter 4 will get into describing all of the alternatives for each of a line's structural components, including the support structures. Here, we describe how a series of spans of wires as found in a transmission line should be understood.

The intention of this chapter is to reveal the behaviour characteristics of suspended cables in a series of spans. These characteristics are hidden from you when your work is taking place within the 'black box' that is computer software. We assure you that line engineers and designers of the pre-computer years were much more likely to have this knowledge and insight at their disposal than you are today, unless you pay attention to this chapter's contents, or something similar. In this sense, this chapter may teach you what grandpa already knew well.

If you read this chapter and are encountering its contents and messages for the first time, we expect that you may not grasp their value as quickly as one reading will take. Take your time, if this is the case, for this is important. It may take a long time (years?) for this information to become second nature to you – to become a readily accessible part of your knowledge. But, once it does become part of your knowledge – second nature to you, then you 'have arrived' and our work is done!

The catenary and the parabola

We begin with a very simple explanation for a very simple formula whose basis for being correct eluded me for years – until I developed this simple explanation. Consider the question: *What is the tension in a long cable suspended from one of its ends?* Figure 3.1(a) illustrates, top left. At the bottom end of the cable, the tension is zero. At the top end, the tension T equals the weight of the cable or $w \times L$, where w is the unit weight of the cable. At any point along the cable, the tension $T = w \times D$, where D is the distance from the point of interest down to the bottom of the cable.

Figure 3.1 Equation 3.1 illustrated

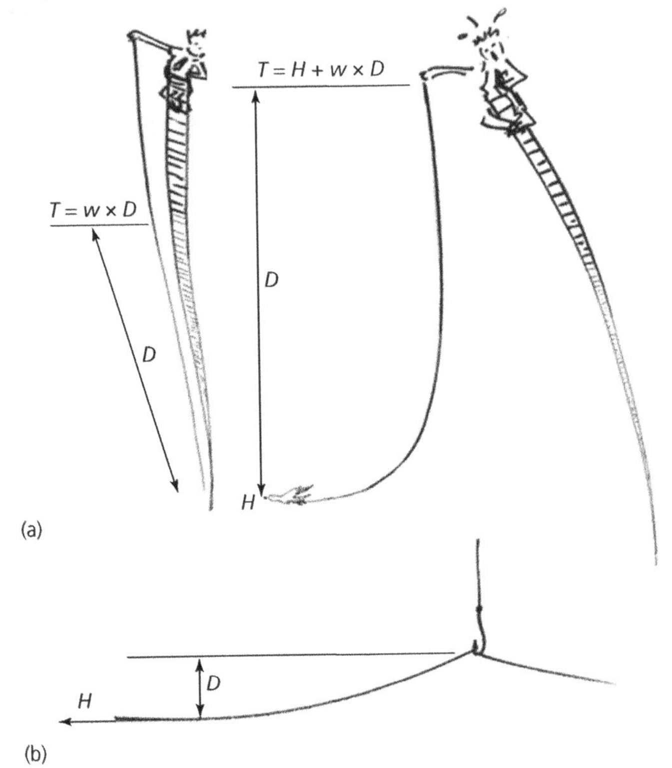

Now, in the top right of Figure 3.1(a), a birdie grabs the bottom end of the cable and flies away with it pulling with a horizontal force of *H*. Consider the question: *What is the tension in a long cable suspended from one of its ends and pulled horizontally by a force* H *at its bottom end?* The answer is simply

$$T = H + w \times D \tag{3.1}$$

Now, let's display this in a fashion better resembling a transmission line cable. See Figure 3.1(b). Here, *H* gets quite large compared with *D*, but nothing has otherwise changed.

We will see later that the hyperbolic version of this simple, simple equation is not something so conducive to a hand calculation. With this simple equation, and knowledge of the vertical distance between a point of interest along a cable and the low point or 'belly' of the cable, you know the tension at the point of interest relative to the value *H*.

Dang, this is simple stuff!

The shape of a 'span' of cable – a cable with horizontal tension put into it – of uniform unit weight and subjected to gravity is called a catenary. The proper equation for a catenary is mathematically precise but cumbersome to use. More importantly, the formula's nature blinds you from an understanding of a span of cable's behaviour that the formula's parabolic approximation provides very easily. A catenary very closely approximates the shape of a parabola under most useful circumstances – but not all.

The catenary is a hyperbolic function:

$$y = \cosh(x)$$

where

$$\cosh(x) = (e^{x} + e^{-x})/2$$

If you expand or 'scale up' this formula by a constant factor C, then it becomes

$$C \times y = C \times \cosh(C \times x/C)$$

From this, we get the hyperbolic sag formula:

$$\text{sag} = C \times [\cosh(x/C) - 1] \qquad (3.2)$$

The equivalent parabolic sag function is

$$y = x^2/2 + 1$$

Understanding that x is half of a level span, we get the parabolic sag formula

$$\text{sag} = \text{span}^2/8C \qquad (3.3)$$

The parabolic function above is the first term in the expanded quadratic expression of the 'cosh(x)' function above. This first term provides an excellent and sufficient approximation, while the remaining terms develop the adjustment to match the two results more precisely.

The difference in shape between the hyperbolic curve and the parabolic curve is very small, and is caused by the parabolic function incorrectly employing a constant weight per *horizontal* unit length of the cable in the span, whereas the catenary function correctly employs weight per unit length *along* the cable. The latter is correct, and the error in the parabola equation is small but increasingly understates the actual unit weight of the cable as its slope increases. When the span has a large slope at the span's ends near the attachment point(s), as in very long or loose spans or along most of the length of steeply inclined spans, the parabolic formula understates the wire weight and therefore understates the sag.

Figure 3.2 The catenary equation

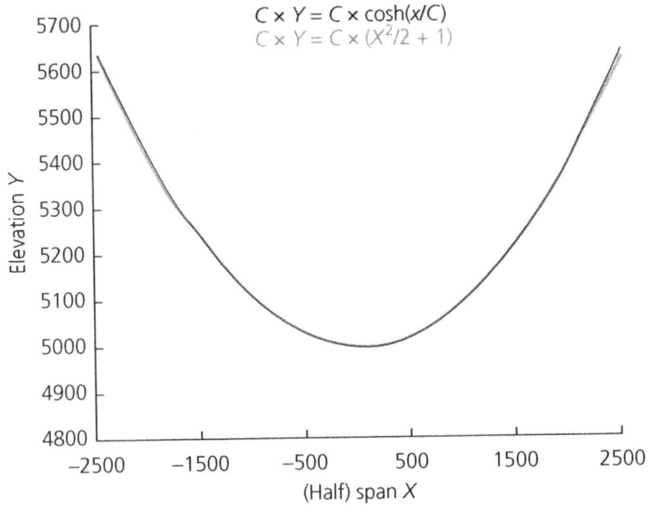

Figure 3.2 illustrates the difference between the hyperbolic equation and the parabolic equation that expresses sag in spans of wire. In the figure, both plots are expanded by the factor C. A large slope produces a difference in the length of cable per length of span compared with the unit length of span itself. Put sample values into the equations and see the sag differences that result.

For example, for a flat 300 m span with $C = 1500$ m, the parabolic sag at mid-span is 7.50 m. The correct hyperbolic sag is 7.5062 m – a 6 mm or 0.08% error in a reasonably long span. For a 1000 m span, the error is 0.77 m (or 2.5 ft). There, the difference begins to matter, but a span of this length at this tension is not realistic. Change the 1000 m span's tension via a more realistic C value of 3000 m, and the parabolic formula's error reduces to 96 mm (or 4 in.). Not bad!

The errors introduced by the use of the parabolic method will be insignificant in practical work when spans are less than a kilometre and when spans are even approximately level. It is possible that the errors may be problematic at much larger spans under most circumstances. With steeply inclined spans or in the inclined sections of very long spans, the difference between constant weight per unit length and per unit of horizontal component of that unit length can produce measurable differences, but measurable differences do not always equate to problematic differences since extraordinary spans often have much better than minimum clearances.

The catenary constant, C

Let's spend a moment on the parameter C. When you expand the entire hyperbolic equation plot up by a factor C, then C is graphically the distance between the low point on the curve where $x =$ zero and the 'origin' where $y = 0$ – a distance that is otherwise unity. This information is useless except for visual people who require a 'place for everything'. That's where C belongs, visually – below the low point of the catenary. Its unit is *length*, and it is called the catenary constant. C is derived by Nigol and Barrett (1980), where we find

$$C = H/w \tag{3.4}$$

that is, C, the 'catenary constant', is the *horizontal tension per unit weight.*

Thus, our selection of C as our expansion factor was not arbitrary but useful. The point to note is that to define C is to define the depth or the 'belly' of the catenary – the wire's *shape*. So, for a given span, specifying C is the same as specifying the sag. In the old days, the shape and sag of a span was represented by a plastic *template* used for drawing the span of the conductor on paper for engineering reasons. Most of those templates were labelled as span- and conductor-specific. Too bad, because all each really represented a C value applicable to any conductor, any span and any situation if their H/w value was shared.

If you speak of the tension in a series of spans, or for that matter in a single span, in terms of C, every good engineer in the room will know whether the wire is loose or tight. If you speak of it in terms of tension (kN or lb), they will not know unless they know what the wire is, and even then they will translate it to C. Think and speak in terms of C.

Inserting H/w for C into Equation 3.3, we get

$$\text{sag} = w \times \text{span}^2/8H \tag{3.5}$$

On display by this simple parabolic equation but not readily seen in the hyperbolic versions are the following:

- Sag is inversely proportional to the horizontal component of tension. Doubling the tension in a span will halve the sag.
- Sag is directly proportional to unit weight. Doubling the unit weight will double the tension, provided that the sag does not change or double the sag if the tension does not change.
- Sag increases with span^2, provided that H and w (i.e. C) do not change.

These simple and direct relationships are enlightening for understanding wires in spans, albeit never exactly true. The reason for the lack of exactness in these relationships is explained later.

Slack

Slack in a span of wire is defined as the difference between the arc length of the span of wire and the straight line distance between the wire's attachment points. Understanding slack is essential, for it lies at the heart of the behaviour of a span of wire. In the hyperbolic world, for level span

$$\text{arclength} = 2C \times \sinh(x/C) \tag{3.6}$$

Then, slack *in a level span* is

$$\text{slack} = 2[C \times \sinh(x/C)] - \text{span} \tag{3.7}$$

where x is the horizontal distance from the span's low point to a point on the wire. When x is at the support point at the end of a level span, span $= 2x$.

The parabolic formula for slack *in a level span* is easier to use and makes certain understandings more clear:

$$\text{slack} = 8/3 \times \text{sag}^2/\text{span} \tag{3.8}$$

or – even more usefully! – in terms of C

$$\text{slack} = \text{span}^3/24C^2 \tag{3.9}$$

The important point to see from Equation 3.9 is that 'slack' is a function of span^3. If the span doubles, the slack goes up by a factor of 8. Remember this exponential relationship.

To continue, take the first derivative of slack with respect to sag, and get

$$\delta\,\text{sag} = (3 \times \text{span})/(16 \times \text{sag}) \times \delta\,\text{slack} \tag{3.10}$$

In other words, $\delta\,\text{sag} = f(\delta\,\text{slack})$. Since a change in slack (a change in the length of wire in the span due to any reason such as creep or temperature change) is a function of the span, then

$$\delta\,\text{slack} = f(\text{span}) \tag{3.11}$$

Substituting that into Equation 3.10, δ sag in terms of C becomes

$$\delta\,\text{sag} = (3C/2 \times \text{span}) \times f(\text{span}) \tag{3.12}$$

In Equation 3.12, we can cancel span from the equation, and see that a change in sag due to a change in temperature or creep is independent of span. This explanation may be very obscure but the relationship is very evident on 'old' sag–tension charts such as that in Figure 3.3. The 'span' notation along the bottom (x) axis is actually the ruling or independent spans, not spans within a ruling span section of line that have interactions. The expression to remember and understand is this:

> A change in sag caused by a change in strain is independent of the span length

or

> A change in sag caused by temperature change or creep or elastic deformation is the same, regardless of the span length.

Figure 3.3 Sag and tension versus design spans

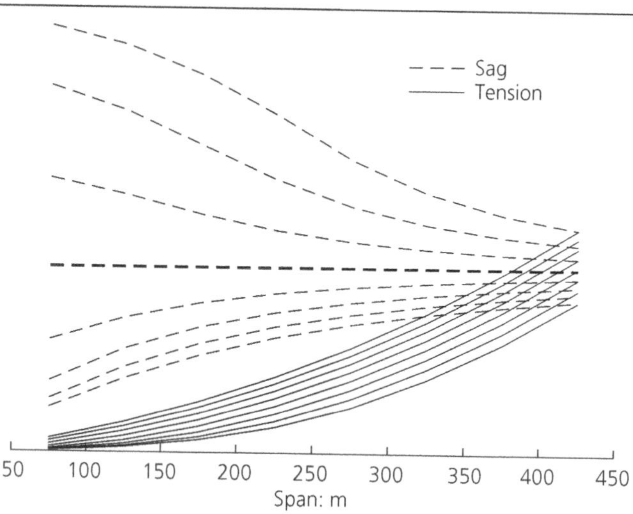

Unless your intuitive abilities are much better than mine, this is not intuitive.

One of the sad side-effects of this computerised world is that charts like that in Figure 3.3 don't get produced anymore. Some information that is presented so well via such graphics is lost, and our understanding of the subject matter displayed is actually set back from what our predecessors understood so well.

Figure 3.3 is a plot of a series of independent sag–tension calculations made at every span length noted on the horizontal axis. The sags are the solid lines rising to the right, increasing with span. The topmost line is for 100°C, and the bottom sag line is for −40°C. The corresponding tensions are the dashed lines that flare out to widely spread values to the left. The topmost tension line is for −40°C, and the bottom tension line is for 100°C. There are no sag or tension values shown because they are not important to the conversation.

The plots of sags and tensions in Figure 3.3 are based on all spans being tensioned to a common value at 20°C with no interference from ice or wind loads. There are no overriding tension limits to complicate the plots. In Figure 3.3, we see that the change in sag caused by a change in temperature (a change in length, therefore a change in strain) is the same regardless of the span. The lines are parallel, but the constant difference disintegrates as the span becomes short. The span range on the horizontal axis is 50 m to 600 m. At the short spans, the elasticity of the wire plays an increasingly significant role, and distorts the relationship. We explore the nature of this disintegration below.

A useful note

Equation 3.10 is most useful for determining the sensitivity of spans of different lengths to a small change in slack. For example, this information allows you to determine quickly the sag change caused by changing the length of a deadend insulator assembly or being wrong with a measured item's length. Deadend insulator assemblies are basically synonymous with strain insulator assemblies. 'Strain' is more correct than the jargon word 'deadend'. It should not be used if accurate sag changes are required and if the slack change is large relative to the existing slack. For example, introduce a 10 mm length (tiny!) into a small span of 60 m that has a C of 1000 m and a sag of 0.45 m. The existing slack in the span is $[60^3/(24 \times 1000^2)] \times 9$ mm: tiny!

The sag change will be

$$(3 \times 60/16 \times 0.45) \times 0.01 = 0.25 \text{ m (10 in.)}$$

Recall that the sag is only 0.45 m before the slack change, so the sag has been increased 55% and the tension will therefore want to decrease 55% according to Equation 3.5. This is a radical change to the span caused by a very tiny length change (10 mm). The large tension reduction will elastically shorten the span, and the calculated sag increase will be less than calculated.

Consider the same 10 mm of slack taken out of or introduced into a longer 350 m span with a C of 1500 m. The sag is $[350^2/(8 \times 1500] \times 10.21$ m. The change in sag is

$$(3 \times 350/16 \times 10.21) \times 0.01 = 0.064 \text{ m (2.5 in.)}$$

In this case, the same introduction of slack affects the 10.21 m sag by only 6.4 mm (0.6%). The change to the tension is therefore equally negligible. If the objective is to introduce a significant amount of slack, relative to the existing slack in a span and to determine with precision the resulting sag, the following relationship should be used:

$$\text{slack}[2] = \text{slack}[1] + \delta\,\text{slack}$$

and, from Equation 3.8,

$$8/3 \times \text{sag}[2]^2/\text{span} = 8/3 \times \text{sag}[1]^2/\text{span} + \delta\,\text{slack}$$

leading to

$$\text{sag}[2] = \sqrt{}(\text{sag}[1]^2 + 3/8 \times \text{span} \times \delta\,\text{slack}) \qquad (3.13)$$

We recently reviewed the installed sags on a small project and found that they were all high – some by a small amount and one by about 0.8 m. Any sag that was under the target meant a tension above the target and we sadly left no margin against the support structures' strengths that the owner would understand as OK. So, the question was, 'How much length must be added into each span to bring the sags to the intended values?'

The answer was tiny amounts. Even a single insulator bell would solve the big sag error. There was plenty of clearance under the conductors, so that would work. But, the other thing we noticed was that the slack error was so tiny – it was much smaller than the contractor created when they installed the spans. In other words, we were asking a lot – maybe the impossible – to have the contractor get it perfect. Consider that: your role in getting an installation installed properly, based on the requirements that you specify. Give the contractor room to do what they must to get it done.

Ruling span in principle

We come to a subject that is most important to understand. Let's weigh in on the subject of ruling spans.

In July 2010, I made a 3-day visit to Brian's home. When with Brian, it is impossible for even a fraction of an hour to go by without the discussion being diverted to the engineering of transmission lines. So, over dinner with his wife, Pat, in tow on the second day, the conversation led me to pose a question to Brian. It was this, 'If the ruling span method and clipping offsets had not been developed many decades ago – in other words, was unknown to us – how do you imagine that we would conduct our business today in their absence?'

The following morning, I returned to Brian's home after a comfortable evening and night at the local inn. He answered the door in a near panic. He had risen at 5 a.m., as usual and was having great difficulty waiting on my 10 a.m. arrival, as planned, because he had had another of his frequent and precious revelations. He had already written the following …

'Yesterday, within idle chatter, Peter mused on what our TL world would be like if Paul Winkleman had not created the workable math of the ruling span method.

'I mused a moment, and said someone else would have had to because the TL systems as we now know and build them would be impossible without the RS method, and then I lapsed into silence, a silence that floated around in my brain and was still there when I awoke quite early this morning.

'Yes, the RS method that I was fortunate by extreme chance to learn from Paul Winkleman himself back in 1950 is the tool that makes possible the present method of building TLs, but the RS method is probably the most misunderstood tool in all the many facets of the work of a TL engineer.

'It is a method of several steps that is necessary to build a TL of more than a single span.'

Brian and I on that day agreed to the fact that it is grossly impractical – to the point of near impossible – either to design or construct a transmission line without the ruling span method in your arsenal of tools. The ruling span is not a handy tool, as it is so often described, it is an essential tool.

Figure 3.4 Equation 3.14 illustrated

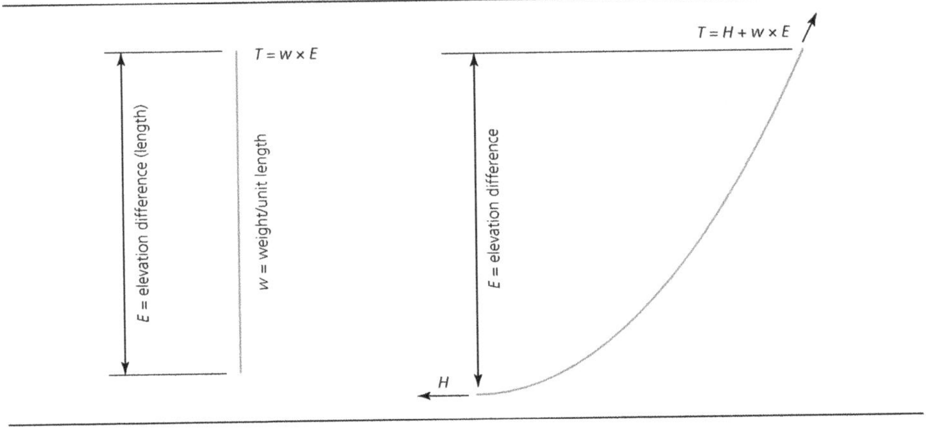

Recall and review the cable hanging vertically from one of its ends. If that cable is 100 m long and weighs 1 kg/m, the tension at its bottom end will be zero and the tension at its top end will be its own weight – 100 kg.

A supplemental tension, whether vertically applied or not, applied to the bottom end of such a suspended cable, causing it to adopt a catenary shape does not affect this formula. Thus, the tension anywhere in the cable suspended in a span, per Figure 3.1, remains as per Equation 3.1. This leads to a most simple and essential formula for understanding transmission line conductors and cables between structures, as Figure 3.4 illustrates:

$$T_{(\text{at support})} = H + D \times w \tag{3.14}$$

where H is the tension in the cable at its low point and D is the elevation difference between the span's low point and its end point being addressed.

In words, Equation 3.14 says: *the tension in the cable at its span's end equals the horizontal component of tension in the span (i.e. the tension at its belly or low point) plus the elevation difference between the low point and the end point times the unit weight of the cable*. This is, of course, the same point made with Equation 3.1. Now, when the cable is installed into a series of spans using travellers or sheaves, the tension in the cable is necessarily equal on either side of each traveller, due to the sheave's lack of resistance to rotation. Each traveller will rotate until the tension in the cable on each side is equal. If the low point of each span is at a unique elevation due to uneven span lengths or terrain profile, then the following must result.

The cable in the traveller between two spans with their low points at different elevations will enter and exit the traveller at different slopes. Since the tension in the cable is equal on both sides of the traveller, the horizontal component of that tension in the two spans

Figure 3.5 Unbalances at sheaves

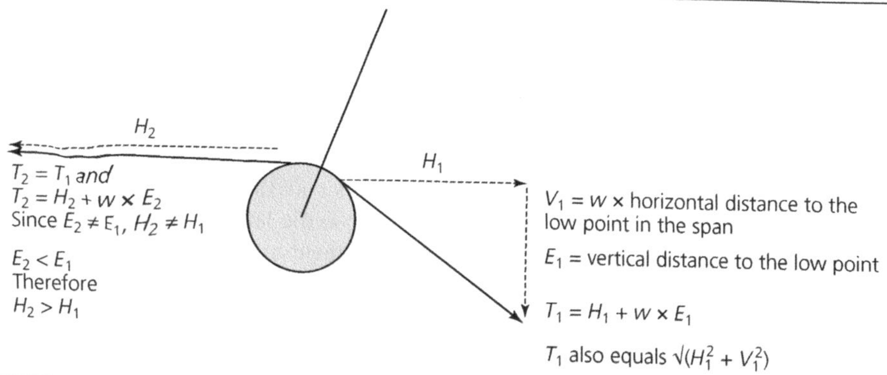

cannot be equal, as Figure 3.5 illustrates. So, the unavoidable fact regarding a cable strung with travellers through a series of non-level spans and/or spans of uneven length is this: *the cable will have a unique horizontal component of tension in every span, when in travellers.*

The ruling span method is composed of steps you take that change this natural situation to create an equal horizontal tension in all spans in a series of spans when they would not naturally occur. We call it the application of clipping offsets. This is explained later. Absent these steps, these tensions are not equal for the reason described above. Thus, the question:

> How would transmission lines be designed and strung in the absence of the ruling span method – in other words, in the absence of an equal horizontal component of tension in each span – in the absence of a common catenary constant C in all spans?

A visit to a parallel universe

Let's visit a universe where things are . . . different! In this universe, there is no greed or hatred but, more importantly, the transmission line engineers do not have the ruling span method at their disposal. We will watch how they design and construct transmission lines in its absence. We will travel slowly so that you do not get lost and left behind. We do not want to leave you in this parallel place because those who love you may miss you.

In this parallel universe, the engineers are designing a line running down a long slope with many spans. According to Equation 3.14, the catenary constant C for each span down the slope will decrease when the wire is in sheaves. As the cable runs down the hill, the cable's tension and, therefore, the C value decreases in the lower elevation spans, and it is lowest at the bottom of the hill. We have a bit of a complicated situation on hand. The larger sag in the lower-elevation, looser spans needs addressing to get the conductors off the ground. Alternatives when these clearance conditions become problematic are to make the spans shorter or to use taller support structures or to isolate the looser spans with deadend structures and increase their tension up to a new, higher and more reasonable value. According to Figure 3.5, all of the horizontal components of tension at each support structure are unequal.

Since these engineers have no knowledge of the ruling span or of applying clipping offsets to develop the rational response to its calculations, they are going to place the support clamps at the end of the suspension insulators either at the location of the out-of-plumb traveller or in a plumb location. If they choose the latter, it will be to no avail because the unbalanced horizontal tensions will move their insulator attachment towards the higher span, regardless.

The end result for their design exercise will be (1) high tensions at the higher parts of the profile and lower tensions, larger sags at the lower parts mitigated by many structures creating short spans or by inserted deadend structures used to reinstate rational tensions, and (2) many of the suspension structures along the way will have inclined suspension insulator strings, indicating the longitudinal load unbalances at these supports. All of these features of the design outcome may add considerable expense compared with what we do in our universe.

This is to say nothing of the complexity of managing the design process where the engineer must track the unique tensions and sags in each span as the line's structures are positioned. Yuk! So, we shout out to these poor folks through our joining wormhole, 'There is a much better way!' But we cannot be heard. Let's go home and leave these people to their difficult work.

We should note that this situation is faced by ski-lift system design engineers, where their cable is always in sheaves. In their business, the tension in the cable at the top of the mountain is several thousands of kilograms greater than it is at the base of the mountain. Notice too that many of their towers that support long spans with large elevation differences on the uphill and downhill sides are tilted forward towards the bisector of the bite of the cable. This mitigates the effects of the longitudinal load unbalance at the tower. So, it is clearly not impossible to design systems this way, but the complexities are great, negating the concepts of tower families and sag templates.

In our universe, we have discovered, thanks to the insights and guidance of Paul Winkelman of the Bonneville Power Administration (Winkelman, 1960), that converting the natural and unavoidable array of horizontal tensions inherent to a series of spans in travellers into a single, common horizontal tension in all of those spans has incredible value to the cost of the line that we are designing and building and to the simplification of our work.

Flash forward to present day. PLS-CADD is perhaps the most popular software package in the world for designing transmission lines. It offers two 'modes' for analysing the wires between support structures: the ruling span method and its finite element (FE) method. Be clear that the FE mode cannot be initiated without first establishing a condition in which the length of cable in each span is declared as 'known'. This is done by one of two methods. First, you can let the ruling span method accomplish its intended task of creating a common horizontal tension in all spans under a specified condition and then 'locking in'. Second, you can very accurately survey the spans and set each span's shape (catenary) to the survey's findings and then 'lock in'. Of course, this second method can only be used if the line is already constructed, so it is useless as a design tool.

For new designs, the required approach is to effectively assume the ruling span method was put in play by way of clipping offsets if necessary to establish the base case. In other words, the FE mode is not an alternative method to the ruling span method, but is only a further refinement based on unverifiable assumptions – that is, based on work yet to be done in the field during construction or ignored for lack of interest, understanding or time.

Since the construction process creates close-to-equal H values in all spans either by clipping offset efforts or by the assumption that they are unnecessary, the finely tuned list of lengths of cables that you see in the software is an approximation of reality. It is followed by an approximation of behaviour thereafter as well. Except for extreme conditions of line geometry, the FE mode is quite a cumbersome and unnecessary tool for design and analysis.

Clipping offsets – making the ruling span method work

'Clipping offsets' is the term used to describe the length of wire that must be moved from one span to the next at a support structure during the clipping in or installation of the cable's support clamps. The exact definition of the measured length of wire, its calculation of length (offset) and the rationale for its use/application is well described in the well-known paper by Winkelman (1960). This invaluable paper, once possessed, must never be lost.

The assumption used to make clipping offset calculations is: *the sum of the slacks in sheaves equals the sum of slacks in clamps*. Since the 'sum of the slacks' is the complete length of the wire installed (the sum of the spans' arc lengths) (i.e. a constant) minus the sum of the point-to-point span lengths (i.e. another constant), then it cannot change. So, the clipping offset process requires that you put the desired length of wire out there to begin and then move it around to undo the natural tension unbalances.

There is a way to estimate the need for offsets. The percentage difference in tension H between the two spans while in travellers equals E/C as a percentage, where E is the elevation difference between the spans' bellies:

$$\Delta H = E/C\ (\%) \tag{3.15}$$

For example, if $E = 100$ m and $C = 1500$ m, then the difference in H between the spans will be 100/1500 or 6.7%, with the upper span C being something greater than 1500 m and the C of the lower span being less than 1500 m. Their exact values will depend on the spans beyond.

A close read of Winkelman's 1960 paper will tell you that the clipping offset calculation depends on several things. First, the calculation applies between the installation equipment set-up locations and not between the line's permanent deadend structures that will eventually bracket the line section in question but may be at different locations. These equipment and deadend locations may be the same, but this is not always the case. This means that offset calculations made without the equipment set-up locations properly known will be completely wrong. In other words, offset calculations cannot be published before the set-up locations are known.

We also point you to IEEE standard 524, which describes the method for tying stringing sections together at suspension structure spans when clipping offsets are being applied (IEEE, 2003). It's a process, and, surprisingly, some contractors of good reputation don't understand or believe it. It can be an ordeal to watch an engineer try to teach a contractor how to string wire. Rather ... watch a contractor learn from an engineer!

> A colleague was running a project, and his well-respected contractor did not believe the offsets could be correct at 8 ft. A conference call to another contractor got the answer to the question, 'What are the largest clipping offsets that you have installed?' Not knowing why the question was being asked, he answered, '32 ft'. My colleague's contractor learned something that day.

Second, the offsets calculated are distances that must be measured along the installed wire *before* the wire is moved into its final clamped position at any of the section's support points. This creates a significant cost for using offsets because the work of clipping in cannot start until all positions are marked. *NOTE*: This offers an opportunity for some aspiring 'Winkelman' to write a progressive offset program that will permit simultaneous marking and clipping in.

It is possible to minimise the need for and the cost of offsets by judicious positioning of stringing equipment, essentially by separating line sections that are at different elevations. In addition, the engineer must establish criteria for the need to apply offsets. Acceptable limits on sag errors and tension variations will vary with voltage and the roughness of the terrain, and will also depend on the tolerances that have been built into parameters such as tower strengths, ground clearance and conductor tension limits.

The Winkelman method is based on the parabolic equations, and brings with it the accompanying approximations. When a traveller sits at the end of a long insulator string and there is a tension difference on either side of the traveller due to uneven sag belly elevations on either side, the traveller will roll towards the higher elevation belly. The Winkelman calculation has no interest in where the traveller sits on the wire. The nature and position of the insulator string and traveller is of third-order importance to the calculation.

Finally, if offsets are omitted when needed, the result will not only be incorrect final sags and ground clearances, unequal tensions and inclined insulator strings but the sagging crew will waste time trying to achieve the impossible. Therefore, even if the line conditions are such that offsets may not need to be applied, the sagging crew may need to know the adjustments that must be made for sighting in. In other words, provide the contractor with the 'sag in sheaves' values according to the clipping offset calculation results but don't follow through with its minor adjustments.

That is worth repeating: *when sagging in steep terrain where the tension values* H *in spans is variable, sag in to the 'sag in sheaves' values, not to the final values that are based on* H *being equal in all spans – because they are not equal!* To do otherwise is to chase your tail.

Ruling span in detail

It has already been noted that short and long spans react differently to changes in temperature, as they also do with creep or ice loads or any influence that loads or strains (changes the length) of the cable. That seems to be an unruly situation. But, there is a very practical solution called the ruling span. A series of spans of random lengths supported on suspension insulator strings will collectively behave as if they were all of some intermediate span length – a span designated as the ruling span. The first mention of the ruling span seems to be by Thayer (1924).

Think about that. What a neat trick! A cable placed in travellers while spanning between a series of supports – such as a transmission line's conductors during stringing or a cable on a ski-lift system – will position itself so that the tension in the cable on either side of each traveller (wheel) is equal, forcing the horizontal component of tension H in every span to be unequal unless the spans are all equal in length and the support points are all at the same elevation.

But, if that same cable is taken out of the travellers and placed in a clamp that is suspended from a reasonably long string, the horizontal component of tension in every span becomes equal. To do this, the grip of every clamp must hold the necessarily unequal tensions on either side of it. Yup . . . that's a neat trick! Handy too, because having H common to all spans makes the engineering calculations so much easier, as we recognised from our visit to that other universe.

Our very easy calculation is made available to us by the existence of a span that can represent all of the actual spans in our calculation. It turns out that there is such a span, and we call it the ruling span. Again, a series of spans of random lengths supported on suspension insulator strings will collectively behave as if they were all of some intermediate span length – a span designated as the ruling span.

The ruling span (RS) formula for relatively level spans is

$$\mathrm{RS} = \surd(\textstyle\sum \mathrm{spans}^3/\sum \mathrm{spans}) \tag{3.16}$$

Note: $\sum \mathrm{spans}^3 \neq (\sum \mathrm{spans})^3$

For inclined spans, the formula can be embellished to

$$\mathrm{RS} = \surd(\textstyle\sum \mathrm{span}_i^4/C_i/\sum C_i) \tag{3.17}$$

where C is the chord length (straight line) between span ends.

Equation 3.16 can be quite incorrect for inclined spans. For inclined spans, resort to Equation 3.17. We are going to explore the fact that the horizontal tensions in each of the spans in a series of unequal spans wants to change, to become unequal if ever made equal – over time and pretty much all the time. The amount of cable that has to be fed from one span into the adjacent span to equalise their differences is quite small, and that small amount of cable is fed back and forth between spans by very small movements (inclinations) of the suspension units (insulators) at the structure support points.

The horizontal tensions throughout the series of spans would all *always* be the same and exactly equal to the tension of this theoretical ruling span value of H if the insulator strings were infinitely long and permitted unrestricted back-and-forth adjustment of the clamp positions. That is to say: if the insulators could feed the slack between adjacent spans without resistance as the strain changes due to temperature and time take place.

However, insulator strings are not of infinite length, and it follows that the suspended system will rarely be exactly as predicted by the ruling span method. Fortunately, in all but the most extreme conditions, the discrepancies between the theoretical and the actual sags and tensions will be of no consequence, well within the tolerances of surveying, sagging-in, creep estimations and so forth. In other words, why bring the sags on the engineer's table to a very precise accuracy when the formulas used to manage the wire's change with time and temperature, and so on, and the field activities cannot be brought to a comparable accuracy?

There are numerous technical papers dedicated to the accuracy and suitability of the ruling span method. Go ahead and read some of these, but then please move on. The papers tend to be authored by vendors of alternative methods, academics or souls lost in the useless pursuit of accuracies that can never be achieved. Revisit the first and fourth mantras written in this book's preface.

Furthermore, longitudinal loads caused by the inclination of the insulator strings will always be negligible with respect to usual structural strengths. It is reasonable to explore the limits and approximations of the ruling span concept with a simple two-span example: spans of 150 m and 400 m, which, if joined by a tower with suspension strings, would comprise a series with

$$\text{RS} = \surd[(150^3 + 400^3)/550] = 350 \text{ m}$$

CASE I: Consider the three spans of 150 m, 400 m and 350 m as separate deadend spans isolated from each other. For ease of computation, assume H at 15°C is 1875 kg and w is 1.5 kg/m, making C equal to 1250 m by Equation 3.4. For each span, we calculate the sag and the slack by Equations 3.3 and 3.9 (Table 3.1).

It is interesting to note that the total slack of the 150 and 400 m spans is 0.090 + 1.707 = 1.797 m, producing a *rate of slack* of 1.797/550 = 0.00327 m/m. The *rate of slack* of the 350 m ruling span is identical: 1.143/350 = 0.00327 m/m. And that is the point!

Table 3.1 Exploring the ruling span

Spans	150 m	400 m	350 m (ruling span)
Sag at $C = 1250$ m	2.25 m	16.00 m	12.25 m
Slack	0.090 m	1.707 m	1.143 m

Thus, the basis for the ruling span equation can be understood: *the ruling span is a span that has a rate of slack equal to the rate of slack of the entire series of spans.* This might have been evident from Equation 3.16 for the ruling span, which sums the spans and slack, because this equation is a function of span^3. It is no coincidence.

With an increase of temperature from 15 to 80°C, a sag tension program would show new values for the three spans (Table 3.2).

Once more it is noted that the rates of slack remain equal ((0.116 + 1.928)/550 = 1.337/350), but the C and therefore H values are no longer equal. A deadend or strain type of tower and insulator assembly inserted between the 150 and 400 m spans would have to resist these inequalities and unbalances. If we were to join this 150 m span to the 400 m span with a long suspension insulator that would equalise H between them, we could understand the sags and tensions in each span using a single calculation on the ruling span of 350 m.

CASE II: Convert the support structure between the 150 m span and the 400 m span from a deadend to a suspension tower with suspension insulator assemblies. This creates a two-span series with a ruling span equal to 350 m. All will be in balance if the conductor is installed at 15°C and with the sags of Table 3.1 applied as above, with clipping offsets and if the conductor remains at 15°C and the conductor does not lengthen with creep over time. However, if the temperature rises to 80°C, there is going to have to be movement of the clamp on the suspension string out of the short span and into the long span in order to equalise the now 386 kg difference in the horizontal tensions in the two spans, as seen in Table 3.2.

Both spans would be brought close to the C of 1155 m if the clamps were to move about 0.06 m into the long span, a calculation easily made by using Equations 3.3 and 3.10. Try it! This clamp movement will incline the insulator string and produce a resistance to the

Table 3.2 Further ruling span exploration

Spans	150 m	400 m	350 m (ruling span)
Sag at 80°C	3.06 m	17.00 m	13.25 m
Slack	0.166 m	1.928 m	1.337 m
C	919 m	1176 m	1155 m
H if $w = 1.5$ kg/m	1379 kg	1765 kg	1733 kg

balancing of forces approximately equal to

unbalance = movement × w × weight span/string length

That is simple vector geometry at the insulator string. If the 150 and 400 m spans of the example are level spans, they have a weight span of 150/2 + 400/2 = 275 m at the intermediate support structure. If the string length is 2.2 m, then the restraining force is approximately 0.06 × 1.5 × 275/2.2 = 11 kg.

Let's look at these span interactions in another way. Equation 3.12 told us that a change in sag caused by temperature change or creep or elastic deformation is the same, regardless of the span length. But, we see that the sags changed to 3.06 – 2.25 = 0.81 m and 17.00 – 16.00 = 1.00 m, respectively, in the short and long spans when isolated from each other. The ruling span (350 m) sag changed by 13.25 – 12.25 = 1.00 m. These are not equal changes in sag. What is going on?

In the short span, the temperature change caused a sag change from 2.25 to 3.06 m (Table 3.1 to Table 3.2) – a change of 36%. When the sag increases that much, Equation 3.3 requires that the tension decrease proportionally – 36%. In the long span, the sag changed by a greater amount – 1.00 m – but this is a 6% increase in sag requiring only a 6% decrease in tension. This temperature increase lowered the tension in both spans but the decrease wants to be much greater in the shorter span. Table 3.2 reflects this, but not exactly.

Our simple equations ignore one fact of the conductors and cables. They ignore their elasticity. They are only correct if the cable is infinitely stiff – it does not stretch with increased tension or shrink with decreased tension. A 36% decrease in tension causes the cable to shrink in terms of strain much more than does a 6% decrease as experienced by the longer span. As sag increases in a cable and the tension decreases, the cable shortens. Due to the elasticity, the sag increase is less than it would be and the tension decrease is less than it would be compared with an infinitely stiff cable. This is why the changes in sag due to strain (temperature) change get smaller and smaller as the span gets shorter in Figure 3.3.

Let's beat this dying horse one more time. When H is common between spans of different lengths, the sags are related by the ratio of the span lengths squared:

$$\text{sag}_1 = \text{sag}_2 \times (\text{span}_1/\text{span}_2)^2 \tag{3.18}$$

So, when joined into a series of spans they are represented by a ruling span equal to 350 m. With the temperature rise, the ruling span sag became 13.25 m – up from the 12.25 m of Table 3.1. With C held common in both spans by the small movement of the suspension insulator between, the sags in the short and long spans become 2.43 and 17.31 m, respectively, according to Equation 3.18 – not 3.06 and 17.00 m.

But then, the inclination of the insulator strings upon the onset of extreme, high temperatures (when ground clearances are critical) will produce a modest increase in sag in the long spans over and above the sags computed by the ruling span method. Fortunately, it is very difficult to find a situation where these inaccuracies will exceed a few centimetres at most.

With spans of varying lengths, the insulator strings will be vertical only at the time of sagging-in (assuming a precise sagging-in procedure) and again if shrinkage due to cold temperature just equals the accumulated creep.

With an increase in temperature (or creep), the strings will swing into spans longer than the ruling span and out of spans shorter than the ruling span. This illustrates what is happening throughout a line section carried on suspension structures.

As the conductor length changes when the temperature rises and falls, as creep continues to take place and as wind and ice come and go, the suspension clamps move backwards and forwards in an attempt to equalise tensions between adjacent spans. This tension equalising by the suspension insulator strings (with almost negligible loads transmitted to the structures) is the reason why transmission lines perform so efficiently and so reasonably predictably according to our RS method calculations.

The results depend on more than just the span ratio and are most sensitive to string length and to the weight span on the intermediate structure because any error implicit to the ruling span method is directly related to the angle of inclination of the insulator string and the load in the string. That is to say, the ruling span method works with increasing imperfection as the suspension strings incline more and carry more weight (i.e. they support larger unbalances in H). The exercise could be repeated with spans of 50 and 400 m (a very large span ratio of 8/1), but the sag error will in fact be reduced. This is because with a span of 50 m and a C of 1250 m, the sag is only about 0.6 m, and the slack at 15°C is about 0.00333 m (3.3 mm!). It is as if the two structures behave as one when they are this close together. The span between is so short that the wires behave as stiff or rigid members effectively uniting the two structures into one from a transfer-of-slack point of view.

Furthermore, the *change of sag/change of slack* ratio for such a short span is about 38 : 1, with the result that a very, very small movement of the clamp will produce a very quick adjustment of and a quick balancing of tensions. There is no possibility of the clamp moving more than a fraction of an inch, and thus no significant inclination of the string. With only a small inclination of the suspension string, there is no possibility of significant horizontal force being introduced that will invalidate the ruling-span-based sags.

Long spans and short spans

The availability of comprehensive computer programs for solving all kinds of transmission line engineering problems has led to considerable discussion questioning the adequacy of the ruling span method and to examine other specifics such as using an FE calculation method in its place or at least to consider the need for limiting adjacent span ratios when the ruling span method is used, because not to limit span ratios invites unacceptable errors.

Some of this questioning results from an uncertainty or lack of understanding of the ruling span method, although there is some, but very little, legitimate concern about errors or approximations involved in applying the ruling span method. Small discrepancies have been found in some studies, and have been turned into unnecessarily conservative design criteria. Some of these criteria appear as adjacent span ratio limits. Unjustified application of span ratio limits can have a devastating effect on the efficient spotting of structures.

Engineers of one major utility had studied the issue of the ratio of the ruling span to the average span. Their work demonstrated that with a six-span series of 3 × 700 ft (210 m) and then three of about 2200 ft (670 m), a temperature rise of 120°F (49°C) will find excessive sag in one of the short spans of 11% or about 17 in. (43 cm). The ratio of the ruling span to the average span (RS/AVE) is 1.323, and this is considered to be important and a severe ratio. The insulator string length was about 9 ft (3 m), and the terrain was flat. Of great interest is the fact that the greatest error was not found at the short span adjacent to the long spans where the adjacent span ratio was about 3/1.

This artificially created scenario was designed to maximise the ruling span problem and find a possible extreme limit to the problem. However, if we adjust our own simple two-span series from 150 and 400 m to 150 and 450 m, we will have an average span of 300 m, and a ruling span of 397 m for the same, supposedly dangerous, ratio of RS/AVE of 1.32. Reworking our example with the spans and a 3 m insulator string length would find very small differences with the ruling span sag values, the short span error being almost exactly the same as the error found previously.

There is no doubt that the ability to equalise tensions when the conductor length increases (under high temperature and/or creep) will be slightly impaired, with short-span sags always being greater than those predicted by the ruling span method. The short spans will always behave as slightly shorter than the ruling span while the long spans will behave as slightly longer than the ruling span.

The subject becomes more complex if we realise that the short-span sags at some temperature such as 160°F (71°C) depend not only on the creep – which adds to the temperature rise problem – on the string length and on the weight span at the point but also on the temperature rise calculated from the temperature at which the line was sagged. The attempt to do something precise about the overall problem given these realities would require knowledge of the temperature at which the line is going to be sagged before the structure spotting can be done. In other words, accommodating such a calculation is impossible since the necessary data does not exist when you need it.

The second difficulty is that a control such as the proposed limit on the RS/AVE ratio can be applied only after a spotting is made and the average and ruling span determined and the weight spans are known at the structures at each end of the short span(s) in question.

Being aware, at this point, of the complexities of the subject on one hand and the relatively modest errors involved on the other, we suggest that the whole subject is of academic interest only, and has no impact on the real world of line engineering. Of great importance in the real world of line engineering is that all restrictions on span ratios be ignored unless someone can justify otherwise. So far, they have not.

We often get it into our heads that mixing long spans with short spans is bad. We noted that short spans are sensitive and easily subjected to tension changes whereas long spans are not. The longer and more flexible your insulator strings, the more equality of tension you will achieve and the more your line will behave (sag) like the ruling span method says it will.

Fortunately, it takes extreme line geometry for loads and sags of any damaging characteristics to develop. Typical tolerances of surveying, sagging-in, creep calculations, etc., override the calculable sag errors and tension unbalances created by long- and short-span combinations.

If you understand the topic thus far, you'll recognise that short spans, high tension and short insulators can unfortunately combine when all are present to produce conditions more like the 'extreme geometry' noted above. The extra ingredient necessary to cause a problem is something that will generate a tension change. Barring the breakage of something, a large tension change typically can only come from large ice loads, large creep or large temperature changes. This is why the issue of high-temperature line operation has been such an important topic in recent years.

Before the 1970s, very few utilities felt compelled to heat their lines sufficiently for this to be an issue. By the turn of the century, running lines hot as an intentional capability had become reasonably standard fare. The boundaries of the playing field have, in effect, been moved, and premises used to establish our 20th-century assumptions do need revisiting when that happens. But, the exercises provided here are meant to guide you in that review, and we expect you will find that some points of change that you identify are of academic interest but are otherwise not worthy of inserting into your actual day's work. Know when to do the fancy work and know when not to bother.

We would ask you too to understand that long-span and short-span combinations are not inherently a problem. When you understand that the combination of short spans, big tension changes and short insulators are what it takes to aggravate the problems of combining short spans with long spans, you should be able to translate that into a central point.

Short spans and high tensions create very small slack values for the spans, and short insulators inhibit the ability to easily transfer small amounts of conductor between spans to equalise the rapidly changing tension unbalances between the spans. It should be easy to see that the use of overhead ground wire clamps and post insulators pretty much denies the slack transfers and tension balancing that the ruling span method assumes. The central point to understand is this: the ability to transfer slack is the issue. The question becomes: *Can you transfer slack and is it there to be transferred?*

The unit curve

> When I first got into this business, there were no computers available to me and calculations were made by hand or maybe on a modest calculator. When I wanted a sag–tension calculation done, I sent data to a vendor (Alcoa or Alcan, I don't remember which), and they returned a printout response some days later for a fee of $600 per run. How times have changed! It was not an exercise easily done or funded, so I explored a different path and discovered the Thomas chart in the *Standard Handbook for Electrical Engineers* – 'Fink And Beatty' (the latest edition is the 16th, 2013). At the time, I remember it being also called the 'unit curve'.

In Figure 3.6, the unit curve plots stress versus strain for a cable suspended as a catenary shape between support points at its end. The key feature of the plot is that as the strain value gets smaller and smaller, the effect on stress is exponentially greater and greater.

Recall that the rate of slack is expressed as length/length (unitless), as is strain. Therefore, Figure 3.6 plots stress against the rate of slack. If you have a specific cable and span in mind, the plot's units can be converted from stress versus strain to tension versus slack, and the design-specific relationship remains identical.

For any span, the tension and sag change across the range of temperature applicable to the location and the conductor's use. This means that the conductor's rate of slack

Figure 3.6 The unit curve

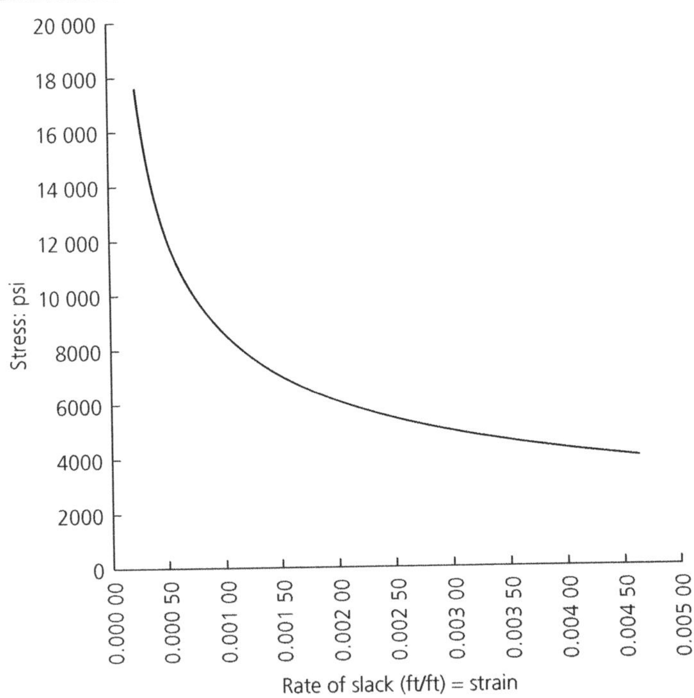

(strain) changes within a specific range – being smallest at the coldest temperature and largest at the conductor's hot, design, temperature.

The point of seeing the tension–slack relationship of a cable in a span in this form is to have a tool for recognising when your design can get into trouble by having a volatile sag–tension relationship. When sags are small, either by way of the spans being short or the tensions being very high, the slack and rate of slack will be small, and your design will reside towards the left side of the plot. There, small changes in slack and the rate of slack result in large changes in tension. There, a small error in sag during installation sets up a tension quite different than intended.

If you 'blow' the installation sag by 10 cm when the sag itself is only 1 m, the error in sag and tension is 10%. If you 'blow' the sag by the same 10 cm when the sag is 5 m, the error is only 2%. For any design exercise (project), you can easily and should plot Figure 3.6 for the project's range of temperatures to see visually what you are creating. This is easy to do with values available from a sag–tension calculation covering the full range of conductor temperatures. Be a bit conservative by plotting the initial values. Without trying to imply an exact requirement, your design should reside within the safe zone shown in the horizontal (solid-line) box to the right in Figure 3.7, and stay out of the vertical (dashed-line) box in the upper left of the figure. A design in the vertical box

Figure 3.7 Safe tension zones

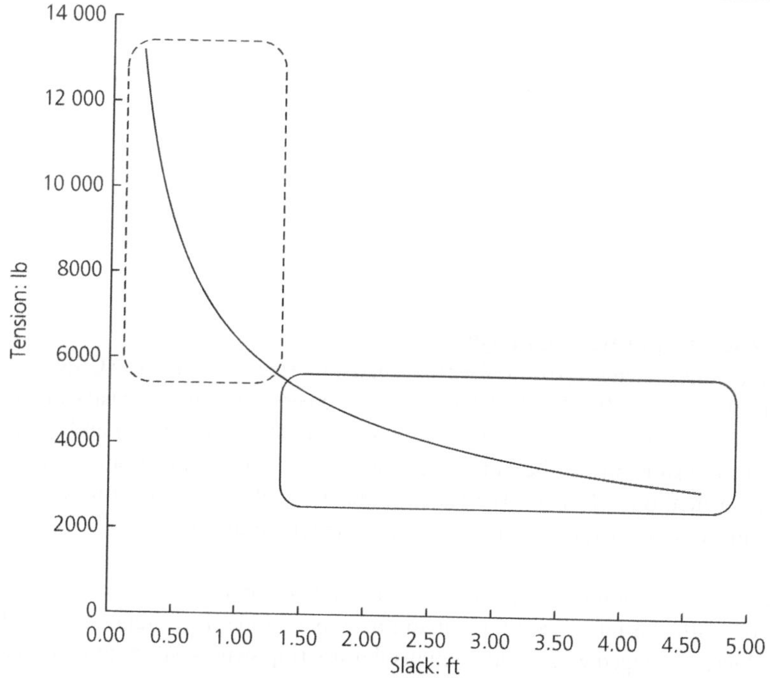

invites unwanted surprises in tension. The discerning boundary between what will work well and what will invite trouble can be defined by the slope of the curve.

It should be recognised that lower tensions on short spans offer great protection against the troubles noted here and at no meaningful cost. This is because the sags on short spans are small, and increasing them by large percentages to reduce tension significantly and to increase slack leaves you with sags that are still small. The added costs associated with the slightly taller structures needed to accommodate the larger sags and lower tensions are modest compared with the potential cost of the trouble avoided.

Look at this subject another way. When you have a design in the vertical (dashed-line) box of Figure 3.7, you have a design that resides on the far left side of Figure 3.3 where sags hardly change and the tensions are very volatile (out of control). A design in the horizontal (solid-line) box to the right in Figure 3.7 is the same as having a design in the right half of Figure 3.3 where sags are significant and under control, as are the tensions. A manageable change in sag causes only a modest change in tension.

> Several years ago, I investigated a distribution line (short spans) line failure. The line, although designed with very low tensions as measured by the percentage rated tensile strength had minuscule slack in its very short spans. It was designed to −30°F (−34°C). On the first night when the temperature dropped to −42°C, many of its deadend hardware clamps failed under tension. The shrinkage of the conductors with dropping temperature was a change in length comparable to the slack in the spans. Sending slack to zero invites a tension increase to infinity. That is impossible, and when the sag and slack are all gone, the tension change must come from the conductor's elasticity. The tension skyrocketed that night. There is real value in understanding your design's nature in the terms expressed by this 'tension–slack' relationship.

A close read of the tension limit footnotes in the US and Canadian safety codes shows that it has been understood for a very long time that high tensions on very short spans are very problematic, but the limits in these codes alone will not protect you from the possible problems. It will be good for you to understand this very well.

Creep and temperature change

The unit change in length (strain) for 50°C on a 45/7 steel-reinforced (ACSR) conductor is about $50 \times 20.7 \times 10^{-6} = 0.00104$ m/m. The creep strain for 10 years on the same 45/7 ACSR conductor is approximately 0.001 m/m. Thus, both creep and temperature rise produce distortions in the same direction and of similar magnitude. In theory, the ruling span method will invite the fewest unbalances if the temperature at sagging is about midway in the expected range of operating temperatures.

However, the inclusion of creep will, in effect, shift the balance between shrinkage and expansion with the inference that sagging-in is best done at a higher temperature. If the conductor is sagged when cold, then insulator string swings will be at their maximum when high temperature adds to the creep effect. In practice, we have no control over the sagging-in temperature, and the finer points of the behaviour of a series of spans as described here are not in our control. This describes one more approximation that we must make and accept in our work.

Wind and weight spans

Loads on wires, including their own weight, are the major contributor to loads on structures. Wire loads are applied to the structure through the wire attachment points. Lateral (wind) loads are applied to a wire's 'wind' span and vertical (ice and wire weight) loads are applied to its 'weight' span.

The wind span h is defined as the sum of the two half-spans on either side of a structure or, restated, as half the sum of the two attached spans. This definition is only useful when we apply the most typical load to a structure – that being wind applied uniformly and perpendicular to the entirety of both spans attached to the subject structure. Wind span is a fixed value until you move a structure (change the span lengths). We also note that we routinely consider the wind span to be a value independent of the shape or sag of the conductor (i.e. independent of C, the catenary constant). Should we consider the crazy notion that the wind may blow on the actual shape of the wires and in complex forms rather than perpendicular and uniformly along the entire span? Should we revisit the subject in more complex detail? That would be crazy – well, actually just difficult!

The weight span v is the distance between the low points of the spans on either side of the structure. The distance can be negative, meaning that the low point is outside of the actual span, and the span imposes uplift on the structure. When the weight span is negative, the wire is putting an upward force on that structure. If the terrain is flat and all supports are of equal height, the weight span v will equal h because the low point will coincide with mid-span.

But, the weight span is not fixed on sloped spans. It varies with changes in sag due to creep, temperature, wind and ice load changes. If the structure under study supports the conductor at a higher point than the adjacent support points, the low points are beyond the mid-spans so that v is greater than h. Conversely, if the support points for a structure in question are below the adjacent structure support points, the low points are closer to this structure, and v is less than h. Both of these conditions pose special problems because the position of the low points will shift with a change in C values; that is, the weight span v on these higher and lower structures is very much a function of temperature, loading and wind.

When doing tower spotting, a designer generally looks at more than one sag 'curve' (C value). At the very least, you look at the maximum temperature and/or iced conductor curves for ground clearance checks and at the 'cold' curve for uplift checks. There is a very handy relationship that makes this often onerous chore of manually calculating various C values very easy.

Without explanation, here is the formula and an example. Assume you have the hot curve plotted, and the curve is defined by a catenary constant $C = 5000$ ft. At a structure of concern, you can calculate the wind span h as the average of the two adjacent spans. Say, $h = 1400$ ft. Also, you can see the two low points on the 'hot' curve, and literally measure the weight span as, say, $v = 1000$ ft.

The question is, 'What is the weight span at some (any) other condition (sag)?' The relationship between C and v is

$$\frac{C_1}{C_2} = \frac{h - v_1}{h - v_2} \qquad (3.19)$$

where C_1 and v_1 are for the known condition, and C_2 and v_2 are for the condition with unknown weight span v_2. Therefore

$$v_2 = h - \frac{C_2}{C_1} \times (h - v_1)$$

Using the example numbers, and assuming $C_2 = 6500$ ft, as obtained from a sag–tension calculation:

$$v_2 = 1400 - 6500/5000 \times (1400 - 1000) = 880 \text{ ft}$$

The computer design program **TLCADD** used this formula and was the basis for the program's easy management of many load cases (varied C values). This formula lets you avoid plotting all of the curves. Instead, plot the hot curve and calculate the weight span values for all other conditions. But take care . . .

The ability to calculate the change in v under the condition of applied transverse wind that moves the conductors out of the vertical plane has only been practical since the advent of 3D line models in computer programs. The low point on inclined spans of wire moves along the span away from the higher attachment point when the wire is blown laterally by wind further and further out of the vertical plane.

Structures with attachment points above its two adjacent structures (i.e. structures that already have the larger weight spans) end up with even larger weight spans to carry during 'blowout' conditions than is calculated assuming the wire is in the vertical plane by Equation 3.19. Conversely, structures with low v values find their small weight spans become even smaller under blowout conditions. The use of pre-3D computer 'templates' ignored this feature, as did the 2D computer program TLCADD. The 3D computer program PLS-CADD allows the option of including or ignoring the fact. Test the difference there.

If you ignore the effect of blowout on the weight span due to choice or lack of analysis tools, please be conservative with structures on high points of the line carrying large weight spans and with the blowout and electrical clearance calculations at structures carrying small weight spans. Since insulator swing is greatly affected by the wind-to-weight span ratio – high wind combined with small weight creates huge swings – designers often limit the wind-to-weight span ratio as a structure design constraint to control conductor swing towards the structure causing flashovers. In rough terrain, this limitation can be devastating to the cost of the line.

Never use a support structure design with wind-to-weight span ratios efficient for a flat ground design when doing a line layout for a rough terrain design. The rough terrain demands much greater weight spans for a given wind span (actual spans) than does a flat ground design to be efficient. Failure here leads to too many towers and lots of weights or deadends to mitigate extreme high and low wind-to-weight span ratios.

Another way to manage the wind-to-weight span ratio in steep terrain is to force the sag low points to stay within their own span under all loading conditions by lowering tensions. This will often lead to fewer deadends without much cost increase because there is often excess clearance space available in this type of terrain.

Measuring sag

There are three ways to sag a cable suspended between support structures: measure its tension, measure its wave speed when a wave is invoked and measure the sag directly. Since tension and sag are in a direct relationship per Equation 3.5, measuring tension can be equivalent to measuring the sag. All that is required is for the dynamometer used to measure the tension to be reasonably accurate.

> We once reviewed the sagging reports submitted by an inspector on a job that was being sagged by a dynamometer. Oddly, every report showed that every section of the line was sagged perfectly 'to the pound' noted in the sag tables. Suspicious indeed! Did it have anything to do with the fact that it was raining nearly every day, all day? So, we went out to check the sags by surveying the sags. We found them to be in error by up to 30%. It seemed that (1) the inspector had no interest in getting wet every day, so he sat in his truck and recorded perfectly fictitious results and (2) the dynamometer was not working well at all. In fact, it seemed to be stuck on a single reading as if a stone was lodged within somehow. The work got revisited.
>
> On another project that was sagged by dynamometer, the contractor proved that his work was good by taking a photograph of the dynamometer as its needle showed the desired tension. That of course does not mean anything at all with respect to the actual tension in the cable, since the dyno could be way out of calibration, and the photo just shows the wrong tension. Silly!

Measuring sags by measuring the tension with a dynamometer requires that the dynamometer is calibrated well at the very least, and that requires attention and work – not faith. It also requires placing the dynamometer in a location that measures the tension provided for construction.

The stop-watch method involves jerking the cable at one end of the span and measuring the time for the wave to travel along the span. If the span is short or strung to a high tension, the time is short, and several runs of the wave along the span are timed and the overall time is divided by the number of times that the wave travelled back and forth. The formula for the sag is

$$\text{sag} = 48.3(t/2N)^2 \text{ (inches)} \tag{3.20}$$

where t is the time in seconds and N is the number of return waves.

Some people swear by the stop-watch method, but you should review the sensitivities of the answer to the input and the need for the accuracy. If you have a design that hugs the dashed-line zone of Figure 3.7, then this indirect method of measuring sag may be less suitable than others. It does seem that, if the lay of the land allows, the best way to ensure the best (most accurate) results is for the sag to be measured directly, since this can typically be done with an error of less than a few centimetres, and a few centimetres usually translates into only a few percentage points of sag and tension error on reasonably long spans.

There are three sags that we speak of for a span of cable: the mid-span sag, the maximum sag and the sag at the point where a line parallel to a line drawing between the spans end point attachments touches the cable. Figure 3.8 illustrates. On a level span with the end point attachments at the same level, these three sags are all one and the same. Otherwise, they are all different from each other. As spans become increasingly non-level or 'sloped', it is necessary to pay attention because measuring the wrong one will lead increasingly to errors.

With a bit of thought, you can see that none of these sags can be readily identified in the field by the crew charged with measuring it. They will find it handy to measure some other sag that is based on one of these values. The **PLS-CADD** software produces the

Figure 3.8 Various sag definitions

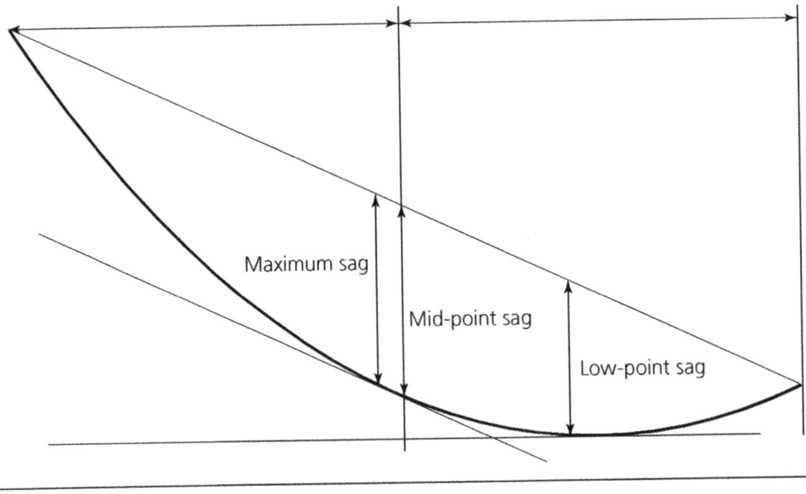

mid-span sag as its output value. Unless the spans are fairly level, this is a sag that cannot be discerned in the field, and it needs translation to something the field can use. Knowing that the presently popular computer program PLS-CADD delivers the midpoint sag, you can see that a translation is required in steep terrain. Winkelman (1960) can help you do that.

A somewhat obscure formula of possible use for checking sags on existing lines requires the visual sighting of a line tangent to the wire's sag belly onto the two supporting structures. Not level, just tangent to ... The sighting line strikes each structure a distance of h_1 and h_2 below the wire's attachment points. The sag is calculated as follows:

$$\text{sag} = [(\sqrt{h_1} + \sqrt{h_2})/2]^2 \qquad (3.21)$$

This little formula requires that you climb only one structure – the easy one – to estimate or establish a sag in the span. Cool, eh? Just trying to help! It is only really useful if the sag is large due to the span being of reasonable length. Set up at the base of one support, with the sight line being a calculable distance via the drawings below the cable attachment point above. The equation calculates the distance below the attachment point at the other support structure to the sight line. Climb to that point on that structure to provide the sight line target. This only works if the sag is big enough so you are not above the top of that structure.

Sag–tension calculations

Recall the quote in our preface on evolving conditions – the yin and yang of things. In the 1970s and earlier, in North America, one had to have Alcoa's (Aluminum Company of America) sag–tension program SAG10 or one's own program, or had to forward line data to Alcoa or another conductor provider to have them run a (single) sag–tension run for a hefty fee. It was tedious to the extreme compared with the 1990s and beyond. If an engineer was not working for an experienced company or self-educated to the max on the subject of bi-metallic conductor behaviour, dependency was necessary, tricky, slow and rather expensive.

> In 1985, I recall saying to a friend, 'I can't wait until my PC has at least 512k of memory so that I can run my own sag–tension calculations on it.' This is amazing to me because, in my head, 1985 was just 'the other day'.
>
> I recall, from the same era, that I once input an unusual constraint – not an unacceptable constraint, just a rarely used one – into a version of a sag–tension program. In the output, I got negative sags. Awesome, if true, I thought. Now my towers can be very short and very far apart. Turns out that it was not true, and the program was 'not so wonderful'.

SAG10 was developed by very bright minds at Alcoa back when that company sold electrical conductors. They retained the serving and development of the program for several decades after they stopped selling overhead electrical conductors. In 2008, Alcoa sold the software to another conductor manufacturer, Southwire. In a world where there are competitors to Southwire that sell unique products that compete with their products, it should be necessary to watch for lack of product coverage and perhaps lack of accurate representations of competitor products in the program going forward – not through malice but through lack of concern or understanding of the nature of competitors' products.

In the 1980s, Alcoa's SAG10 program became readily accessible on personal computers and even got embedded in the line design program of the day – TLCADD. In the 1990s, its identical 'engine' became embedded in the very successful and widely used line design program PLS-CADD. By way of this embedment, sag and tension calculations became trivial exercises that occurred instantly anytime you wanted during the line design process for PLS-CADD users.

It also meant that the calculations became a 'black box', and too few engineers can any longer tell you what the calculation actually is. Trust became blind trust, and a bad answer became nearly impossible for a less than wise engineer to recognise. High-quality (valuable) results are not intrinsically linked to computerised methods of working. The computer calculations do not yet flag poor choices.

Additionally, the embedded method is simply a method that mimics real behaviour by a particular and unchangeable set of assumptions. If a better, more realistic calculation is developed, it will need to be the embedded method, replacing the current embedment. Such a change will have lots of opposition. Sadly, that better calculation does exist, and has since about 1980.

Sag–tension calculation methods

There is a wide range of methods or forms of calculations to determine the sags and tensions of conductors as temperature, ice loads and wind speeds change. CIGRÉ (2007) describes these, and we are not going to reiterate these descriptions here.

We will compare results that the various methods produce and most importantly, we will talk about limitations that various methods have with the new types of conductors that have come onto the market in the last 10 years.

As with any calculation method that scientists or engineers develop, it is usual that the boundaries within which the solution is presumed to be and is sought are defined by the subject matter (products) of the day. Some of the modern conductors that are now on the market have characteristics outside of long-standing boundaries. The result is that a calculation method developed prior to their existence may not represent them well. This is the case with some sag–tension calculation methods for some conductor types.

We dare say that if you are using a combination of calculation method and conductor type that we describe here, you are probably not getting a result that is either close to correct or close to an efficient use of the conductor. CIGRÉ (2007) describes three approaches to modelling the stress–strain properties of conductors.

It is vitally important to understand the differences and shortcomings of each model, with a focus on the actual behaviour of the conductor being modelled. Any model that simplifies a non-homogeneous (i.e. core material different from the outer strand material) conductor with a homogeneous model will misrepresent the conductor's behaviour in some way. This is common practice in many countries. The inaccuracies resulting from this homogeneous modelling of non-homogeneous conductors will result in significant errors when the conductor is far from homogeneous, with core material behaviour much different from the outer strand behaviour. Most high-temperature low-sag (HTLS) conductors fall into this category.

SAG10 is a computerised mimic of a graphical method known as the Varney method (Aluminum Company of Canada, 1950). The method was the first to use curves in lieu of straight-line approximations of the three moduli of elasticity of interest. The curves are derived from standardised stress–strain test procedures The essence of the method is to determine the load sharing between the aluminium (or copper) and the core material. If you consider the standard stress–strain (or, more precisely, the load–strain) plot as per Figure 3.9, you must understand that the two materials plot basically through the origin of the plot at the temperature when the test was conducted. At any other temperature, the length of each material changes in accordance with its coefficient of thermal expansion. Thus, at a temperature other than the test temperature, the lengths of the two materials change at different rates, and the load sharing between them changes accordingly. Tension sharing at other temperatures is determined in this method by shifting the core and aluminium plots away from their test temperature locations by the strain amounts caused by the temperature change between the test temperature

Figure 3.9 Stress–strain curves for a 2.18 cm (26/7) ACSR conductor. (From Nigol and Barrett, 1980)

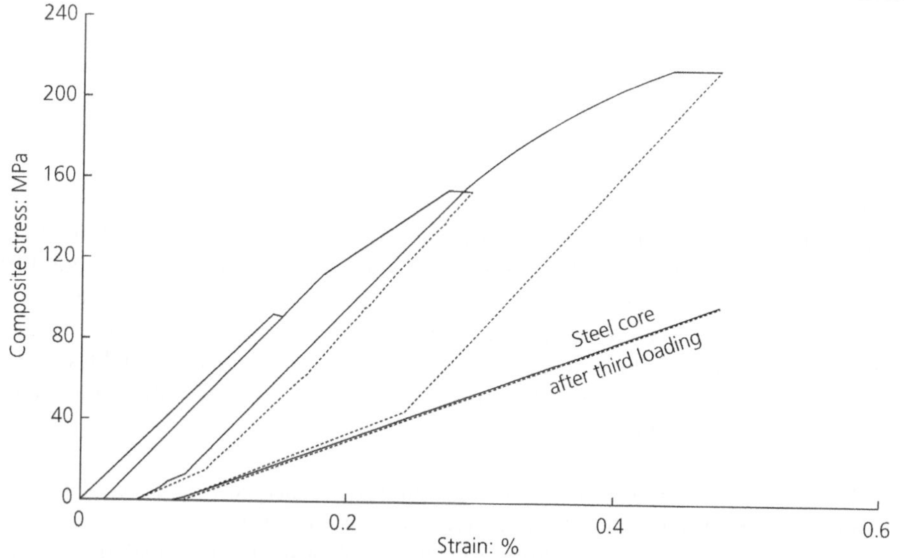

and the temperature of interest. You can imagine that this would be really tedious to accomplish by hand – but yippee for computers!

> Personally, all of that plot sliding leaves me cold, and I admit that I trust that it works. After all, 'five trillion flies can't be wrong' they say as we watch them feed on ... poop. Sadly or gladly, I entered the business at a time and location where another very different sag–tension calculation program was being introduced to the business. From it, I got my education on how two-material conductors work.

The Canadian Electrical Association (CEA) undertook a study in 1978 to develop a sag–tension computer program that would more correctly predict the sags that appear to occur in bi-metallic (usually aluminium and steel) conductors at high temperatures. The idea of running conductors to new high temperature limits began in the 1970s in response to the growing difficulty in being able to find rights-of-way for new lines. It was becoming necessary to push more amperes through the existing lines, thus increasing their operating temperatures and sags to new maximums. Some line owners were seeing that the sags at previously unseen temperatures were larger than expected or predicted by the existing calculation methods.

Generally speaking, calculation methods are developed in response to a need, and that need can be defined as having boundaries to its range of interest. So, very often, these boundaries get crossed because the range of interest changes over time. All too often, the calculation method fails to represent reality outside of its original boundaries. Such is the case with these sag–tension calculation methods. Watch for it elsewhere.

The CEA produced a three-part report (Nigol and Barrett, 1980–1982). The computer program is called STESS and sadly may disappear into the history books because it is not embedded into a line design software program where the Varney mimic holds the high ground as an incidental calculation. We highly recommend a read of Nigol and Barrett (1980–1982), especially Part II as it provides a great deal of insight into the behaviour of bi-metallic conductors such as ACSR conductors, and that behaviour's impact on sags and tensions.

Stress–strain curves

Figure 3.9 has interesting information to show us. The light line with small, hollow circles along its length laying above all those other lines is the path plotted by the laboratory exercise of applying load to 30%, 40%, 50% and 70% in intervals, holding the tension at these loads for a prescribed time, relaxing the load back to near zero and then repeating the exercise to the next higher load value. This loading pattern and test methodology is defined by an American Society for Testing and Materials (ASTM) standard, or some other nation's equivalent.

The load is held for a period of time at each 'limit' then released before reapplying and loading to the next higher limit. The exercise stops at 70% (or higher with some non-ASTM standards) because the laboratory folks don't want to risk breaking the wire and smashing all of their equipment. Over time and between standards, the tensions at which the holds are made have changed.

The smooth, heavy line is drawn through the series of 'after having been held' points, and is the industry's standard source for *initial* stress–strain values. These laboratory tests actually measure load and strain, not stress and strain. Some results are presented as stress versus strain, and some are presented as load versus strain. There is nothing to distinguish the two for our purposes, and the expression 'stress–strain' is accepted as applying to either. The pattern that tends to generally repeat itself on the release and reloading parts of the exercise provides the industry's source for *final* stress–strain data.

When combined with separate creep test data, the wire's entire stress–strain and creep characteristics are defined for the purposes of making sag–tension calculations. In Figure 3.9, the wire's *initial* stress–strain relationship is represented by the bold, upper curved line. *Initial* is defined as the stress–strain relationship when the wire is subjected to a load for the first time, and after the required hold times of generally 1 h. *Final* is defined as the stress–strain relationship for all loads less than the largest load seen to date.

The heavy *initial* line is drawn through the 'after having been held points', since it is reasonable to assume that a real installation of wire will be subjected to loads more slowly than the rate of application in the laboratory. If laboratory loads were applied slowly and at a uniform rate, the initial curve would track below the uppermost plot created by a fast application of load. It is the 'after having been held for an hour or so' curve that is universally used for initial condition sag–tension calculations.

Keep in mind that this curve will fall further and further away from the fast application curve as the load is applied more slowly, albeit at a much decreased rate over time. Still, the initial curve is a decent approximation, but only an approximation, of a wire's condition at the time of being sagged and clamped. The actual time that a conductor is held in the travellers varies at a bunch of tensions for some varied period of time. The tensions can also be looser than the eventual sagging in tension or not, and the time can vary considerably from an hour or so to a few days. Yet, some of us hang onto the third significant figure with our decision-making.

> Some years ago, a contractor in Texas asked if they could leave the conductors in the travellers over the weekend. The conductors were not clipped in yet, but they really wanted to get away for the weekend. The owner allowed. Over that weekend, an ice storm hit the line with the conductors in travellers, before clipping in. On Monday, I was asked if an adjustment needed to be made to the sagging tables because the conductors may have moved off their initial values.
>
> I used the program STESS to answer the question, because that program has a time component, which means that the load cases are entered as a history of events with a timeline. This was a type 13 conductor, 1590 kcmil (800 mm^2), and the effect of the time and ice load was trivial. I said, 'carry on as is'.

There is another reason not to attach great accuracy to these curves. The initial and final curves that are published by the industry as representing a conductor type are created from interpolations from or averaging of many tests. They are not the results of single tests. All tests of a type of wire do not plot identically with each other. There is some scatter to the results. So, your wire should be expected to behave somewhat like the published data defines. Do not expect exactness.

There is much less exactness to the initial curve data than to the final data. The initial data depend on the behaviour of the core material and the aluminium stranded onto it acting in combination and at a time before the strands have settled into their final positions due to applied tension and due to factory equipment settings and conditions, etc. Much of a line's design decisions are based on the conductors' final sag–tension results. Unless you pre-stress (pre-tension) the conductors to very high loads during installation, the final condition values depend on the initial data's separation from the final data. You are almost always held hostage by the initial data. Did we already say, 'don't expect exactness'? Yes, we did – twice!

Let's talk more about *initials* and *finals*. Follow the path of the plot in Figure 3.9 as load is applied, released and reapplied. From this, you can see that the initial condition is only in play when the conductor is seeing a load *for the first time*. At all other times the behaviour of the conductor's sag–tension relationship lies along the final data, with that data connected to the initial data plot at the maximum load seen to date.

Suppose that a conductor is strung onto a series of structures and for some reason sees a tension greater than the tension that occurs when the conductor is finally clamped in. This can occur due to intentional pre-stressing or by some action taken by the stringing crew or because the line was left overnight in travellers, the temperature drops overnight and the tension rose in response. If this occurs, the sagging table information should correctly come from the initial curve for tensions above that seen prior to clamping in and from the *final* curve for tension below the maximum tension seen. Ask yourself where your sagging tables are getting their information. It is not likely to be from the sources we describe as correct. Before you panic, remember that not all discrepancies that exist in principle are a problem in reality.

Final sag–tension data has two very specific and independent sources. Recall that we have said that the initial curve data is developed from the 'after being held for an hour or so' points on the stress–strain plot. And that, if the time held was very short, the initial plot would be higher on the plot and further from the final plot data. Or, if held longer would be lower on the plot and closer to the final data. The creep curve plots lower than the initial data because it is meant to represent the case when the conductor was held at that tension for 10 years, never mind 1 h.

The *after-creep* condition case for a *final* sag–tension result assumes simply that nothing has ever happened to the conductor except that it was subjected to one single temperature, ice and wind condition for 10 years. The environmental condition that you should use for the *after-creep* calculation is the mean, windless and iceless temperature that you think the conductor will experience most often over the 10 years. If the line will be lightly loaded in the long run, and if the sun rarely shines on it, this will be the average annual temperature of the location. If the sun shines a lot or if the line is often heavily loaded, this temperature will be higher than the average annual ambient. Here again, we are having to make an estimate of a 10-year event that the output of a computer program will express to the nearest 1/100th of a metre, and tension to the nearest newton.

If computers were not so darn productive when asked, any reasonable manager would fire them for being so disruptive to the practicality of the work at hand!

The *after-load* condition for a *final* sag–tension result ignores the issue of creep and takes the tension to the declared value established by the described weather condition – usually one with a low temperature, high wind or heavy ice, and offers all results off the final (after load) plot data. The program SAG10 looks at both *after-creep* and *after-load* final calculations, and provides the output of the one that caused the greater sag increases. If the program's output report says that 'creep is not a factor', then the results you see are those of the large applied load. If the program declares that 'creep is a factor', then the after-load tension failed to develop sags greater than those developed by creep. That program does not simultaneously provide results for both the after-creep and after-load conditions. Too bad!

PLS-CADD, on the other hand, offers sets of both *after-creep* and *after-load* final results, and you get to review the data and make your own decisions.

Strain summation

The discussions above on sag–tension calculation methods and stress–strain curves illustrate that our industry has at its disposal linear methods and non-linear methods for doing the calculations. Clearly, the actual stress–strain relationship that conductors exhibit is non-linear, and the linear methods are going to provide only rough approximations of the actual behaviour compared with the non-linear methods, and their approximations can be very poor for certain conductor types or under certain conditions. The following discussion is based on the non-linear methods in order to enhance your understanding of conductor behaviour. If you happen to be working in an environment that uses a linear method, you might try changing that for the benefit of your work's quality.

The CEA program name STESS (Nigol and Barrett, 1980) is an acronym of 'Sag-Tension Evaluation by Strain Summation'. The program's logic is worth discussing because it provides a level of understanding of conductors that is otherwise hard to find.

By definition, the *amount* of each material that makes up a conductor (the mass) in a span between two clamps (structures) is constant. Some conductors are made of one material only (copper or aluminium), but many are made with two materials – the aluminium (or copper or even steel) conductor and a core material (usually steel, but not always). Even though the *amount* of material is always unchanged in a span, the *length* of each material in that span will change for various reasons such as creep, thermal expansion/contraction and elastic (tensile) elongation. Each of these 'change in length' actions is expressible as *strain*. The logic of the program is enlightening and worthy of your study. It is also worth pointing out here the many ways in which conductor stress–strain behaviour will never be calculable to our nemesis – the third decimal place.

We tip our hat to Dr Steve Barrett, recently retired from Ontario Hydro, for this fine work.

The identified mechanisms causing strain changes in the two metals are slack, strand settlement, creep, thermal strain and elastic strain. The list is the same for both metals, except that slack does not apply meaningfully to the core. These five mechanisms are described/defined below to a degree useful to our purposes. For more, read Part II of the report by Nigol and Barrett (1980–1982).

Expressed in words:

> the change in strain (length) for the various reasons in the two materials must be equal to each other

Expressed in math:

> slack + strand settlement + thermal strain + creep strain + elastic strain (all on the aluminium)
>
> must equal
>
> strand settlement + thermal strain + creep strain + elastic strain (all in the steel core)

The equation is solved by seeking the tension in the two metals that equalises the sum of strains under the condition of the moment. One of the factors affecting the creep strain is time. This introduced time as a variable in the sag–tension calculation and opened the door to queries on the conductor's sag–tension relationship at times other than right now and 10 years from now.

NOTE: There is an opportunity for an eager engineer to bring the STESS program up to date with a useful I/O interface and bring its powers to our industry. Sadly, STESS was written in the days of mainframes, FORTRAN language and punch cards. Versions do exist for PCs but the I/O interface is cumbersome.

Slack – the other kind

Look again at Figure 3.9. Notice that the origin point of zero stress does not coincide with zero strain. In fact, the ASTM laboratory exercise 'zeroes' the strain at 8% rated tensile strength (RTS). Our search for accuracy is constantly sabotaged by practical constraint. We therefore don't get good information below 10% RTS or above 70% RTS where the tests stop. Operating a conductor outside the tested range puts you into uncharted territory. Same old theme, isn't it?

In laboratory work, the conductor is pulled to 8% RTS to get the kinks and idiosyncrasies out before attaching the strain gauges. The process hides the details of what

Figure 3.10 Stress–strain plot. (From Nigol and Barrett, 1980)

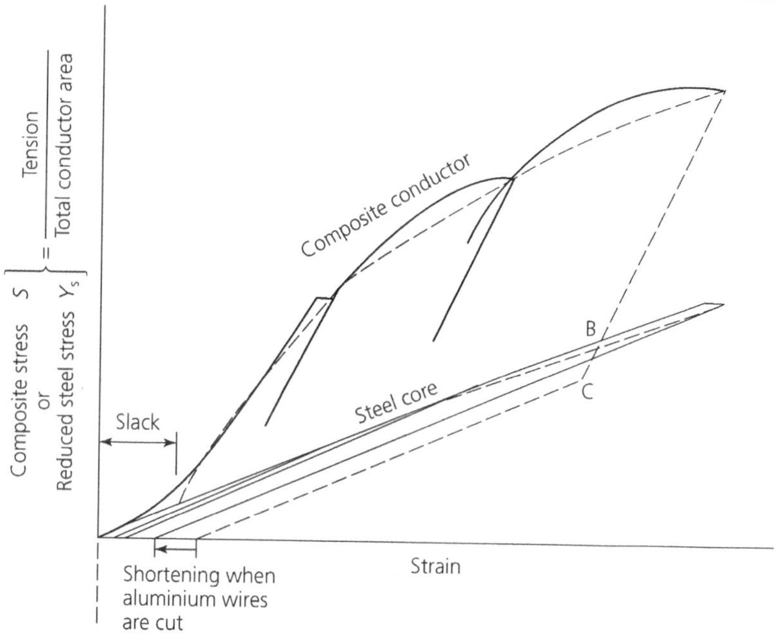

is actually happening in this corner of the stress–strain chart. In fact, stranded wire can have an inherent looseness to its stranding and layers depending on the quality or intentions of the manufacturing process.

Figure 3.10 shows a typical 'hook' or displacement of stress–strain curve from the 0–0 origin of the chart. This displacement is called 'slack' in STESS. This slack is the same word as we have discussed above to refer to the length of wire in a span but has a different meaning here altogether. Slack in a conductor's construction refers to the looseness of the aluminium strands on the core, expressed as strain.

Slack applies only to the aluminium, and carries a zero value as its default in STESS. STESS was written by *research* types at Ontario Hydro's research division, which later became a separate company called Kinectrics. These were bright folks who liked to install the 'capability to explore possibilities' into all of their creations. STESS has features that *production* people would not bother with. This is one of them. Besides, quantified knowledge of this slack is never going to be available to the engineer on a project basis. To the degree that it exists, it messes with the hunt for third significant figure accuracy. Read Chapter 6 to see how this slack bit us on one occasion.

Strand settlement

The laboratory stress–strain test is applied within a time period of a few hours. Since the *initial* stress–strain plot is a curve based on a permanent elongation of the wire as the load is increased, we need to assign a cause for the permanent elongation (deformation). The action generally assigned to the elongation is called *creep*. However, metallic flow creep takes time to occur, and it cannot explain all of the deformation that occurs in the quick test. The elongation is more correctly understood as including a component called strand settlement or – when the stress is in the yielding range – elastic yielding.

All of the strands in a conductor follow a helical path, except for the central straight wire called the king wire. Each layer of strands is laid in the opposite direction to the one below, causing the strands in a layer to cross over the strands of the layer below. As tension is applied to the conductor, the helical strands try to straighten and head inward towards the king wire's straight path. In doing so, they can crush into the layer beneath. The diameter of the helical path of each layer gets shorter as the crushing occurs. The strands first soak up any looseness (slack) inherent to the wire's construction but then can indent into each other at the crossover points, allowing the helical path to become even smaller in diameter. This mechanical action is called strand settlement. Strand settlement is the shortening of the paths of the helical wires under tension, or it is describable as the lengthening (strain increase) of the strands. Strand settlement is irreversible, and only occurs each time a new maximum tension is applied.

The amount of strand settlement that can occur is also affected by the number and nature of the contact areas, by the strands' lay lengths and by the hardness of the metal. Flat, trapezoidal strand contact areas are larger than those created by round strands, thus developing lower stresses at the contact points and allowing less indentation. Harder alloys will indent less than softer alloys. Steel will indent less than aluminium, being a much harder material.

Creep

As noted above, this word in the context of STESS refers only to 'metal flow' creep. Figure 3.11 displays the fact that you get about a third of the conductor lengthening in the first few hours as you do in the first year and as you do for the rest of the life of the wire. Consider Figure 3.11 as a qualitative plot. Figure 3.12 comes from a particular conductor's laboratory creep test, and disagrees. It suggests that the first 24 h of creep account for about 10% of the 10-year creep.

Understand that creep is a complex mechanism that defies accurate calculation but which can be approximated well enough to permit efficient line design and construction. Outside of STESS's separated definitions of creep and strand settlement, we deal with the two matters as if they are a single matter under the incorrect title of 'creep'. Since creep, especially the strand settlement portion, can be highly variable between production runs, it is best to ensure that conductors of a bundled phase all come from the same production run and absolutely from the same production facility. To mix conductors of a bundle is an invitation to problems, as latent and unequal adjustments occur.

Creep as a function of stress and time seems to be a relatively easy concept to understand because, if the stress level is kept constant and the creep is plotted on log–log paper, the result is a straight line. See Figure 3.11.

However, there is difficulty in predicting creep even with this relatively simple relationship with stress and time. CIGRÉ developed a predictor equation for creep:

$$\text{creep} = k \times e^{n \times t} \times \text{stress}^{p} \times {}^{q/\text{stress}^{r}}$$

Figure 3.11 Components of creep. (From Nigol and Barrett, 1980)

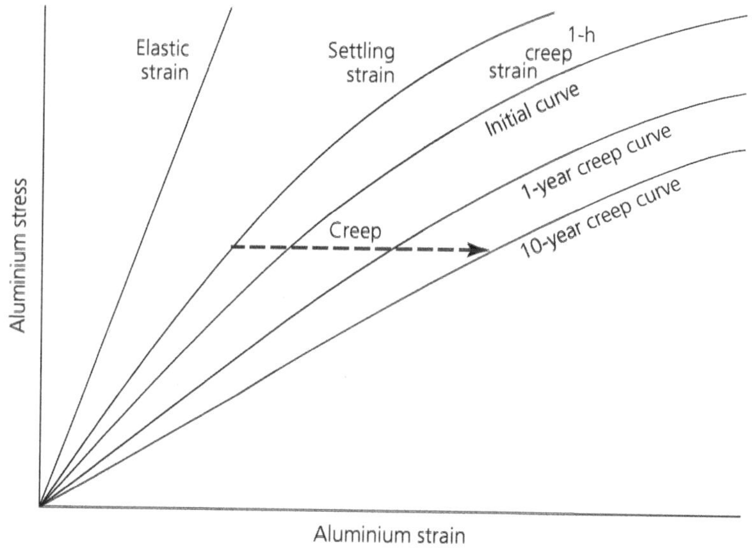

Figure 3.12 Typical creep test plot

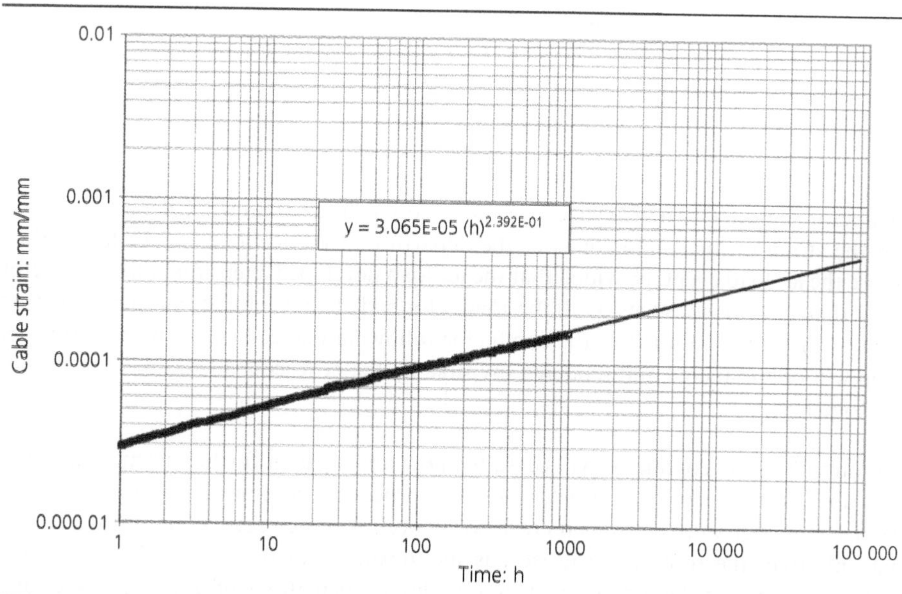

Table 3.3 Values for parameters in the CIGRÉ creep equation. (From CIGRÉ (1977))

Stranding	Steel: %	*k*	*n*	*p*	*q*	*r*
Cast						
18/1	6	1.2	0.023	1.5	0.33	0.13
54/7	13	1.6	0.017	1.4	0.38	0.19
30/7	23	2.2	0.011	1.4	0.18	0.04
Rolled						
48/7	9	3.0	0.010	1.9	0.16	−0.01
54/7	13	1.1	0.017	2.2	0.34	0.21
24/7	13	1.6	0.23	1.9	0.19	0.08
26/7	16	1.9	0.23	1.8	0.23	0.08
12/7	58	0.66	0.11	1.9	0.27	0.15

The values of k, n, p, q and r in this equation vary with strand count ratios and with the type of aluminium: cast or hot rolled; t is the time in hours. Table 3.3, from the CIGRÉ work, illustrates.

The point of displaying these values is to recognise the many anomalies that may result from the different methods used at the several laboratories that produced the data or, more likely, that the conductor samples came from different stranding machines, that lay factors were different and so forth. This is not to fault the quality of the CIGRÉ effort.

All else being equal, creep can be expected to decrease with increasing steel content. Thus, the many anomalies evident in the data of Table 3.3 raise some doubt about the validity of such complex formulas. The CEA goes into great detail to justify the particular creep formula used in STESS. There is more than one creep theory out there to read about and fall in love with. If someday you become an expert in creep, take issue with the formula and change the world to your liking.

One saving aspect of creep is that creep strain is not additive to other strains. A conductor subjected to a heavy ice load may adopt a large strain while the load persists, and the strain in the conductor after the departure of the ice load will be larger than before the ice load event, provided that the ice load developed a tension in the conductor larger than any previous tension. If the conductor had already adopted a creep strain that was less than the strain left in the conductor following the ice load event, the strain of creep will be smothered and rendered invisible by the post-ice-load strain. They do not add together.

Not only that, the conductor will cease to show any signs of further creep strain until such time as the creep strain, if left to its own processes, develops the strain left by the ice load. In other words, it is as if creep strain occurs in the background of load strains and only rules the roost when it exceeds the maximum load strain to that time. It is for this reason that the independent, non-additive calculations of *after-creep* sags and *after-load* final sags and tensions are a legitimate view of conductor behaviour. One or the other will produce the larger final sag, but they do not affect each other.

Having explored some of the details, it is necessary to back off and put matters into perspective for, in practice, sag increases resulting from creep are not very large. Consider the 10-year creep for aluminium (AAC) and aluminium alloy (AAAC) conductors at 90°C at 15% RTS. The calculated creep is just over 1000 μstrain or 0.1%. This is an extreme value for creep because the test conditions were extreme. No one could afford to operate a conductor at 90°C for 10 years due to the high cost of losses it would develop. Nevertheless, it is instructive to use this 0.1% to determine the limits of possible sag adjustment.

A 440 m span of AAAC conductor with $C = 1600$ m would have a sag of 12.5 m (from Equation 3.2) and a δ sag/δ slack ratio of (3 × span)/(16 × sag) = 6/1 (Equation 3.9). Thus, an extreme creep of 0.001 × 400 = 0.4 m would produce a sag change of 0.4 × 6 = 2.4 m. A more typical creep value would be of the order of 250–500 μstrain, producing sag increases of 0.6–1.2 m. To have assumed a creep of 400 μstrain would infer a range of sag error of about 0.3 m, or somewhat less than the buffer or allowance usually inserted to allow for sagging or survey errors. Thus, there is little need for attempting a precision that is not, in any event, possible. This realisation should drive crazy the many line engineers presently engaged in finely tuned clearance analyses to establish publishable line ratings.

There is a useful approximate method of allowing for creep. Creep can be rated in terms of a temperature change, because both temperature increase and creep produce a change in length proportional to length (strain). If the coefficient of thermal expansion of the conductor is, for example, 20.3×10^{-6}/°C for a 45/7 ACSR conductor, then a creep rate of 420 μstrain would be equivalent to $420 \times 10^{-6}/20.3 \times 10^{-6} = 21$°C.

If the actual conductor temperature during sagging is 30°C, the conductor could be installed with sag and tension data for 30 − 21 = 9°C, so that the final after-creep sags and tensions will be set in place, as desired. The use of this method requires only that you know the 10-year creep strain of the conductor in hand.

Thermal strain

Explaining and understanding this issue is relatively easy. Every metal expands and contracts with temperature change according to its coefficient of thermal expansion. These coefficients are considered constants, so the thermal strains in the metals are strictly a direct function of temperature.

A feature of the program STESS that plays into the thermal strain calculation is R – the ratio of the aluminium temperature to the steel core temperature. R allows you to impose a cooler temperature on the wind-cooled aluminium strands compared with the inner core temperature. In detail:

$$R = (T_a - \mathrm{AMB})/(T_s - \mathrm{AMB}) \qquad (3.22)$$

where AMB is the ambient temperature. The default value for *R* is 1.0, making the steel core temperature equal to the aluminium temperature.

This is of interest since more recent adventures into sag and tension and ampacity calculations are trending towards recognising the varied temperature across the radius of a very hot conductor. In fact, it is readily accepted that the temperature through the depth of the aluminium from the surface to the core is not constant. Some folks point out that the temperature of a conductor is not constant within a span or between spans given the wind shadowing of trees and clouds and variability of wind, etc.

Now, a lengthy aside

The proponents of such features of conductor characteristics note that the variances can be quite large. However, it is unlikely that there will be any industry-wide implementation of formulas to incorporate these points into our work so ... these are adventures into the domain of the third significant figure. Have fun!

It is so important to make the distinction that there are subjects that are rightly subjects of academic interest and worthy of academic study but that are not worthy of a role in practical engineering, given today's tools and methodologies for the real work. The plot of an experiment's relationship between a conductor's sag and temperature displayed in Figure 3.13 is a case in point.

Figure 3.13 Sag versus varying temperature plot

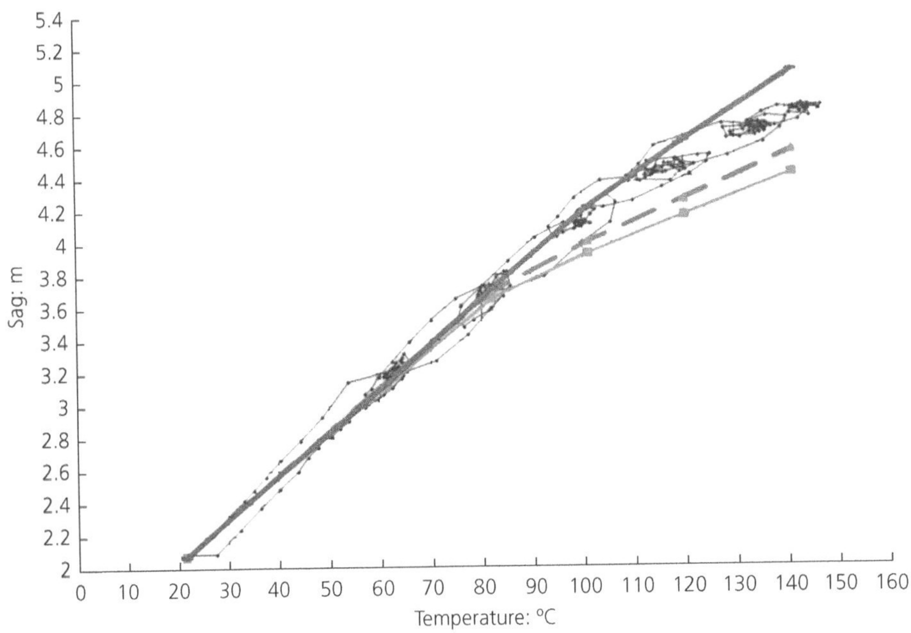

Ignore the several thick, straight lines and observe only the very loopy and squiggly thin line hiding underneath these lines. This line is the plot of sag versus temperature as the temperature is cycled through a rising and falling pattern. It points out that, at least in this experiment, the sag at any temperature can be about 0.3 m different, depending on whether the temperature is rising or falling.

If this sag difference is applicable to all spans in the real world or only some spans, or could be markedly different between different conductor types or for installations of the same conductor but at different spans or tensions, or if we learn which sag applies to the rising temperature and which does not and, finally, if we know whether the temperature is rising, falling or stable when an existing conductor is surveyed, then ... take a deep breath!

If, if, if... then maybe, just maybe, we could predict the sag in a conductor to an accuracy of better than 0.3 m – provided we also solve all of the other mysteries affecting sag. Until then, this and so many other things that we study and discuss endlessly are of academic interest only. Go ahead and study these things and have these discussions. You need to do so to put the actual work that you are charged to execute into a proper context.

Elastic strain

This strain element is also easy to understand. Like most materials, a conductor has an elastic strain response to a stress change within a range of stresses below which plastic yielding takes place. The elastic relationship between the stress and the strain is defined by the modulus of elasticity. Many conductors are made with two materials that have different moduli. This bi-material nature of most conductors is one reason this work is interesting.

We do not speak of a conductor's initial modulus, only of its final modulus – of which each of a conductor's materials has one of its own. The initial modulus that could be spoken of does exist, of course, but it is buried in other concurrent activity when it could be employed in a calculation. To understand it, we will describe the final moduli first.

Figure 3.10 shows a typical and simplified version of a stress–strain curve. The dashed line curving from the bottom left corner of the plot to the top right corner is the initial 'curve'. Because it is a curve, we cannot assign it a simple modulus value to describe its slope. The dashed line then runs in a straight line from the top right position to the point labelled 'B'. That section of the stress–strain plot defines the final modulus of the composite (two-material) conductor by its slope. From point B back towards the bottom left corner of the plot, we have the final core modulus of the conductor. As described above, point B, but more precisely point C, is the knee-point. At stresses and strains below the knee-point, the aluminium has been relaxed by decreasing the tension in the conductor to the point that it carries no more tension and all of the tension in the conductor reverts to the core only. Thus, the core's modulus is the one representing the stress–strain relationship from that point downwards.

As described above, the conductor's stress–strain behaviour is represented by these 'final' moduli (composite and core) whenever the tension in the conductor is lower than it once was at any time in the past. As noted, the stress–strain behaviour is represented by the 'initial curve' only when the conductor is experiencing a tension for the first time.

At the very beginning of the stress–strain plot in Figure 3.10, down in the bottom, left corner of the plot when tension is being applied to the conductor for the first time and the initial curve is the one tracking the behaviour, the slope of the initial curve looks very close to the slope of the composite, final composite plot. This means that the initial modulus of the conductor is the same as the final composite modulus but the strand settlement and, to a lesser degree, the metallic flow creep are also lengthening the conductor with increasing tension, causing the plot to bend to the right, creating the curve and hiding the elastic modulus when tension is applied for the first time.

At some high stress, the aluminium and the core material move beyond the range of elastic behaviour and plastic yielding takes place. The stress at which the plastic yielding becomes unacceptably large is vague and very different between different materials. Where this unacceptable stress value is decided to be is one reason that tensions in a conductor are limited to whatever they are. As new materials are brought into the business and their unacceptable strain limits are seen as quite unlike other products, we find the need to revisit our premises about how limits are set, how calculations are made and so on. More later, when we describe these new conductors.

So ends the discussion on components of strain that feed the nature of sags and tensions in a stranded wire in a span. It can be a complicated subject. Understanding what matters is important, even though understanding the strain components in detail may remain difficult.

High-temperature compression in aluminium

The reason the CEA wrote the sag–tension program STESS was to offer up a tool for understanding and calculating the effect of a phenomenon called aluminium compression. This is the phenomenon in which the aluminium of a two-material conductor goes into compression even though the conductor on the whole is in tension. If the aluminium is in compression, that compression must be offset by additional, opposite and equal tension in the core material. Added tension is added strain, and that causes added sag.

In a span of conductor made of two materials, the length of those two materials between the clamps at the ends of the span must be equal. Pretty obvious, yes? As temperature changes the length of the two metals, they will want to become different lengths from each other, provided that the coefficient of thermal expansion in one material is different from that in the other. When the aluminium of the conductor is paired with steel or carbon fibre or a silica matrix, the coefficients are different.

When the temperature rises, the aluminium will want to be longer than the core material. But, the aluminium is restrained from lengthening relative to the core by the clamps at the end of the spans. This restraint forces the aluminium into compression. Since the overall tension in the span cannot change, the compression in the aluminium must be countered with an equal increase in tension in the core. It is this corresponding tension increase in the core that causes an elastic elongation and sag increase that is labelled 'excess'.

Figure 3.10 illustrates this phenomenon. After a high tension is applied and then released, such as by the heating of the conductor, the sag–tension relationship is represented by the final slope of the dashed line coming down from the initial curve. The stress and strains track downwards with increasing heat (tension relaxation) along this line. The tension in the aluminium reaches zero when the plot intersects the core plot at point B. If you insist that the aluminium cannot take compression or when it actually does not, the composite plot must coincide with the core plot, meaning that all further reduction in tension requires that there is no tension in the aluminium and all of the tension is in the core.

But, the Figure 3.10 plot – a figurative representation of actual plots – shows that at tensions below point B the strain in the aluminium is more than in the core material by a constant amount as the tension decreases. This strain difference is the elongation in the aluminium that is restrained into a compressive force. When you run calculations with STESS, the result is an increase in sags at tensions below that found at point B that are greater than the sag found when the compression is ignored by an amount of about 0.5 m.

Here's a difficulty. The amount of compression that a conductor can support depends on a number of variables such as the number of layers of aluminium and their lay lengths. To begin, a single-layer conductor supports virtually no compression. The idea that the conductor has a limit to the amount of compression that it can support is because, at some point, the long, helical strands of wire collapse under compressive load. This is called birdcaging when it occurs. In multi-layered conductors, the outer layers support the inner layers against collapse. With small single-layer conductors such as 6/1 and 4/3 strandings, there is no outer layer to inhibit compressive collapse.

As a result of this complexity, the CEA said that their laboratory work found the range of compressive capability to be 6–12 MPa, with one test showing 18 MPa (Nigol and Barrett, 1980–1982). The greater this value, the greater the excess sag. Since you, the line design engineer, are never going to know the exact values derivable from the conductor that will show up for your project, you must estimate the value. The CEA suggested 10 MPa (1450 psi).

Of course, this must make you realise that you have no capability whatsoever to know what the hot sag of a conductor will be within about 10–20 cm.

It gets worse! Not everyone believes the compression principle described by the CEA report. Other mechanisms have been described to explain the excess compression.

They do say that the CEA mechanism is in play but it is not capable of explaining the amount of compression taking place. What are we to do? Remember the distinction between science and engineering? Let the scientists study and debate the source of the compression and let them try to improve its measure. In the meantime, we have an approximate value to use, and that is as good as it gets in this business.

Rated tensile strength defined

If a stress–strain laboratory test stops at 70% RTS, then we can deduce that the 'rated tensile strength' of the wire is known before the test begins. That implies, correctly, that the RTS might have little to do with the actual breaking strength of the wire. That is why the employed term is '*rated* tensile strength'. The establishment of the RTS of electrical conductors is defined in ASTM, IEC and other national/international standards, and it is done by calculation, not by laboratory testing.

The purpose of ASTM-like standards is to define ways for purchasers of engineered products to discover by testing and measuring whether the vendor has provided the product that the purchaser believed they were buying. The purpose of ASTM standards is *not* – we repeat, *not* – to help describe limits of use for the product. Usage limits have to be developed elsewhere.

Conductors of different materials have formulas for calculating their RTS that can be quite different from each other. The reason for this is to manage the obvious objective of requiring the product to be a conductor of electricity in concert with an ability to be structurally loaded rationally and safely.

ACSR conductors and their various incarnations that mix the rather standard alloy of aluminium (1350, formerly EC grade) with a steel-stranded core have the following stress–strain characteristic. Grade 1350 aluminium will break at about 1% strain, perhaps a bit higher. Steel, on the other hand will not break until it strains to about 3–4%, although it begins to yield appreciably near 1% strain. There is no value in an ACSR-type conductor if, upon the application of great strain, the core is intact but the aluminium strands have ruptured. Under such a condition, the electricity will be forced to jump to the core, overheat it and melt it to nothing useful. So, the calculation for determining the RTS of ACSR-type conductors is done at 1% strain, assuming that, by doing so, the aluminium remains intact and viable as a conductor of electricity. Any reserve strength of the core steel is of no value.

The RTS of ACSR-type conductors (1350 aluminium combined with a steel stranded core) is the sum of the loads that each aluminium strand can carry at 1% strain plus the sum of the loads that each steel strand can carry at 1% strain. The ability of the steel core to carry more load at higher strains is ignored. In the applicable ASTM standard, this calculated sum is de-rated a bit based on the number of layers of steel and aluminium strands. In the IEC calculation, which is otherwise identical, the stranding de-rating is not applied. The Japanese do something entirely different, but the outcome produces near-identical results for most typical conductors. For conductors that are outside the norm, we have seen the Japanese formula produce quite different results. We suggest that you keep your curiosity antennas fully extended and watch for the differences when necessary. Recall the comment of boundaries that set the usefulness of calculation methods.

Consider ACSS conductors, once referred to as SSAC 'sack' conductors. These conductors are identical to ACSR conductors in that they combine 1350 aluminium strands with steel core strands. However, the aluminium is factory 'cooked' to anneal it. This not only saps its strength to about one-third of the strength in its 1350 alloy state but it also changes its breaking strain from 1% to far more than 10%. Wow! Think taffy. This requires a change in the basic RTS calculation to retain the intention of keeping the conductor useful as a conductor of electricity and make use of component strengths. With ACSS conductors, the full strength of the core up to its yield point can be used without breaking the aluminium.

Thus, the RTS of ACSS conductors is the sum of the breaking strength of the steel strands at the ultimate tensile strength (beyond its yield point) plus the breaking strength of the aluminium at that strain. Even though the core is used to a fuller extent, the result is a significant strength reduction compared with the same wire in its 'pre-cooked' ACSR form. Consider too that the full strength of ACSS conductors requires the development of the larger strain of the steel. This means the development of very large unrecoverable sags are associated with a very high percentage usage of ACSS conductors. To bring the strength of ACSS conductors back to a marketable value, the core steel strength is sometimes increased by substituting a very-high-strength steel alloy in place of the typical ACSR alloy material. This is of value if the constraints of a project's criteria put the conductor up against a %RTS limit, which is not always the case.

Enter the new breed of conductors. In recent years, two new conductors have appeared on the market, and both have non-ferrous cores with unique properties. They have been named ACCR (aluminium conductor composite reinforced) and ACCC (aluminium conductor composite core) conductors. ACCC is actually a registered trademark name for one company's product line: CTC Global of Orange County, California. A standardised, industry-wide acronym for this type of conductor is to date not developed. We have suggested ACCS: aluminium conductor composite supported.

ACCR conductors have a silica oxide/pure aluminium matrix stranded core, and ACCC conductors have a large single 'rod' of carbon fibre surrounded by glass fibre – all impregnated with epoxy resin. The ACCR conductor marries its metallic core to thermally alloyed aluminium, and the ACCC conductor marries its fibre core to annealed, 1350-O aluminium. Both were developed for high-temperature applications. There are other thermally alloyed conductors, some with a mechanical disconnect of sorts between the core and the aluminium. What are we to do with the RTS calculations for these? The answer has not yet been laid down in the ASTM-type standards but is left with the vendors themselves to declare.

It is interesting that this can happen, and it is interesting how they muscle their way through the development of their method selection and the selling of it to the customer base.

Since the intent of the ASTM and other standards is defining the test methods for ensuring product quality and *not* for use as setting usage limits of the product, it is very, very sad indeed that our industry has linked certain usage limits to %RTS values even though there is no consistent relationship between that limit and the performance of the conductor under the action being managed.

Restricting a conductor's tension to a percentage of its (more or less) actual strength by a %RTS rule is useful only for managing safety. Otherwise, the RTS of a conductor does not relate to its actual limitations. We speak primarily of Aeolian vibration damage management. Many regional, national and international standards have for years managed Aeolian vibration fatigue damage by specifying tension limits in %RTS terms. It has become a very entrenched way of speaking, and it is near impossible to make it stop. We have tried! We will bear down on this subject in detail later in Chapter 4, where we discuss different conductor types and Aeolian vibration damage management in some detail.

Pre-stressing

Pre-stressing – also called pre-tensioning – is the act of pulling a new conductor to a tension above its installation tension so that it will behave as if it has already undergone a significant ice-loading event or years of creep and, therefore, does not have to pay the penalties of high, temporary initial tensions that can block you from eventual and more attractive stress limits on the aluminium. Without pre-stressing, the conductor will have a considerable share of that tension carried by the aluminium strands as represented by the stress or load-sharing data provided in the computer program outputs. A view of the load sharing in the *after-creep* or *after-load* parts of such programs' output reports compared with the load sharing in the *initials* part of the output indicates the amount of load and therefore stress that is shed from the aluminium by 10 years of creep or a large loading event.

The effect of offloading the aluminium is to move the knee-point – visible in a 'sag versus temperature' plot for the conductor – to a lower temperature. See Figure 3.14. The effect is also to remove from the conductor's installed condition the high aluminium stresses that invite Aeolian vibration damage as suggested by the AAMT (average annual minimum temperature) plot. In this sag versus temperature plot, a low sag means high tension and stress. This example had the conductor strung to 6000 lb (2720 kg) at 20°C and clipped in. At the time of installation, the knee-point is at about 120°C. Upon the application of 0.5 in. (13 mm) of ice with some wind (the US NESC (National Electrical Safety Code) loading case), the tension that develops stretches the wire and permanently elongates the aluminium, so that upon removal of the load, the knee-point moves to about 55°C. The application of a larger load, in this case: 1 in. (25 mm) of ice, the knee-point is pushed further to about 20°C.

The point of pre-stressing is to apply a tension such as this immediately – as in, not wait for Mother Nature or Father Time to deliver it – and drive the knee-point to a colder temperature and move the wire's tension from the aluminium to the core. Having done that, and seeing that the cold tensions are now reduced from what was going to be acceptable in the initial years of service, you have a choice: tighten the conductor

Figure 3.14 Effect of prestressing

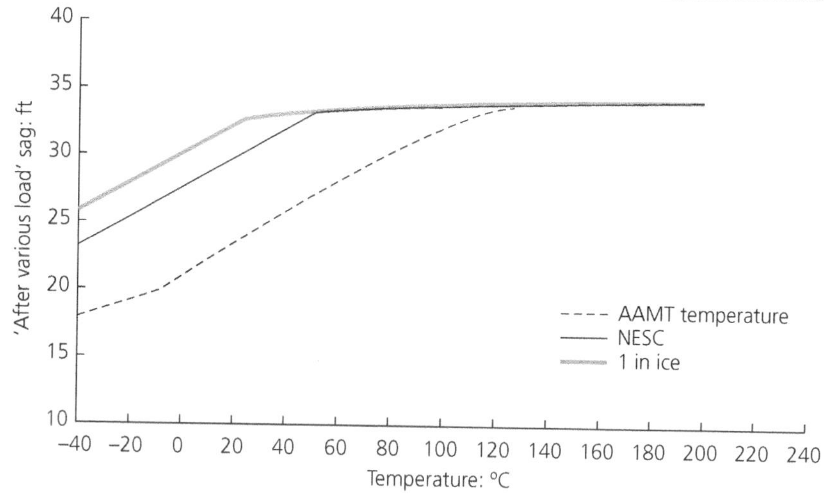

back up to those acceptable tensions and reduce the design sags or embrace the benefits of low or no stress in the aluminium – or a bit of both.

Pre-stressing is done in the field by the stringing crew during the conductor's installation onto the line structures. Certain conductor types are better suited to pre-stressing than others because the exercise is less onerous to perform and/or the reward is greater. Some manufacturers tell us that they always pre-stress ACSS conductors. That is an odd thing to say, because manufacturers don't install conductors. They sell them and make recommendations. They admit that not all installing crews do a good job of pre-stressing. Wouldn't it be nice if the manufacturer would do it in the more easily controlled manufacturing facility as a matter of routine?

To execute a pre-stressing exercise correctly and to produce safe and useful results, the engineer/designer must use the sag–tension programs outside of their intended bounds. Tricky but fun!

So ends our discussion on understanding conductors and any other type of wire when suspended in a span between supports. Wasn't that exhilarating?!

Summary of useful equations

Here, we collect the various equations presented throughout the chapter for your ease of reference.

$$T = H + w \times D \tag{3.1}$$

$$C \times y = C \times \cosh(C \times x/C) \tag{3.2}$$

$$\text{sag} = \text{span}^2/8C \tag{3.3}$$

$$C = H/w \tag{3.4}$$

$$\text{sag} = w \times \text{span}^2/8H \tag{3.5}$$

$$\text{arclength} = 2C \times \sinh(x/C) \tag{3.6}$$

$$\text{slack} = 2[C \times \sinh(x/C)] - \text{span} \tag{3.7}$$

$$\text{slack} = 8/3 \times \text{sag}^2/\text{span} \tag{3.8}$$

$$\text{slack} = \text{span}^3/24C^2 \tag{3.9}$$

$$\delta\,\text{sag} = (3 \times \text{span})/(16 \times \text{sag}) \times \delta\,\text{slack} \tag{3.10}$$

$$\delta\,\text{slack} = f(\text{span}) \tag{3.11}$$

$$\delta\,\text{sag} = (3C/2 \times \text{span}) \times f(\text{span}) \tag{3.12}$$

$$\text{sag}[2] = \surd(\text{sag}[1]^2 + 3/8 \times \text{span} \times \delta\,\text{slack}) \tag{3.13}$$

$$T_{(\text{at support})} = H + D \times w \tag{3.14}$$

$$\Delta H = E/C \tag{3.15}$$

$$\text{RS} = \surd(\textstyle\sum \text{spans}^3/\sum \text{spans}) \tag{3.16}$$

$$\text{RS} = \surd(\textstyle\sum \text{span}_i^4/C_i/\sum C_i) \tag{3.17}$$

$$\text{sag}_1 = \text{sag}_2 \times (\text{span}_1/\text{span}_2)^2 \tag{3.18}$$

$$\frac{C_1}{C_2} = \frac{h - v_1}{h - v_2} \tag{3.19}$$

$$\text{sag} = 48.3(t/2N)^2 \tag{3.20}$$

$$\text{sag} = [(\surd h_1 + \surd h_2)/2]^2 \tag{3.21}$$

$$R = (T_a - \text{AMB})/(T_s - \text{AMB}) \tag{3.22}$$

Here:

T	is the tension in the wire
H	is the horizontal component of the tension T in the wire
D	is sag
w	is the unit weight of the wire
C	is the catenary constant

E	is the elevation difference between two spans
RS	is the ruling span
i	is a subscript denoting inclined spans
h	is the wind span
v	is the weight span
t	is time
N	is the number of return waves for a span
R	is the ratio of the aluminium temperature to the steel core temperature
T_a	is the aluminium temperature
T_s	is the steel temperature
AMB	is the ambient temperature

REFERENCES

Aluminum Company of Canada (1950) *ACSR Graphic Method for Sag–tension Calculations.* Aluminum Company of Canada, Montreal, Canada.

CIGRÉ (2007) *Sag–tension Calculation Methods for Overhead Lines. CIGRÉ Technical Brochure 324.* CIGRÉ, Paris, France.

CIGRÉ (1977) *Conductor Creep Formulations, Technical Brochure 107.* CIGRÉ, Paris, France.

IEEE (Institute of Electrical and Electronics Engineers) (2003) IEEE 524-2003. IEEE guide to the installation of overhead transmission line conductors. IEEE, New York, USA.

Nigol O and Barrett JS (1980–1982) *Development of an Accurate Model of ACSR Conductors for Calculating Sags at High Temperatures.* Canadian Electrical Association, Montreal, Canada. CEA Report 78-93, Parts I–III.

Thayer ES (1924) Computing tension in transmission lines. *Electrical World* **84(2)**: 72–73. (First mention of ruling span.)

Winkelman PF (1960) Sag–tension computations and field measurements of Bonneville Power Administration. *Transactions of the American Institute of Electrical Engineers* **78**: 1532–1547.

Structural Engineering of Transmission Lines
ISBN 978-0-7277-5988-7

http://dx.doi.org/10.1680/setl.59887.087

Chapter 4
A transmission line as a structural entity

It will be noted in Chapter 7, where we examine line failures, that many failures of lines show collapsed or crushed structures but intact wires. What is happening in these frequent scenarios is that, after some sort of initial or 'triggering' event, the wires pull down (destroy) the structures, with or without help from loads applied directly to the structures. The reason for the wires tending not to break during a failure – or, if broken, break first – is that, after stability is lost and potential energy stored in the wires as tension and weight is converted to kinetic energy in the form of large, unbalanced tensions or motion, the usually greater strength and weight of the wires prevails over the weaker structures, and the wires 'win the day', often without damage to themselves.

Now that you have read and understand the nature of wires in spans, this chapter provides discussions on the main structural components of a transmission line. Here, we describe the choices within each component. Then, we describe what the basic differences are that will play a role in your choosing.

Conductors

Conductors are at the heart of the structural design of a transmission line. They are not just an electrical component hanging on a series of structural supports. They are, in fact, the most important structural members of the line. The primary point of this book – aside from its scintillating bedtime reading value – is to instil in you the understanding of this point and to understand the conductors' behaviour in this structural role.

That said, it is also necessary to understand the differences between the plethora of conductor types when it comes to placing them onto a line via the engineering exercise of applying sags and tensions and honouring usage limits. The choices affect cost, electrical characteristics of the circuit and the structural characteristics of the line.

Materials

Around the world, bare, overhead line conductors are defined by different standards. The USA uses American Society for Testing and Materials (ASTM) standards. Canada and European countries use International Electrotechnical Commission (IEC) specifications, sometimes with 'country-specific' deviations. The main IEC specification is IEC 61089. Japan and other countries do their own thing as well.

Table 4.1 lays out the present-day conductor alternatives in US (ASTM) terms. Aluminium options are listed in the first column and the core options are listed in the

Table 4.1 Conductor types

Core material	Aluminium						
	1350-H19	6201-T81	1350-O	TAI	KTAI	ZTAI (UTAI)	XTAI
HS	ACSR ACSR/TW	AACSR	ACSS/GA ACSS/TW	TACSR	KTACSR	ZTACSR GZTACSR	
EHS	ACSR/EHS	AACSR/EHS					
EXHS			ACSS/HS285				
AW	ACSR/AW		ACSS/AW	TACSR/AW	KTACSR/AW		
Galvanised Invar						ZTACIR	
AW Invar							XTACIR
Mischmetal			ACSS/MM ACSS/TW				
Aluminium matrix						ACCR ACCR/TW	
Carbon fibre			ACCC/TW				
None	ASC AAC	AASC AAAC					

remaining columns. The second and third columns show, respectively, the standard 1350 alloy and its stronger counterpart, the hardened 6201-T81 alternative, which offers the same unit weight, thermal expansion and thermal limit but slightly lower conductivity and about double the strength of the 1350 alloy. These aluminium options are not in the high-temperature conductor options because their thermal limit is understood to be 90–100°C under extended use to avoid annealing. Some line operators will let these conductors operate in the short term – measured as a few dozen or a few hundreds of hours over the line's lifetime – to temperatures of 125°C or higher. In other words, they will let a bit of annealing occur, with the understanding that annealing in the field has a very minimal effect on the strength of the conductor, with no meaningful consequences.

All of the other columns in the table are annealed or alloyed aluminium options that allow higher-temperature operation. The first four rows in the table are standard steel core materials (increasing in strength from HS to EXHS) and aluminium clad steel (AW). The AW cores are used extensively in place of a galvanised steel core in corrosive atmospheric conditions, such as near industrial facilities or salt air conditions near the sea, and the aluminium thickness can vary by choice, usually expressed as a percentage of the gross area of the strand (10%, 20%, etc.).

Notice that the very highest strength steel core is used almost exclusively with annealed aluminium for the purpose of replacing some of the strength that is absent due to the aluminium's annealing. We remind you that the value of doing this is not always useful. The remaining five rows of core options are designed to mate with various high-temperature aluminium options.

The discussion here does not try to list all of the specific conductor designations and technical specifications (ASTM, IEC, etc.) by name and number. These things are transient over time and can be found easily on the internet. Simply understand that the world is not a unified place with respect to conductor designation systems, naming conventions and technical specification control. This means that you not only have to learn another language, you also have to learn another designation set when you visit foreign regimes. The information provided here is designed to be generically useful, regardless of where you are sitting on this planet. That said, there are a few new conductor types that have stepped well outside the box of all current conductor types, and these will get some attention.

Each of the aluminium/core combinations used in the high-temperature market has unique thermal limits. The limit is defined by either the core or the aluminium. Annealed aluminium can operate up to 250°C, while the standard galvanising on steel-core strands cannot run this high. To make full use of the aluminium, the steel core must be modified to mischmetal. Mischmetal is best known for being the flint in lighters. It is tolerant of high temperatures.

The carbon fibre/glass fibre/resin core of ACCC is designed for a continuous-use thermal limit of 180°C. This makes less use of the annealed aluminium's thermal capacity, but the conductor has other good attributes that make pushing the core further of limited value.

The last four columns of aluminium options are for alloys with zirconium, to render the aluminium increasingly immune to annealing via heating. Conductors built with these alloys come in varying thermal limits due to varying degrees of alloying. The most well known of these conductors are 3M's ACCR and the 'gap' conductor that employs the mechanical trick of separating the aluminium from the core via the manufacturing process and the installation process. More is said below on these high-temperature conductors.

Type

The 1350 aluminium alloy has a conductivity of about 62% of that of copper. This is expressed as 62% IACS (International Annealed Copper Standard). Annealing the aluminium improves this conductivity a bit to 63% IACS, whereas alloying it to prevent annealing decreases the conductivity to as low as 55% IACS. On a comparative basis, these differences begin to matter when chasing low losses to the finer significant figures. Careful!

All but the hardest alloy (6201) are routinely reshaped from a round strand to a trapezoidal strand (TW). The 6201 alloy is regarded by many manufacturers as too hard to easily reshape. The purpose of the trapezoidal strands is to pack more aluminium into a particular space (diameter). TW conductors come in two forms: with a diameter equal to a round wire size (TWD) and with an aluminium content equal to a round wire size (TW). The latter version has the same area of aluminium (mm^2 or kcmil), and is therefore slightly smaller in diameter than its round wire 'mate'.

The resistance of a conductor is nearly directly related to the amount of aluminium in the conductor, so putting more aluminium into the conductor's diameter reduces its resistance almost proportionally without increasing the wind load that the conductor imparts to support structures.

The natural void ratio (aluminium area to gross area) of round stranded construction is 78%. This means that the aluminium portion of the conductor contains near 22% air within its diameter. Compare 759 kcmil Tern ACSR to its TWD mate, Kettle ACSR/TWD (Table 4.2).

The 45/7-stranded TWD conductor is the same diameter, is 20.5% heavier and has 83% of the resistance. This is compared with a stronger 26/7 conductor in Table 4.3.

The 26/7-stranded TWD conductor is the same diameter, is 20.5% heavier and has 80% of the resistance. The use of TW conductors can be well worth it if the cost of the losses is

Table 4.2 Tern ACSR and Kettle ACSR/TWD

Conductor	Size: kcmil	Diameter: in.	Weight: lb/ft	R_{dc20}: Ω/mile
Tern ACSR	795	1.063	0.8947	0.1143
Kettle ACSR/TWD	957.2	1.060	1.0790	0.0949

Table 4.3 45/7 Drake ACSR and Kettle ACSR/TWD

Conductor	Size: kcmil	Diameter: in.	Weight: lb/ft	R_{dc20}: Ω/mile
Drake ACSR	795	1.108	1.0934	0.1166
Kettle ACSR/TWD	957.2	1.108	1.3180	0.0933

important to you, since the weight of a conductor does not come close to increasing the cost of a project proportional to the weight increase.

Traditional steel cores do little to assist in conducting electricity. This is less true with the aluminium-clad cores because the core's small aluminium content attempts to help out. The higher content of aluminium in the ACCR core also improves the core's contribution to conductivity for that type of conductor. Tempering this contribution is the fact that electricity, being a strange creature, prefers to travel in the outer regions of a conductor. In other words, the nature of the core is not of much interest to the electrons. The fibre and resin core of ACCC conductors is completely non-conductive and, unlike the metal-core materials, is not much of a heat sink either.

Another longstanding conductor type is self-damping (SD) conductors. These conductors have TW strands and a core that is smaller in diameter than the space provided for it. The intent is that the loose aluminium tube can rattle on the core to dampen Aeolian vibration. The SD conductors tend to be used in flat, open terrain where Aeolian vibration activity is most prevalent due to the common, synoptic breezes that occur in such places. We know of one installation in northern Canada where the SD conductor is pulled to a quite high tension with apparent success. We also hear of SD conductors breaking near mid-span because water sits in the airspace at the low point of the spans, corrodes the steel core away and the conductor fails in a time frame that is too short for the conductor to be considered a success. The 'gap' conductors are a fancy version of SD conductors albeit filled with grease so they will not rattle and dissipate any vibration energy.

Another conductor design aimed at addressing Aeolian vibration and allowing a higher tension and lower sag is the T2 conductor. T2 conductors are made up of a pair of regular 1350 ACSR conductors wound around each other on a lay length of about 2.5 m. The non-cylindrical shape of T2 conductors disrupts the vortex shedding inherent to cylindrical conductor types, and the energy imparted into the conductor is shed by axial rotation back and forth, rather than vertical vibration. The vendors will tell you that this allows the conductor to be pulled safely to code limits without concern for fatigue damage. From a project experience, we found this to be not true.

First, the declaration that the conductor can be '*pulled safely to code limits*' is a parochial statement only meant to be accurate for the US National Electrical Safety Code. Even so, it turned out not to be true even in that context. By an oversight not worth noting here, the T2 conductor was tensioned too close to the code limit, and it vibrated like crazy. We are reasonably certain that this transmission line may be the only T2 line in the USA with Aeolian vibration dampers attached to it to keep it under control.

The differences between all of these conductor types competing for your attention can be dramatic on the face of it, and can make you think that you see value in a product accordingly. Be very careful, because nothing short of a fully detailed 'installed cost' comparison using a line's own design criteria and other constraints will reveal the best conductor choice. The results may be surprising to be sure, and you may find yourself faced with having to reconcile and overcome the preconceived notions that you and others on your decision-making team undoubtedly have.

A full blown cost comparison will provide you with an opportunity to see on what basis decisions are made. Pure engineering may not be the basis. Also, understand that the benefits listed by vendors are designed to sell conductors, and while not necessarily untrue – as the story above suggests – they are easily taken out of context. Be careful with understanding vendor declarations. Read the fine print and between the lines. Listen for what they do not say and don't always trust what they do say!

High-temperature conductors

Since the earliest years of the 21st century, several high-tech conductor types designed to allow safe operation at temperatures well above the classic 90–100°C limit that ACSR's 1350 aluminium allows and to be an improvement on the single player in the field for several decades (ACSS) have come into their own. In order of appearance on the commercially viable scene were the thermally alloyed conductors – TAI, ZTAI and XTAI, including the gap conductor GZTAI – followed by ACCR from 3M and, a few years later, ACCC/TW from CTC Global in California.

The sequence of getting a toehold in the market matters, because the competition between these conductor products has generated marketing strategies that fed and affected the information stream reaching potential customers' ears.

> Being in the consulting business has me believe that it is necessary that I and my colleagues understand many things in our business better than the people who would be our clients. Otherwise, the reasons to hire us are seriously diminished. Couple that with the fact that my own interest in the wires on a transmission line exceeds my interest in the other components of a line (structures, hardware, foundations, etc.), and I find it necessary and interesting to understand these conductors in great detail.
>
> In trying to do so, I became well aware that the competitiveness between the vendors of these new conductors is fierce, to say the least. Be aware that every salesman and manufacturer's representative has an agenda that includes beating their competition, regardless of the facts. Some of the information out there is intentionally untrue, and it takes a lot of investigative effort to separate fiction from facts.

> That said, I love some of those guys. Just understand their priorities and understand that they may not understand transmission line engineering as well as you do. Therefore, they may say things that mean little to you or should mean little to you.

TIA, ZTAI and XTAI, and especially the gap version of these conductors, have been used in the UK and Japan, but this may be a fading fact. Transmission lines in both of these countries share features essential to the cost-effective use of gap conductors. The countries are small, and alignments are very crooked, with corners every few structures. As a result, the transmission lines in these countries have deadend (strain) towers at very frequent intervals. Gap conductors require the mechanical separation of the aluminium from its core at frequent intervals, and this is more easily accomplished at deadend structures, where the work is more naturally accomplished. Systems with few deadend structures on them – as throughout North America and countries that are less restrictive with developing new rights-of-way (ROWs) – find the complicated work of installing gap conductors not very attractive.

A gap-type conductor separates the aluminium from the core so that the conductor acts like an aluminium component riding on the core as if the core were a messenger – a common concept in distribution lines at household and industrial voltages. The effect is to make thermal expansion and sag increases at all temperatures above the installation temperature occur at the expansion rate of the steel of the core. In other words, the high thermal expansion rate of the aluminium in the conductor is removed from affecting the high-temperature sags. The trouble with the product line is that the conductor remains as heavy as an ACSR conductor and its rated strength cannot include a contribution from the aluminium.

Next, the 3M Corporation presented ACCR with its lightweight, very strong and less thermally expansive silica aluminium oxide core. 3M's strategy was to offer a conductor that was as easy to understand and handle as an ACSR conductor but offered a higher strength-to-weight ratio and less sag growth at high temperatures. To help the aluminium contribute to the rated tensile strength (RTS) of the conductor, operate at high temperatures and be easy to handle, they selected a strong, thermally alloyed aluminium stranding in either round or trapezoidal stranding options.

The core strands are very strong but brittle, with the feel and breaking behaviour of dry spaghetti strands. Some say that the mix of the very thermally inert silica strands in the core with the surrounding pure aluminium raises a concern for the effects of microscopic strain damage over time due to the two components' very different thermal coefficients of expansion. The conductor also requires great care while running over travellers at high tensions during installation. A traveller radius that is too small runs the risk of cracking the brittle core strands under high tensions. It is not likely that such damage will be revealed during installation if it occurs, because the strength of the aluminium will allow the pulling to continue without conductor failure. Neither of these concerns has developed into a conductor failure to date, and may never do so. Time and/or study will tell. The real hurdle that the ACCR conductors must overcome is cost. The purchase cost of ACCR at this time is about double that of the more competitive alternative, ACCC/TW. As a result of this cost barrier, there are few installations of ACCR of any length. They have been limited to relatively short installations where all alternative solutions were also very expensive, and the options may not have included ACCC/TW due to its timing in coming to market or the owner's lack of trust in that new product.

ACCC/TW conductors offer a list of characteristics that are very attractive to many situations. On an equal-diameter basis to any other conductor type except 'all-aluminium' conductors – which have their own significant limits of use – they offer the best strength-to-weight ratio; equal or lower weight than any alternative; the lowest resistance value; and by far the lowest sag increase with increasing temperature. Their Achilles' heel is the higher elasticity of the very strong composite core that adopts more sag under the application of ice loads than other conductor choices. They also require as much care as any other alternative to install. So, if the purchaser recognises these two drawbacks, there remain many applications where ACCC/TW conductors can be the conductor of choice.

Speaking approximately, when the ice load exceeds the weight of the ACCC conductor, the ice sag will exceed the hot sag and control clearance to ground issues. This is less true as the installed tension increases. Then, the ice load case will rule at lower values. In other words, at large ice thicknesses, the height and therefore the cost of support structures will be controlled by the ice load case, not the thermal sag. *Limits* in this case do not relate to the strength of the conductors, because, at the ice loads shown, the strength of the conductor is not close to being challenged. This is to say that this may be a drawback relative to other conductor types' ice-carrying capabilities, but it is by no means a show-stopping flaw. The drawback manifests as an ice-induced sag that must be accommodated in the design of the line, not as a risk of breaking the conductor.

Even though ACCC/TW conductors were developed for the high-temperature market, studies conducted by the author and others have shown that the conductor is cost-competitive for extra-high-voltage (EHV) – but not hot – applications when (1) the ice load does not greatly exceed the weight of the base conductor and (2) the lifetime operating costs can be high due to high normal operating usage. The conductor is less competitive for low-usage lines if the conductor selection is based on losses being used to offset the capital cost of the work.

ACCC/TW conductors have had a few installation failure events, and each of these has been believably explained by addressable causes. It would appear that the product has no inherent, irreparable flaws. Mostly, the failures can be avoided by recognising that the product requires a new understanding with respect to installation care. You might want to think of these advanced products as Ferraris. If you buy a Ford, you will have no problem figuring out how to drive it safely and how to get value out of it. If you buy a Ferrari, you had better take driving lessons to get the value and be safe. Such are these conductors compared with the ACSR conductors of the world.

Whether these conductors will become the norm of the future requires, as usual, not only acceptance of their calculable technical advantages but also the discerning of fact from fiction with respect to the information reaching the industry's ears. To dismiss them because you will not do your own investigating is to miss out on their advantages.

Mechanical design considerations

Chapter 2 provided a description of the relationships between electrical characteristics and goals and the business of a line's structural engineering. Here, we discuss the mechanical issues in detail.

It should be evident that there is a host of issues to address in the selection and use made of a conductor for a line, and that all of these issues are not uniformly improved with each choice. It is necessary in every case to accept the degradation of certain behaviours in order to gain the desired positive behaviours on matters of more essential need. It is the achievement of a 'best balance' that is called good engineering and good decision-making.

It is wrong to find that balance based on parameters viewed in isolation. It is even wrong to decide on cost alone, even when based on a detailed installed cost and operating cost optimisation. While such a cost evaluation should not be ignored, we suggest that there are other issues of increasing importance such as the acknowledgement of environmentally fatal flaws and the addressing and honouring of sustainability – a buzzword used to address the now well-known subject of climate change as aggravated by our burning of copious quantities of fossil fuels. See Chapter 9.

With that opinion presented, and with all of this said, the best balance is likely to be achieved and qualitatively recognised by which criteria are more closely met by the selection than by competing options. We take the view that a criterion not challenged is to be interpreted as a feature of the selection that is used inefficiently.

The easiest example to describe is that of running one conductor tension to its limit while leaving others unchallenged. For example, a tension limit of a 2000 m catenary constant value might make sense, but if the usage of the conductor's strength under the project's maximum ice load is low, say below 40%, then the conductor may have too much core strength, too much steel. Or, if the conductor seems very attractive by several measures but carries the required current at a temperature well below its capability, is it a poor choice?

Our answer to that last question is 'no', and this illustrates our point that this is a qualitative assessment that we are describing – and this should be kept in mind to keep you thinking.

How tight should you pull a wire?

Code limits

Many national and regional safety codes generally stop you at 50% or 60% of RTS. This limit is tied to a particular load case with meaningful temperature, ice and/or wind load values. These safety code cases may put the idea into your head that 60% or so is an absolute limit for conductor tension, but remember that there are legitimate load cases in which a safety code expresses no interest but in which you should have a great interest

– particularly heavy ice loads. The applicable safety code may force you to keep the tension below some %RTS value for their safety-related zone loading, but you are free to exceed that limit for any other larger ice loading that you have reason to expect during the 40+ years of your line's life.

In fact, these limits have been changing – as they have always done in the long term as the thinking minds in the industry try to improve our approach to the work. When you step back and look at the changes, they are driven by the following ideas:

- simple factors of safety are becoming partitioned into load factors and strength factors
- Aeolian vibration damage mitigation is being better understood and addressed
- reliability-based methods are replacing deterministic methods.

Let's track a few codes. In the USA, the 1961 National Electrical Safety Code (NESC) was issued by the US Department of Commerce, National Bureau of Standards. Part 261.F required tension not to exceed 60% of the ultimate strength under load, 35% of the ultimate strength in the initial unloaded condition and 25% of the same in the final unloaded condition. The latter two conditions are both at 60°F (15°C). The load that limited the tension to 60% was one of three sets of values, depending where the line is located in the continental USA. The values sets were:

- 0.5 in. (12.7 mm) ice + 4 psf (190 Pa) wind at 0°F (−20°C) or
- 0.25 in. (6 mm) ice + 4 psf (190 Pa) wind at 15°F (−10°C) or
- No ice + 9 psf (430 Pa) wind at 30°F (−1°C).

There were small wire exceptions to the values, and the cases with ice also required the addition of a weight value to the resultant unit weight of the conductor – just because. At this time, 500 kV lines were just getting underway, and there was a footnote to the values. It said

> The above limitations are based on the use of recognized methods for avoiding fatigue failures by minimizing chafing and stress concentration. If such practices are not followed, lower tensions should be employed.

In 1984, six years after Canada went metric; the NESC was published by the Institute of Electrical and Electronics Engineers (IEEE) and the American National Standards Institute (ANSI). The tension limit and load values from 1961 remained unchanged, but the metric values were noted as you see them above. Even the vibration damage footnote remained. Also, a new load case was added for extreme wind pressure applied to bare wire at 60°F (15°C). The extreme wind pressure varied from 3 to 8 psf across the country, including Alaska, with the higher values being along the Atlantic coast. The pressures were based on 'fastest-mile' wind speeds, which are reasonably understood as synoptic wind speeds, not gusts. The tension limit on the conductors for the extreme wind case was 80% of the ultimate strength, as far as we can tell. The one change was that ultimate strength became called the rated tensile strength.

In 2012, the NESC had further adjustments. The three load cases noted above remained but the temperatures for the 25% and 35% limit cases were changed from 60°F (15°C) to various lower temperatures, depending on the location in the country. This effectively lowered the tension limit for new lines, since the tension of most designs is limited by one of these 25% or 35% constraints. The 60% and 80% limits for the ice and extreme wind cases remained as they were, but the extreme wind map was completely revised from pressures to wind speeds. The conversion to pressure was now in the hands of the user by way of a formula provided in the code that was a simpler version of the formula in other texts that engineers tend to have at their disposal. So, the results changed from 1984 to 2012.

The notes on Aeolian vibration fatigue damage were also changed to (the comment inside square brackets is mine).

> The initial and final unloaded tension limits may be used at higher temperatures not to exceed 60°F (15°C) [the older temperature limit] if (a) vibration control devices or self-damping conductors are appropriately used, or (b) a qualified engineering study, manufacturer's recommendations, or experience indicates Aeolian vibration damage is not likely to occur

and

> The above limitations may not protect the conductor or facilities from damage due to Aeolian vibration.

What a fascinating shift through the years in the US code! The code has acknowledged for more than 50 years that the bare wire limits may not protect the conductors against fatigue damage or failure. Only recently have changes to the limits trended towards making them more likely to avoid damage, but the caution remains as forceful as ever. All the while, the responsibility has always been the engineer's to get it right. The change in the footnote exists because far too many engineers have fallen into the erroneous habit of believing that any conductor can be pulled to the 25% or 35% limit without risk of Aeolian vibration fatigue damage. As we will see later, there is simply no direct relationship between RTS and the risk of such damage. If you learn nothing else, please learn that!

The Canadian code (Canadian Standards Association (CSA) standard C22.3, No. 1) evolved as well. In 2001 and 2006, the CSA code used the same three tension limits – 60%, 35% and 25% for the same load cases – except that the loaded case limiting tension to 60% of the rated strength employed metric values and a higher wind pressure associated with ice thicknesses comparable to the US values. A footnote to the CSA code said:

> The unloaded tension is intended only as a guide. The unloaded tension may be increased when (a) factors that produce conductor fatigue are at a minimum; (b) suitable steps are taken to suppress fatigue failure; or (c) self-damping conductors are used

and

> The unloaded tension should be decreased where the line is exposed to conditions that can cause conductor fatigue and no steps are taken to suppress vibration.

Then the CSA changed everything in its 2010 edition of the code, as it adopted the reliability-based approach to line design. To begin, the 25% and 35% limits were discarded entirely, and the bare wire tension limit became expressed as a catenary constant limit.

> The initial tension limit at average temperatures during the coldest month (January) should not exceed a catenary parameter of 2000 m for single conductor spans properly equipped with vibration dampers. In the case of bundled conductors, the catenary parameter may be increased to 2200 m.

Note:

> This limit does not apply to special conductors such as self-damping conductors where different limits may be used in accordance with past experience and appropriate studies.

A table is then provided, offering recommended catenary parameter limits that decrease from the 2000 m maximum down to 1300 m as the span shortens from 400 to 100 m. The footnote to the table says:

> The reduction of the catenary value in relation to the span is based on the principle that lower tensions are safer than higher ones from the point of view of Aeolian vibrations. This reduction will not affect the sag significantly but will reduce loads on angle structures.

Where have you heard that before?

The 2010 CSA standard goes on to say that:

> The final tension limit after creep or permanent stretch due to ice and wind loads should not exceed 70% to 80% UTS. A value of 75% UTS may be used and has been applied to many Canadian lines.

So, isn't that interesting? The Canadians have resorted to saying 'should', 'may' and 'recommended', meaning that it is all up to you. The IEC standard 60826 that is used by so much of the world says even less. In Part 7.3.1, Table 16 applies to all conductor and ground wire types and says:

> for Damage Limit load cases, limit the conductor tension to the lowest of a) vibration limit [unspecified], the tension that initiates infringement on clearances [as loose as some constraint that might exist allows], or 75% of the rated strength

(typical range in 70% to 80%). For Failure Limit load cases, limit the wire tension to rupture (100% of rated strength).

This IEC document was championed by an engineer with a long history of employment at Hydro Quebec in Canada. It is very reasonable to assume that his view of the subject of Aeolian vibration management is very studied and well understood, and he was likely to believe that anyone worth their salt also knows the subject well enough not to need guidance from this IEC document.

We will get deeper into the Aeolian vibration management subject below so that you can be about as educated on the subject as this champion of the IEC, the Canadian code and the US code all require you to be.

Respecting rated tensile strength

Manufacturer stress–strain tests run up to 70% and stop, in part for concern of damaging equipment. It seems clear that the sag–tension relationship up to 70% RTS is understood, and beyond that it may not be. We have watched testing take place to higher tensions, sometimes repeatedly. While the purpose of doing so may vary, we have watched the conversation about the high-tension results be a debate about the conductor's behaviour at these loads and, therefore, the impact that this behaviour should have on sags and maximum usage expressed in %RTS terms. When you allow a limit much beyond 70% for any load case, you are basically in uncharted and unstable territory, and you can consider yourself to be on your own.

Consider as well that the popular computer programs PLS-CADD and SAG10 solve the problem of the poorly extrapolating fourth-order polynomial equations that are widely used in the industry by truncating the formulas natural plotting beyond 0.5% strain. They do this because the fourth-order formulas do not trace the stress–strain plots properly beyond about that point for most conductor types. The programs calculate the slope of the plot at 0.5% strain, and project that slope forwards for all greater strains. This leads to errors at strains above 0.5% strain if the actual stress–strain relationship is not straight as assumed. Some relationships are straight beyond that point, but some are not. It is another good reason to hold the strains of conductors to reasonable limits and to believe that, otherwise, you may well be in poorly charted territory.

Since ACSR conductors reach their RTS in the neighbourhood of 1% strain, this approach to managing these formulas does not create much risk, provided the strains and loads are held to about 70% for any load case. It happens that the ACSS and the ACCC conductors particularly do strain well above the 0.5% point with higher tensions, but their stress–strain plots are linear beyond that point, and the computer calculation assumption works out quite well.

There seems to be a practice in some countries to presume the loading of conductors to more than 80% repeatedly, and expect the conductor to behave as calculated. We consider it a risky practice if the conductor's load sharing between the core and aluminium is expected to remain predictable and if either of the parts, particularly the aluminium, is expected to remain structurally viable. We suggest that there is simply not enough information about a conductor's nature after repeated high loads such as this to permit such high-tension use and expect predictable results. If you set limits above about 70%, you ought to presume permanent wire strains other than are represented by sag–tension calculations or that full-blown breakage can occur, and part of your design process should include dealing with the consequences.

'Years ago' was a simpler time, and inventive minds have added complexity to our present-day situation. Our claim that %RTS limits are inappropriate for guiding engineering decisions would be a weaker claim in those past years, but no longer. In the beginning – as it were – the Good Engineer created ACSR and not much more to replace copper as the conductor of choice. On day 8 of conductor creation – as it were – along came alloys for the aluminium and the core. Along came intentional annealing and, eventually, along came non-ferrous and non-metallic core materials. We are boxed into a corner with our once-upon-a-time %RTS-based tension limit rules, and seem to have difficulty escaping them however inappropriate they have become.

The rated strength of ACSR is based on a formula that leaves the conductor in a basically usable condition after experiencing a very high tension in the range of 70–90% of its RTS. The aluminium will – probably – not have broken but it may have adopted a sag beyond the acceptable, although the stiffness of the steel core limits the permanent strain increase in the aluminium to a workable amount. Consider ACSS where its naturally lower RTS value depends on the fuller use of the core with respect to encroaching on its tensile strength limit (i.e. well beyond its yield strength), coupled with the knowledge that the taffy-like aluminium strands will not break but can be strained as far as the core is able to take it. Running that type of conductor to a high %RTS value requires the understanding that you are doing something very different to its stress–strain relationship and the two metals' load-sharing capabilities than you are to an ACSR.

Consider the new carbon-fibre core conductors (ACCC). ACCC has the same taffy-like aluminium but a core that is purely elastic. The core is more elastic and stronger than steel. Its RTS is calculated basically as the breaking strength of the core plus a fairly low tensile strength contribution from the aluminium. The breaking strain of the core is at 2% – twice the strain of ACSR's RTS strain. The RTS of 795 Drake ACSR is 31 500 lb (140 kN). The RTS of its ACCC counterpart is 41 000 lb (182 kN). The RTS of ACCC is highly dependent on the core strength.

Running both conductor types to 20% RTS puts their tensions at 6300 lb (28 kN) and 8200 lb (36.5 kN), respectively – quite different from each other! They both weigh the same at very near 1 lb/ft (1.5 kg/m). Thus, this line of thinking places them at catenary constant (C) values that are equally different. At 60% RTS, the strain in ACSR is about 0.4% and the strain in ACCC is 1.1% – again quite different, with differing consequences. There is nothing wrong with either conductor's characteristics. They are simply very different from each other. Since we will discover that certain conductor behaviours relate better to parameters such as C and 'strain' rather than %RTS, thinking and working in terms of %RTS is to be on a misguided adventure.

Consider an all-aluminium conductor (ASC or AAC) and its alloyed or partially alloyed mate (AASC, AAAC and ACAR). Each of these conductor choices of identical aluminium quantity and stranding are constructed with the same material from certain points of view – unit weight and thermal expansion but with very different strength properties. An ACSS conductor is simply a factory-cooked ACSR conductor. They are identical, as described above for AAC and AAAC, and also very different, as described above. If these non-identical twins are managed by %RTS rules when Aeolian vibration magnitudes and damage propensity has nothing to do with %RTS, then you are not in control of understanding your conductor's Aeolian vibration behaviour.

Before leaving the subject, you must refer to the discussion on hardware, where we describe another very important limit on conductor tensions. That is the subject of splices and terminating connections. These are not all they try to be.

Managing Aeolian vibration

This topic is exciting and of paramount importance. Aeolian vibration is that high-frequency, small-amplitude, usually vertical motion caused by the alternating pressure on a wire when it 'sheds' wind vortices from its top and bottom leeward 'corners'.

Aeolian vibration is a very important subject to understand because the avoidance of fatigue damage to the conductor is done by one of two actions. The first is if tension is held below a threshold and the second is if vibration dampers are attached to the conductor in quantities dependent on the chosen tension threshold. Very often, the tension limit chosen to manage vibration damage is the tension limit that overrules all others. If not for concern for Aeolian vibration fatigue damage, most conductors could be safely pulled tighter than they are.

It is useful to think of this in terms of 'energy in and energy out'. The wind is putting energy into the wire, and the wire sheds that energy by vibrating. The energy dissipation achieved by vibrating is done not only by motion but by the internal rattling or rubbing between strands that the motion creates. When the strands of the conductor are tightly held together, generally by higher tension in the conductor, the strands lock down on each other, and this rattling and rubbing is reduced, and the energy dissipation is reduced. If the conductor cannot dissipate enough of the energy input by the wind by its own self-damping capabilities, then it needs assistance from attached dampers to keep the vibration amplitude and frequency of occurrence below a threshold for the material to survive against fatigue failure. Thus, the basic relationship between tension and energy dissipation and vibration-induced damage is easy to understand, but very difficult to quantify.

Here's our take on the history of vibration control. From the beginning (early 20th century), line designers knew that conductor vibration and fatigue failure was a problem, and they sat down to decide how to control it. They had a dilemma. Based on a lot of data, the vibrating seemed to be a function of '%RTS' *and* of the ratio 'tension/mass'. They decided, largely for the sake of ease of use, to go with the %RTS choice. Well, they blew it, basically!

Back then, most or all ACSR wires were natural strands, meaning the diameters of the steel core wires were the same as the diameters of the aluminium strands. Conductor design had not become complicated yet. This means that the only stranding families were likely to be 6/1, 18/1, 12/7, 30/7 or 54/7. Some were not heavily used if the steel content was quite small.

The 30/7 stranding has a heavy steel content at 39.65% by weight. In later years, other stranding combinations were created by using different aluminium and steel core strand diameters. This allowed a variety of steel core contents – as defined by 'percentage of area' – to be made available, and the performance of the '%RTS' rule became less effective in controlling vibration because new data plotted in new places, and the %RTS relationship with the vibration weakened.

One of the greatest clues that 'tension/mass' (T/m) is the valid method is the presence of this term in vibration frequency formulas used throughout the world. The Bonneville Power Administration did a study some years ago in which it discovered it could string ASC (AAC in the USA) for their 500 kV system to about 33% RTS without creating vibration problems (Catchpole, 1996). The essence of its work was that different conductor types showed similar vibration behaviour when sagged the same; that is, were sagged to the same T/m value, which you will recall is H/w (Equation 3.4). By Equation 3.14, H is practically equal to T, and w (weight) is m (mass) with a constant gravity factor thrown in.

A page from Catchpole (1996), and more specifically from an old Italian damper manufacturer (Salvi in Italy, Fargo in North America) catalogue, displays big dots that indicate a condition beyond which a damper arrangement is required. The Salvi page shows that for all wires noted, at a span of around 500 m, the damper is required for tensions above where the dot is placed. At each round dot – where damper needs and therefore vibration problems are the same – the %RTS is different but the T/m value is constant. Some players in the industry saw that the T/m basis for managing Aeolian vibration has more validity than the %RTS basis.

So, why is this exciting? It appears to be clear that if vibration is a function of T/m, then it is also related to C. We have already noted that C *is* sag for a given span length. Here's the claim being made:

> For any span of wire, the vibration performance is the same for any type of wire, when installed in a span at the same sag.
>
> I had a problem with that simple declaration. It seemed incomplete. I prefer to say that wires of any size or stranding configuration (excepting self-damping types) that are sagged to the same catenary constant may exhibit a common propensity to vibrate but the damage that may occur due to that vibration is not necessarily equal among the wire types nor is the self-damping.
>
> Several years ago, we had the opportunity to 'reverse engineer' Alcoa's damper recommendations. For a few years and until the sale of SAG10 to Southwire,

> Alcoa embedded their damper recommendation program (Vibrec) into SAG10. Southwire removed it because they don't sell Alcoa's dampers. Their recommendations are a function of C alright but the value of C was *not* a constant. When I asked the member of the CIGRÉ group that produced Alcoa's work to explain the very different declarations that existed between CIGRÉ's work and Alcoa's work, I eventually understood his answer to be, 'They are both right.'

Let's wear this out because it is important. The following are quotes from a host of sources.

> If the H/w parameter is in fact the governing design parameter regarding vibration performance, then it will follow that conductors of any type but with the same sags should have the same vibration performance.
>
> If the H/w parameter does govern vibration performance, a high strength ACSR ... will have the same sag as an all aluminum stranding.
>
> Suggested limits (should) not be used unless verified from other sources.
>
> Work done to date suggests that ... bare conductors of any commonly used type will not suffer fatigue damage with H/w values up to about 4500 ft (1370 m).

All of these quotes were offered up by Brian in his work. The first quote says *should.* The third quote says you must verify by other sources. These are red flags. However, the second and fourth quotes dare to make quantified claims:

> If reliable experience on existing lines proves that a (higher) T/m value did not cause fatigue damage, it is quite reasonable to use the same T/m value for a similar line, with the same conductor and running on similar terrain and wind conditions.
>
> It is instead quite dangerous to generalize such experience for other conductors, or different terrain and different wind exposure conditions.

These quotes are from Rudolpho Claren, the godfather of Italian leadership on the subject and Brian's friend. While Claren was held up as a proponent of the constant C value as a viable expression for managing Aeolian vibration damage, these two quotes do not actually support it as a simple matter to manage.

> A ... review of technical literature at BPA convinced us that, in theory at least, H/w appears to be the more relevant parameter ... to be used as a basis for tensioning conductors to control Aeolian vibration damage.

This quote is from Peter Catchpole. We suggest that there is hesitation in the words 'appears to be more relevant'. This is simply an acknowledgement that the idea trumps the %RTS method but does not declare that 'this is the simple Holy Grail answer'.

> The [undamped] safe limit may be as much as 1400 m. However, there is not enough experience cases ... to determine where ... the limit should fall. One can only conclude it is somewhere in this interval [between 1000 and 1400 m].
>
> Caution is compulsory ... vibration-induced fatigue of conductors is a problem of highly complex and highly random nature ... the wind, the sole cause of Aeolian vibration is random ... the conductor responds ... by displaying markedly different amplitudes and frequencies ... Conductor self-damping ... is far from a constant ... which depends on loading history ... Fatigue endurance ... is another random variable.
>
> Obviously, simplification is required to overcome complexities. Some conservatism is needed to counterbalance uncertainties.

These are the caveats embedded in *CIGRÉ Technical Brochure 273* on safe tension limits (CIGRÉ, 2005). They are saying, in other words, that *this is the best we can come up with given that we were determined to offer some form of guidance to the industry*. CIGRÉ went on to say that:

> Much of the poor experience that resulted from applying the recommendations [of the 1953–1962 EDS Panel – the panel that had originally recommended that %RTS be the method of choice] was the result of ignoring the Panel's reservations [about the applicability of their recommendations].

In other words, be careful what you say because you will be misunderstood. All of these quotes, when read critically, led to the understanding:

> Wires of any size or stranding configuration (excepting self-damping types) that are sagged to the same catenary constant may exhibit a common propensity to vibrate but the damage that may occur due to that vibration may not be equal among the wire types.

CIGRÉ Technical Brochure 273 declared safe tension limits for single conductors (not bundles with spacers). The tension limits are expressed as C values (constants) for a wide range of wire types. It offers four values for C, depending on the windiness and exposure to wind of the conductor's location. Their work is offered as plots of C versus the factor Ld/m, where L is the span, d is the conductor diameter and m is the conductor's unit mass (weight).

The reverse engineering of Alcoa's Vibrec program produced quite different results (Catchpole, 2006). One individual, Chuck Rawlins, was party to both productions, and his claim was that both the CIGRÉ answer and the Alcoa answer were right. This claim reveals very much about the subject at hand. Chuck is absolutely no slouch in this field, and we laud his knowledge, contribution to the industry and his all-round great guyness.

The original Alcoa work and the CIGRÉ work were both based on gathered field data. The data allowed plotting on a *C* versus *Ld*/*m* chart with the accompanying information that the installation was incurring Aeolian vibration damage or it was not. It is useful to understand these components of the plot's variables. Diameter and mass are features of the conductor being strung. Span is a feature of its application, and is a set of fixed values for a line. Thus, '*Ld*/*m*' is a statement of the conductor's nature expressed as an inverted unit of density. *C* is made up of tension that varies with weather and use and mass. The placement of *m* on both axes can be cancelled out, so we have a plot of the conductor installation's fixed nature against its varying usage.

On such a plot, the installations that were unsuccessful tended to plot to the right side of the chart at higher *C* (tension) values or high on the chart at higher *L* (longer span) values. Field data is crude by nature, so the plot appears to show a general but crude separation of successful installations to the lower left and unsuccessful installations towards the upper right, where the *C* value is high and/or the span length is long.

A line drawn between the successful and unsuccessful points on the plot amounts to a declaration of the *C* versus *Ld*/*m* relationship that can be used to manage the vibration damage. CIGRÉ chose to draw a vertical line separating the two. We do not know if it believed that this was the best representation of the physics in play or just a simple but approximate representation. We expect the latter based on its quotes above. It was perhaps driven by the more than 20-year-old push within the industry to use *C* as a single value that is independent of span length or conductor type for addressing the subject – a notion that we and it acknowledge as suspect.

Alcoa did not draw a vertical (single *C* value) line through its plot of similar data. The plot is not even close to vertical. Their curves, as shown in Figure 4.1 for one size of ACSR, honour the idea that span length and/or diameter affect the vibration, that all conductor sizes and types do not vibrate or damage equally. In other words, they were more receptive to acknowledging the complexities of the energy in versus energy out equation in their work.

To say that both systems of separating what works from what does not work means that the understanding of the subject is crude and that while both formulas for separating the workable from the unworkable are decent, perhaps neither is efficient. The use of both systems and taking the worse case as guidance may be overly prudent and expensive. The use of both systems and taking either case as acceptable would be acceptable and more efficient. In other words, we like the intelligence of the Alcoa calculation but suggest that CIGRÉ values may be used as minimum values for longer spans of lighter wire.

With all of that said, you will see that any damper vendor's recommendation tends to come with a disclaimer of responsibility. Great! The very best guidance should be

Figure 4.1 The catenary constant versus *Ld/m*

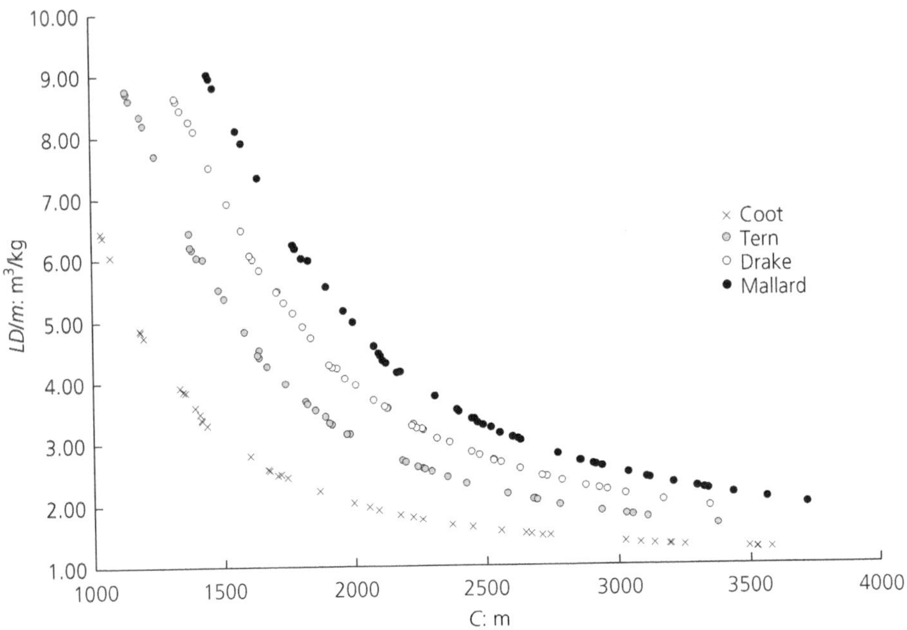

tempered by your knowledge of and experience with other lines in the area or similar terrain with a similar conductor. To be fair to all, we must distinguish to what subject each party is addressing their remarks. When you seek a threshold below which no dampers are required, you are addressing the subject of a conductor's own self-damping. When you are seeking a threshold below which a non-zero number of dampers work successfully to manage the situation, you are addressing the subject of damper characteristics. It is the former subject that CIGRÉ and most of this topic was addressing. It is the latter that the vendors' disclaimer addresses. Since vendors' dampers are unique products with unique capabilities, their recommendations cannot be transferred to a competitor's product.

It is a well-understood feature of ferrous metals that they exhibit a stress threshold below which all the vibration cycling possible will not damage the material. This threshold does not exist for non-ferrous metals such as aluminium. With aluminium, fatigue failure can occur at any stress level, however low. However, the lower the stress, the more cycles required to do damage.

In the cycling–stress relationship, we could control damage by either limiting the cycles or limiting the stress. The choice is available when the wire is ferrous (steel), but when the metal is aluminium, we must limit the cycles. There are spans of steel cable in existence where the damage problem is controlled by limiting the stress in the wire. With aluminium strands in a wire (conductor), we must use self-damping or dampers to limit the amplitude of the stress cycling to an acceptable level. There has been some effort put into determining a suitable stress level in aluminium conductors to avoid Aeolian vibration damage. In principle, this level seems elusive since it must be linked to the cycling frequency and amplitude.

Ultimately, on the issue of determining the need for vibration dampers on a project, the line engineer solicits a 'recommendation' from a damper manufacturer. This is necessary because each product has unique damping capabilities that are unknown to the engineer. The manufacturer is the only one to know its dampers' capabilities. The dissipation side of the energy equation is the manufacturer's to quantify. In fact, each manufacturer uses different data from the line designer to make its analysis. It is not an exact science! Who do you trust?

As the US, Canadian and IEC standards/codes all indicate, successfully mitigating Aeolian vibration damage is up to you, the engineer. We have shown you that the subject is poorly understood, and that solutions are even more poorly specified by various guides. You would likely be remiss in believing in any one of them implicitly and ignoring the others. Consider it all, pay attention to the debate among the experts, watch for facilities such as the one you are designing/analysing to see what has been working and what has not, and then make good judgement calls for your project. This is why you get the big pay cheque!

We discuss elsewhere the point that the capital cost of a project can be affected by the tension limit put on the line's conductors, and that tensions are most often controlled by the tension limit designed to manage Aeolian vibration damage. Since we have painted the picture that the tension limit designed to manage the damage caused by Aeolian vibration is a very poorly understood value, does it make any sense at all that you seek accuracies and depend on accuracies in your design work that far exceed this very important choice? Of course not . . . except to appease the person with the money who is paying your bills and does not understand what you understand. Carry on!

Conductor selection

We have already stated that a genuinely, optimised conductor selection is typically made via a detailed installed cost assessment along with accommodation of environmental constraints and of issues that are not well represented by the classic, detailed cost assessment. It is not necessary to run the assessment on an exact model of a line. In fact, it may be argued that one should not, because every such detailed assessment may lead to a different conductor selection, and any large organisation will not want a whole set of one-off conductors hanging on their line structures.

Many selection exercises are actually limited to exploring the choices within a long-standing short list of conductors considered as the 'standards' for the organisation. If such a list of conductors has choices in which the aluminium content varies by more than 30–40% between adjacent sizes and there is only one core size for each, then the exercise has been pretty much sabotaged, and the choice can be made quite easily by a less than exhaustive exercise.

> We came across an organisation that was forced by a system operator's decree (law?) to execute a line optimisation for every major (not trivial) project. A line optimisation is essentially another term for conductor optimisation, in that the support structure choice and structure design really need to be done in concert with the conductor selection to find a cost-effective solution, because the two items so clearly affect each other.
>
> The frustration with the exercise was twofold, and one of the reasons just makes one chuckle. First, they had only one structure family to work with, and they had only five conductors to choose from – tiny to large – but they wanted an exercise that was exhaustive to the max. The second frustration was that I later met an engineer from a company that was subjected to the same law, and he said, 'Ah, we just blow that off.' People are interesting to say the least.

However the exercise is to be run – exhaustively or on the back of a napkin – the work is actually very important. So, we will discuss some basics and guiding principles and give a description of a valuable comparative process that is independent of exact project details. Then, we will discuss what is important and what is not, as usual. Finally, we will forward the subject to Chapter 9, where we describe a whole new paradigm for conductor selection.

The conductor selection/optimisation process involves two parts – capital costs accrued during the installation of the conductors and the operating costs accrued during the operating life of the conductors. If greater amounts of capital cost monies are reasonably well spent, they will tend to cause lower operating costs. This means that, as capital costs rise, the operating costs tend to decrease, as illustrated in Figure 4.2 several pages on. This is most recognisably the case when the additional cost is spent at least in part on putting more aluminium into the conductors, because this single act reduces electrical losses practically in proportion to the amount added. The classic exercise is one of finding the conductor that provides the lowest sum of capital cost plus the present worth of a selected number of years of operating costs.

As noted, not all conductor selection efforts are exhaustive. Some organisations are so focused on capital costs – the outflow of money today – that they ignore the counter-balancing operating cost savings. In the end, they are likely to have made a less than optimal choice, but, in such a corporate culture, it is because few will care about that.

Keep in mind that there are two conductor characteristics to chase down within the conductor selection exercise. The two things relate to the two parts that make up most conductors: the aluminium and the structural core. First, you are looking for the optimum amount of aluminium to put in place to cost-effectively transmit power with all coincident issues such as noises, fields and desired impedance levels addressed. Second, you are looking for the right amount of structural support within the conductor (i.e. its core content) needed to support the loads imposed by ice, wind and cold temperatures.

Figure 4.2 Net present value (US $) cost versus conductor size (kcmil)

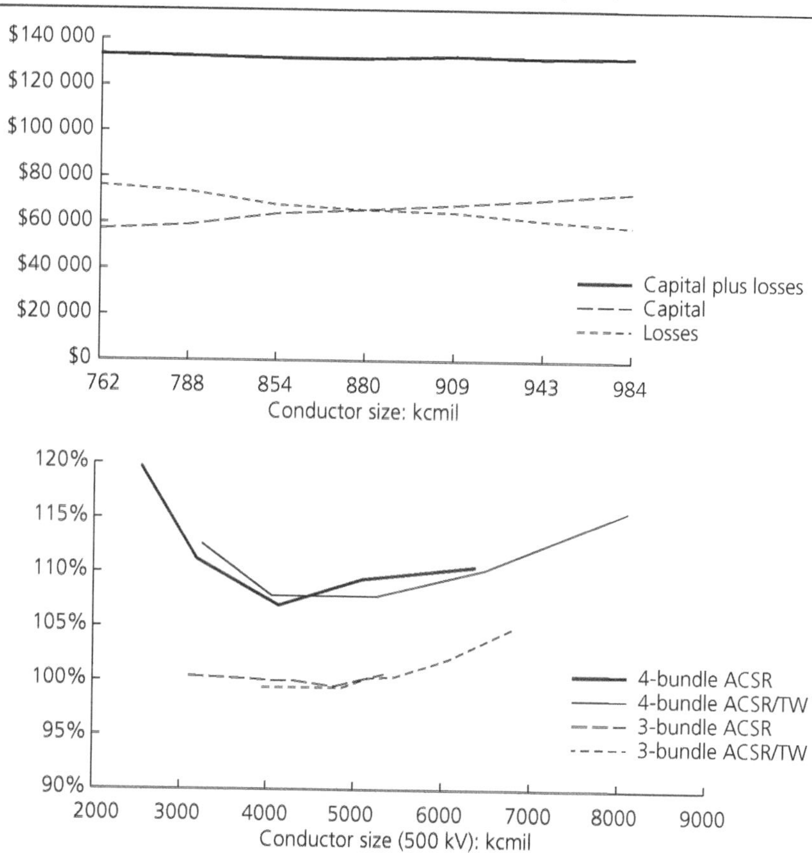

These two searches cannot be run independently, because not only does the core material support the conductor against these loads, it also greatly affects the rate of sag increase that occurs with increasing amperes delivery. This is to say that the core impacts the hot sag and limits the amperes that can be delivered. Said in these terms, you can see that the core choice plays a role in deciding the capacity of the conductor and the cost of the design.

If a planning department of an enterprise selects the conductor for a line, it will not likely get this right because it tends to comprise electrical engineers with little experience of core selection for structural reasons.

Capital costs

The calculated capital costs for installing a set of conductors could include everything that is done to get them installed to make the study valid. For a new line, this will include the costs to run the route selection process, get permits, pay legal costs and so on. Then, there are the costs for engineering, procurement, construction management and company overheads. Some of these costs are very difficult to predict, and a cost assessment that tries to include them will be unwieldy and fraught with inaccuracies to be sure. Although we just said that the 'capital costs... could include...', we want to make it clear that the 'capital costs... need not include...'. It is much easier and more accurate to make the assessment using only the components of cost that will vary with the conductor choice. It is *more* accurate because we can ignore the more variable costs noted above that are common to all conductor choices.

Thereafter, a sensitivity assessment can and should be conducted to reveal whether the cost relationship between a conductor choice and any cost component is sensitive or insensitive. Components of cost can be revealed to be insensitive to the conductor choice. When this happens, you have two choices. Complete insensitivity allows you to ignore the component and move forwards with the exercise. Weak insensitivity allows you to be less concerned with the accuracy of the cost component's value and to focus on the cost components that matter. Either way, the exercise is simplified, and focus on the sensitive relationships will yield better results for less effort. And do not forget the all-important choice of the very elusive chosen tension limit that must make sense. It would be easy to pull a conductor really tight to reduce sag and save a lot of money, but pulling a conductor tight without thought or knowledge of the many consequences invites real disaster.

Some of the items affected by and affecting the capital cost and viability of a conductor selection are:

- environmental loading criteria and conductor tension limits
- ROW width-constraining clearances and workspace
- noise, corona, electric field limits
- ground profile and alignment
- span capabilities and usage
- maintenance practices
- construction method constraints
- structure preferences and geometries
- cash flow profile of capital costs on drawn-out schedules.

Here is a simplified calculation process that develops comparative capital costs between conductor alternatives on a spreadsheet platform.

Step 1. Subject all conductor choices to a sag–tension calculation with common span, load cases and tension constraint principles. Select a reasonable design span for the line's circumstances. Use the same ice and wind cases for all conductors but truncate the maximum operating temperature for each, based on the ampacity goal of the project, if there is one. Otherwise, make good use of each conductor to get its capacity well utilised.
The discussion above on managing vibration-induced damage to the conductors suggests one of two rational principles for a tension limit. Either subject all conductors to the same C limit or subject each to a

unique limit that intends to produce a reasonably equal risk to vibration damage. For example, the Alcoa recommendations specify span limits for various numbers of dampers placed on the spans. For different wire types and sizes, these limits are also different. Select a C limit for each conductor choice that places each at the midpoint of its own declared span range for, say, one damper per span.

Step 2. The output values of interest from Step 1 include the maximum (design) sag and the maximum (design) tension. The maximum sag will be developed by each conductor's maximum temperature requirement or by a large ice load, if such a load case exists in your exercise. Use the larger value.

Step 3. You now have a list of conductor options that all hang between structures of common span length and height. Designate one choice as the base case. To rationalise the options, you need to correct the ground clearance differences as represented by each conductor's maximum sag. Rather than make structure height adjustments, retain their heights and adjust the spans by the formula derived from Equation 3.18, in which sag_1 and $span_1$ represent the base case conductor:

$$span_2 = \sqrt{(sag_2/sag_1)} \times span_1 \tag{4.1}$$

By Equation 4.1, the conductors with sags larger than the sag of the base case will have their spans shortened, and vice versa.

Step 4. Calculate the wind-sourced transverse load on each structure as a function of each conductor's diameter and now unique span length. Check the highest wind case and any iced conductor case with concurrent wind, and apply the desired/necessary load factors. You now have a list of maximum transverse loads for each conductor option applied to structures of a common height.

Step 5. Calculate the factored design tensions with load factors included. This is the maximum tension that will design the deadend and corner structures. You now have a list of longitudinal design loads for your structures. Realise that, from this point forwards, you are dealing with two structure sets of interest – those designed by wind and those designed by wire tension. Any transmission line is populated with a combination of these two types of structures, and they are not always clearly distinguished from each other. It is not necessary to make a clear distinction at this time, as we will see. Be approximate.

Step 6. There is a simple formula to estimate the weight difference between two structures based on the different loads applied to them. It is

$$Wt_2 = Wt_1 \times (T_2/T_1)^k \tag{4.2}$$

where $k = 0.25$–0.70 and is dependent on the structure type and T is the horizontal load applied to the structure by the attached wires.

The equation says that the weight ratio between two structures equals the applied load ratio raised to the power of k. Clearly, this simple formula will work best when the two structures are equal in height and the applied loads are not too different from each other and of a common nature. This is why we have developed comparisons of varied spans that retain structures of equal height.

For the wind-designed structures and separately for the tension-designed structures, calculate weight differences based on the design wind load and the design tension load, respectively, using Equation 4.2, and express the differences with the weight of the base case structure as 1.0. Some structures will weigh 0.9 and others will weight 1.11, as example values. Do this for the range of k values unless you have data that reveal a more accurate value for k.

Step 7. You now have a list of relative structure weights in the two useful categories for each conductor option – wind-designed and tension-designed structures. From here, assemble the important capital costs. Include in the cost calculation mechanism the ability to declare the installed cost of a wind-designed (suspension) structure, and then express the cost of the tension-designed structures (deadends, corners and structures designed by a broken wire case) with a factor that is a multiple of the cost of the suspension structure. We suggest a multiplier on the suspension tower cost of between 3 and 5 for towers designed by conductor tension.

List the number of structures per unit distance as kilometre/span length (or mile/span length, if you prefer) for that conductor, as determined in Step 3. Set up and declare the percentage of wind-designed structures in the line. The percentage of tension-designed structures is the remaining percentage. A very straight line with a 'not very imposing' containment criteria on all structures will have a very high percentage of wind-designed structures – 90% or so. A line with many corners and/or with a severe broken wire load case applied to all suspension structures will have a lower wind-designed structure percentage – say 60%.

Step 8. Thus, the installed cost of all structures is developed as:

$$\text{installed suspension structure cost/km} = \text{weight} \times \text{suspension cost} \times \text{structures/km} \times \text{percentage of suspension structures on the line}$$

$$\text{installed tension structure cost/km} = \text{weight} \times \text{suspension tower cost} \times \text{tension structure cost factor} \times \text{structures/km} \times \text{percentage of tension structures on the line.}$$

Be sure to include mobilisation and foundations, etc., in the cost – a truly complete installed cost but expressed in the unit weight value.

Step 9. Add in the conductor and fittings purchase and installation costs per kilometre for all conductor options. These should be gathered from the manufacturer and a reputable contractor, respectively. Take care to solicit values of common definition

Step 10. You now can list the capital costs for all conductor options relative to the chosen base case conductor. For the cost difference to be meaningful, effort must be made to include all differentiating cost components. Note that the cost differences between conductor choices are in actual dollars (or euros, yen, etc.) and are not expressible as a percentage of the capital cost for the line because all costs are not included in the exercise, as we described above.

If you have set up the exercise well, you have the ability to easily revise the variables that we have described: *k*, the suspension structure cost, the strain structure factor and the percentage of wind-designed structures. The key to satisfying yourself that the exercise provides valid results is to run each of these variables through a range of 'as small as reasonable' to 'as large as reasonable', and record the effects on the outcome. Independently plot the sensitivity of the cost differences between the conductor options to the range of each variable.

This sensitivity exercise will reveal the importance of getting each variable right, and you may be surprised to see that it does not matter if some of these variables are not accurately determined because the outcome is either fairly immune to a variable's selection or you will see a clear trend and you can fit your project into that trend. Once you have the variables set to where you are content and the capital cost differences are on display, you now have values to add to the operating costs for each conductor option so that you can seek the absolutely best conductor in the whole world.

Operating costs

The items that are or may be of importance for this calculation are:

- the number of years of operation used to accumulate costs
- the cost of money (interest rate) over those years
- the expected increases in cost of power in all of its incarnations
- the loading profile of the circuit over those years
- the line losses based on the conductors' resistance and capacitance
- line and ROW maintenance costs.

There are some basic trends with these items as they affect the outcome. For example, if the normal condition load on the line is low, it will be very difficult to recover the capital investment by counting on savings on losses. This situation drives the favoured conductor choice to the less expensive conductors – mostly as defined by purchasing cost and less so by the impact of the structure, foundation and installation costs, as these are all relatively constant costs. If a very high N-1 capacity is coupled with a low normal condition load, the attraction to a high-temperature conductor increases. But, their costs increase relative to a standard round wire ACSR, so some normal load flow may be useful to cover the higher cost.

I recently worked with a client who, when I asked what the price of power from his facility was, answered, 'Things were simple once but the lawyers got together, and now we sell four different colours of electrons depending on the time of year, reservoir level, generators on line and whatever else. At the end of the day, the price didn't really change, but the lawyers were happy with themselves.'

The capability of conductor types to deliver a good amount of power at a good cost is as follows, in the general order of purchasing cost – lowest to highest:

1 ACSR/TW
2 ACSR
3 ACSS/TW
4 ACSS
5 ACCC/TW
6 ACCR/TW
7 ACCR.

The capability of a conductor type to lower the capital cost of other line components such as the support structures and foundations provided the constraints of existing facilities are not a significant factor (best to worst, using the numbers from the list above) is:

5 ACCC/TW
7 ACCR
6 ACCR/TW
4 ACSS
3 ACSS/TW
2 ACSR
1 ACSR/TW.

You can see that the two lists are nearly opposing forces. The best conductor is not likely to be intuitively obvious and is highly dependent on the normal load and N-1 load ratio and on the impact that your choice can have on the cost of structures and related items.

Does all of that seem daunting? A simple formula to identify the optimum aluminium content is offered. When we came upon this simple formula, we were told by an economist, 'I don't understand it but it works.' Cool!

$$\text{optimal aluminium area} = \sqrt{(\text{PWL} \times A/k)} \qquad (4.3)$$

where:

- A is an estimate of the needed area of a single wire (phase or subconductor in a phase)
- PWL is the calculated present worth of losses for amperes/phase/km
- k is the cost per extra kcmil of three phases of the conductor plus support costs/km. If each phase is a bundle of four conductors, k is then the cost of adding 1 kcmil to 12 wires. Why does everyone want to use k – this is not the k of Equation 4.2!

This formula for optimal area merely represents the least sum of the present worth of losses plus those capital costs directly related to the conductor size. These capital costs include the proportion or ratio of the structure cost needed to support the extra conductor; that is, for an extra \$1 of aluminium, something like \$0.33 is required in extra support costs.

As an example, if $A = 795$ kcmil and if the estimated **PWL** for three phases of 795 kcmil is US \$60 000/km, and $k =$ US \$96/kcmil per three-phase km (based on a conductor cost of US \$72 per three-phase kcmil/km, and a support ratio of 33% for a total of US \$96), then the optimal area is

$$\sqrt{(60\,000 \times 795/96)} = 705 \text{ kcmil}$$

To verify by Kelvin's law:

related line costs are 705 × \$96 = \$67 680/km

PWLs are \$44 000 × 795/705 = **\$67 680**/km (these should be equal)

total cost = \$135 350/km

It is useful to explore the sensitivity of this selection process to errors in the input data. It is assumed that the data of the above example are correct; that is, PWL = \$60 000 for 795 kcmil and $k = \$72\ (1 + 0.33)$ per kcmil per three-phase km. Assume the PWL had been estimated too high by 25% at \$75 000/km instead of \$60 000. The calculated A would be 788 kcmil.

If the line were to be built with 788 kcmil conductors, then

related line costs are 788 × \$96 = \$75 648/km

PWLs are \$44 000 × 795/788 = **\$60 532**/km

total cost = \$136 190/km

Thus, the error of +25% in PWLs will waste only \$1007/km or 0.7% of the total related cost. Similarly, examples have been worked to demonstrate the extra total line costs incurred if the PWLs are estimated at −25% or if the support cost ratio is mistaken by +25%, +100%, −25% or even −50%.

The results are as shown in the sensitivity analysis in Table 4.4, which indicates the line cost errors. The table shows the capital costs rising with increasing aluminium area (kcmil) and the operating costs (losses) falling with the same. But, the point to notice is the near-constant sum of these two cost components. Across a conductor range from 4000 to 6000 kcmil (i.e. 2000 kcmil ± 20%), the cost of the installation is effectively constant.

Figure 4.2 shows data extracted from a real conductor optimisation study in Canada (2006) for a 500 kV line. The four lines plot the net present value (NPV) of the 'capital cost plus

Table 4.4 Optimal aluminium area sensitivity analysis

Case	PWL: US $/km	Support cost: %	Optimal size: kcmil	Capital: US $/km	PWL: US $/km	Total cost: US $/km	Penalty: +%
I	73 000	33	880	65 975	65 975	131 950	Base
II	93 000	33	984	73 775	58 999	132 774	0.62
III	54 750	33	762	57 150	76 161	133 311	1.04
IV	73 000	41	854	64 080	67 957	132 037	0.07
V	73 000	66	788	59 100	73 648	132 748	0.60
VI	73 000	25	909	86 175	63 845	132 020	0.05
VII	73 000	16	943	70 725	61 543	132 268	0.24

the operating cost (losses)' against 20 conductor options ranging from three-bundle and four-bundle options with round wire ACSR and their ACSR/TW counterparts. For the three-bundle options, the cost of the project is constant within 3% for a size range from 3000 to 6000 kcmil, regardless of the wire type (round or TW alternatives). For the four-bundle options, the cost of the project is constant within 3% for a size range from 3500 to 6500 kcmil.

This figure shows that the selection of a precisely optimal conductor for a project is not very sensitive to the aluminium content of the choice. Thus, the selection of the aluminium quantity is not effective in reducing the NPV of the 'capital plus losses' cost of the project, but is effective in determining the balance between these two cost components. We say more on this point in Chapter 9 on sustainable development. Figure 4.2 warrants two further comments.

This study showed that the three-bundle options were all about 7–8% less expensive than the four-bundle options. We would prefer that you do not draw this conclusion from this figure in the belief that it can be applied to your projects. Things vary with time and place, so we expect the reasons for this gap in this analysis, while hidden from us, may also be unique to that project.

Notice too that the TW plots are shifted one step to the right. The data behind the charts is such that each point on the TW line represents the TW version of the round-wire ACSR that is on the companion line one step to the left. For example, the top left end of the '4-bundle ACSR' line is 636 Grosbeak conductor and the top left end of the companion '4-bundle ACSR/TW' line is Grosbeak/TW. And so on. The second conductor in the list was Drake so the plot shows that the NPV cost of 4-bundle Grosbeak TW is identical to the cost of 4-bundle Drake ACSR. Until the conductors get quite large and the capital costs are rising significantly, the TW conductors are offering a competitive alternative to the round wire conductors. This point too is taken up in Chapter 9.

Structures

The selection of the appropriate type and strength of supporting structures is a critical step towards arriving at an efficient, low-cost and durable line. Read that sentence again. The selection has more to affect than cost. It affects efficiency and durability. In fact, cost may be the least important of these on the face of it, because missing the mark badly on efficiency and durability will translate into cost some day.

There is no best general overall choice, but each line project probably has its own best solution. In fact, a long line that traverses different terrain and land uses will have a different best solution for each different portion of its length. It is not by accident that there is no such thing as a 'one structure type fits all' solution out there, although this is only part of the reason.

The available options go from wood pole structures that can be used effectively at lower voltages where there is economical local supply to the more exotic and usually much more expensive steel tubular structures that are sometimes required and justified for close-up visual benefits, albeit sometimes without aesthetic success.

The process of selecting your preferred structure type is frequently limited or restricted to making a few minor adjustments to existing standard designs, limitations made necessary by some overriding single imperative such as having to build on very narrow ROWs or building in urban areas where aesthetics control. At the other extreme is the line to be built in the countryside where the line engineer has only the pure challenge of matching their techniques and art to the demands of the terrain and the forces of nature. If you have never had this opportunity, trust us ... *this* is the more rewarding engineering exercise of the two, due simply to its engineering purity. Love to be spoiled!

Following is a discussion of the options available for a high-voltage line design with a listing of some of the issues that should be assessed before reaching a decision.

A very general comment

Transmission line structures serve no purpose whatsoever except to:

- hold the highly electrified conductors far enough above the ground and other surfaces so as to allow the safe activities that we all must and choose to perform in their proximity to occur with an acceptable level of safety
- hold the conductors in a position from grounded items and from each other to allow the air around them to provide sufficient insulation so as to allow an acceptable quality of operation, and finally
- be works of art on our landscapes and in our neighbourhoods for us to enjoy or, at the very least, not be offended by them.

This last purpose is wishful but not meant to be entirely facetious. If the entire human race is either connected to electricity or would like to be (with a few exceptions), then why is it that transmission lines – most notably, their support structures – are so disliked by nearly everyone?

It is a frustration, is it not that we have chosen to be in an engineering field where what we produce is so disliked. Maybe, I take this a bit too much to heart

> and, if so, it is because more than one of my friends have over the years come to understand and have told me that I am 'not an engineer but a frustrated architect' and that I am 'a transmission architect'. Perhaps this aesthetics bugaboo is mine to bear.

It would behove our industry to do two things regarding our transmission lines. First, we must stop putting structures that people think are ugly in their neighbourhoods. We are good at erecting ugly structures and unnecessarily so. Second, we should execute a campaign that argues against the notion that our lines are a blight on the landscape. Why have we let the public and their representatives win the aesthetics argument? In this section, we explore structure functions, materials, types and styles, and will offer facts and opinions on better ways of doing things than we often see done. We will pick up this aesthetics argument again in Chapter 8 on projects.

Technical functions

Remember that the operational functions of transmission line structures are to provide acceptable levels of safety and power delivery security, and to please – or at least not offend – the eye. Their technical functions are to keep the conductors at a distance from various things, including each other, and adequately to resist the structural loads that the conductors impart on them for various reasons. In this later role, we make a number of distinctions.

Conductors cannot be held in the air above the ground without adopting tension within. So, wires in a span not only have weight and are pushed down additionally by ice or snow or construction activity and are pushed laterally by wind but they are in tension according to the formulas offered in Chapter 3. In this environment, support structures fall into two functional categories, albeit messily. Here, we want you to understand structures in this functional context.

There are structures that do not terminate the line permanently or temporarily or are not used to change the wires' direction and there are those that do. Structures that do not terminate the line permanently or temporarily or are not used to change the wires' direction are not affected by the tension in the wires. Conversely, structures that do terminate the line permanently or temporarily or are used to change the wires' direction are affected by the tension in the wires.

The magnitude of the effect of the wires' tension on the support structures increases with the magnitude of the direction change and the chosen circumstances for termination duty. That is strangely worded, but it is a clean definition of function that is easily translatable into practical terms.

Deadend or strain structures

A structure used for permanent line termination at the end of a line is called a *deadend* or *strain* structure. Such a structure is always understood to be capable of supporting the highest tension of all of the attached wires as is expected to occur during the line's construction, maintenance and operation periods. The electrified wires will be terminated on the structure with insulators that, due to the tension in the wires, will be oriented in line with the wire tensions. Deadend structures are clearly necessary at the end of a line, and that structure might take the form of a substation strain structure. Deadend structures can also be placed along the line at logical locations. These logical locations are where large angles are turned, uplift is to be managed or where a tension change is advantageous. These latter two reasons for a deadend's use may include no line direction change at all.

A deadend at the end of a line may be subjected to tension loads applied constantly to its one face. A deadend structure set within the line may be normally in a nearly balanced load condition but is expected to be capable of carrying all of the tension from one direction only under any circumstance, however rare.

False deadends

A main point to understand about deadend or strain structures is that they are generally understood to be capable of supporting the largest conductor tensions that the line was designed to experience. A deadend or strain structure is easily recognisable by the strain insulator assemblies placed on it to connect the conductors to it. A deadend or strain insulator assembly is distinguished from a suspension insulator assembly by the simple feature that it is oriented in the direction of conductors whereas suspension insulators are oriented in the vector sum of the line tensions attached to the structure from the two directions.

There are very few structures in use where the insulator assemblies are deadends or strain assemblies but are connected to a structure that is unable to support the maximum loads that the engineer plans for. They exist but they are dangerous. This is not academic. *If it looks like a deadend structure, a line crew may believe that it can carry the big loads of a deadend. Therefore, it must.*

There are tower designs that use strain insulators to turn a modest line angle. This is done because the alternative is to have suspension insulators that are swung laterally towards the tower and away from the tower, and some organisations don't like this for some reason. Strain insulator sets can also be put on a light duty tower simply to allow a change in conductor without employing a mid-span splice. Some people do that. These uses do not require a strong tower.

When the tower looks just like a tower that can support all the tensions ever expected based on the appearance of the insulator assemblies on the tower but the tower's strength can be understood only by the most experienced eyes, and that strength is actually not there, you invite real trouble. People have been hurt and have died for this reason. More is said on this in the next chapter.

Corner structures

Corner structure need not be deadends. It might be clearer to say that they need not be strain structures. It is often reasonable to turn a corner with suspension insulator assemblies. When this is done, the structure is called a running corner. The ability to turn a corner with suspension insulator assemblies depends on the type or shape of the structure to which they are attached. We have seen running corners greater than 90°.

So, the label 'deadend' or 'strain' structure is a function label. The 'corner' structure label is simply a geometry label.

There is one nuance to the corner structure definition. That is with large vertical angles. A tower that turns no significant line angle in a lateral sense but that carries the conductor through a large vertical angle change is somewhat defined by the wire tensions. This is a fact, but since vertical loads on towers are not very expensive to support compared with supporting horizontal forces, there is little impact from this fact.

Conditional terminations

While deadends placed within the line are intended to carry small tension unbalances routinely and maximum tension unbalances under rare and problematic circumstances, we refer here to another function often labelled as 'failure containment'. While deadends designed for maximum load unbalances can be used to stop cascading failures, it is often regarded as acceptable that the tensions in play during a cascade failure are not the maximum tensions, and containment can be successfully achieved often enough by structures that are capable of withstanding conductor tensions that are less than the deadend structure design tension. In other words, a cascade failure can be contained by a type of structure that has sufficient strength to do so but only under some circumstances, not all circumstances. So, some containment structures use suspension insulator assemblies (not deadend assemblies) and suspension strength towers (not full-capacity deadend towers).

It is the case that selected condition containment structures are very often beefed up suspension structures. It is a matter of risk assessment and choosing to spend less on containment in the belief that doing so is economically sufficient. If you want to make this sort of risk assessment wisely, then you are well advised to be very sure that you understand the effect of slack in starting and ending cascade failures.

Suspension structures and tangent structures

After all the fancy work is done by a small number of structures at the line's ends and corners, special locations on a radical profile and selected other locations, you get to traverse most of the line's length with structures that do little but carry the conductor through light angles, both vertical and horizontal and through no line angle at all. These structures all use suspension insulator assemblies, and those that turn no line angle are called tangent structures. The 'suspension' label is a functional label, and the 'tangent' label is another geometry label.

Note that a tangent structure can be a suspension or strain structure type, just as an angle structure can also be a suspension or strain structure type. Still, it is usual that the vast majority of structures on a line are suspension towers with little or no line angle to support. As this becomes more and more true, the economics of a line design requires more and more attention to the design of these structures.

Wire system structures and support system structures

Now, we will see where these functional and geometric structure types fall within the useful pair of functional categories that we announced above. A wire under tension that changes direction imparts a force in the plane of the wires equal to

$$\text{force} = 2 \times \text{tension} \times \sin[(\text{angle turned})/2]$$

When the direction change is horizontal and is through a line angle of θ, the horizontal (transverse) force T is as below:

$$T = 2H \times \sin(\theta/2) \tag{4.4}$$

In this equation the horizontal component of the wire tension H is the only part of the tension that feeds the horizontal component of transverse load T. The direction of the vector T is horizontal and along the bisector of the angle turned. When the angle turned is vertical, as caused by gravity and perhaps supplemented by adjacent support points that are out of vertical alignment, the vertical force V is as follows:

$$V = T_{\text{back span}} \times \sin(\theta_b) + T_{\text{ahead span}} \times \sin(\theta_a) \tag{4.5}$$

where

- T is the tension in the wire (not the same as T in Equation 4.2 – lousy symbol control in this business!)
- θ_b is the angle of the wire from the horizon in the back span (+ is below the horizon)
- θ_a is the angle of the wire from the horizon in the ahead span

From Equation 4.5, the sum of θ_a and θ_b is the angle turned through the support point, and is often called the departure angle. Suspension clamps must be shaped to accommodate this turned angle of the wire to avoid damaging the conductor where it bends sharply at the ends of the support.

It is described that we have two categories of support structures – those designed by the wire tensions and those designed by wind loads. For small line angles – say up to about 10° – the effects of transverse load on structures developed by the wire tension are modest, and for most spans where the vertical departure angle of the wires through the structure's support points is modest, this separation of functions and labels is useful. But, the boundary between the two functions is not perfectly clear. It is a transitional concept. When we talk about useful load cases for structures and line failures, we will make much more use of this set of definitions and take the concept to another level for the purpose of understanding lines.

Here's a preview. We have declared that the conductors and other wires that span between the support structures should be understood as being the most powerful structural elements of a transmission line. In fact, a transmission line should be understood not as a series of structures with wires spanning between but as one very long structure that is primarily made of wires. With that in mind, we separate that very long structure into two components: the *wire system* and the *support system*. The support structures that

Figure 4.3 The wire and support systems

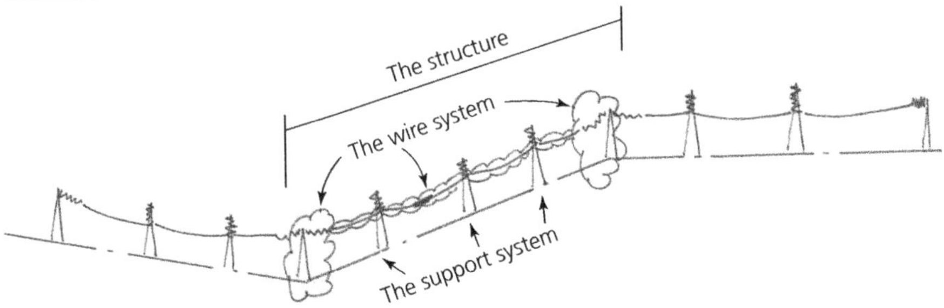

are designed by or heavily affected by wire tension are considered as part of the wire system. The structures that are basically unaffected by wire tension and designed primarily by the wind are part of the support system. This is the foundation for understanding line failure mechanisms and understanding how to respond to failures with your design activities. Let's be clear on this.

The wire system

If we understand a transmission line *structure* as all support structures, spans of wires, hardware and foundations that could cover many kilometres of distance between and including the deadend support structures that establish the limits of the *structure*, then the parts of that structure that are designed primarily by the tension in the wires are the wire system. Figure 4.3 illustrates.

The support system

All of the support structures, hardware assemblies and foundations in the potentially very long transmission line *structure* defined above that are defined by the wind and not by the tension in the wires make up the part of the *structure* called the support system.

In the wire system, there are all of the wires in the spans between structures, support structures that are designed by wire tension, including the hardware assemblies on those structures and the foundations under them, as shown encapsulated in Figure 4.3. In the support system, there are all of the support structures designed by wind forces blowing across the line, including the insulator assemblies loaded primarily by gravity, and including the foundations under these structures.

The complexity is that some support structures fall in between these two simply defined groups. They straddle the fence, as it were. For example, structures at small line angle changes tend to drift into membership in the wire system as the line angle increases. The same can be said for tangent structures that carry very large vertical loads or have many of their components designed by a broken wire case.

When we talk about and consider failures and failure prevention in our design work, recognising this very long *structure* and distinguishing these two systems takes on value.

Materials

Structure materials vary widely – and why not. We humans keep trying new things, and we favour things at hand or things that fall apart at the slowest rate. Things at hand and the rate at which things fall apart are not constants across the globe. All preferred choices are local. So far, we have not seen line structures made of mud, china or cardboard! We list the reasonable possibilities in some logical manner.

Wood

Wood comes in enormous varieties around the world. We have hardwoods and softwoods, and the variety in strength, weight, growth rates, decay rates and straightness of the grain and limbs is truly all over the map. Few species seem to be ideal for transmission line poles and sawn braces or arms. Those of us in North America live in one of the lucky parts of the planet in this regard. Here, the best species are light-weight softwoods that are found in many regions and grow reasonably quickly. We also have a climate that is reasonably kind to the wood once installed as a pole. This is not the case everywhere.

Many parts of the world have fewer attractive species from which to choose. They may have a lesser quantity of trees from which to choose. Their environment may be less kind. And so on. There may also be competition for the material that raises the cost or causes theft.

Where useful wood species and quantities do exist, wood does make a good transmission line structure material option. To enhance its lifespan against decay, preservatives are typically applied to either the butt that will be placed in the soil or to the entire pole. Wood is easily sawn for creating shapes useful for spars and braces. It can be glued into larger sizes. It can be drilled and climbed with tools and spurs without much ordeal or much harm to it.

The unique problem with wood compared with all other material options discussed here is that it is a naturally created material, and quality control is not so good. Local species of value are offered as options, and these have various strengths and decay rates. Old-growth trees tend to offer different strength properties than second- or later-growth trees. The base of a tree is not the same strength as the upper parts of the same tree.

Finally, trees are used for many other purposes. If someone takes your perfect tree first, to carve it into a totem pole or into a bunch of sawn timbers for roof beams – that is too bad. It takes time to replace the perfect tree. You won't want to wait! The cost of wood poles is related to supply and demand, and the industry constantly believes that the supply is diminishing. Thus, the rise of the alternative materials described below to replace wood poles that have historically been economical provided as wood. In the past, most poles were smaller than desired today, because the voltages were lower. Wood is sold by volume, and the volume rises exponentially with length. Wood remains a very attractive material for small poles except where environmental constraints apply. As poles become tall, the economics of some alternatives increase rapidly, and at some point wood poles are not a sensible option.

In North America, the warmth and humidity of the atmosphere is modest in all but the southern parts of the USA. Wood poles do quite well, but their weaknesses include woodpeckers and rotting of the base as a function of the amount of air that can access the soil in which they are set. Wood poles last longer in clays than in sands, where the moisture changes seasonally. Woodpeckers are a mobile problem. Areas that have been without the problem for decades can find this is not a permanent condition. Their attack rate can be impressive.

In our preface, we allude to the fact that one of the few constants that you should expect is change. Statistical analysis – at least as we use it in this industry – assumes that data from the past will predict the future. Ha! In recent years, the North American winters have warmed enough not to kill pine beetles and the like with winter temperatures. Our evergreen forests came under attack and were being wiped out in huge swathes. We see dead trees to the horizon in every direction in the West. All of these dead trees attract ants, and they are not discriminating between the dead trees and the power poles in the same area. This is to jump ahead one step, but we are being told that steel today is not the quality that it was two decades ago, and problems are surfacing because of it. Beware the assumption of the constant!

Steel

Would it be silly to say that steel has been with us since the Iron Age? Steel is a well-developed, man-made material that has wide use due to the varieties available at excellent performance-to-cost ratios. It can be shaped into virtually any form and engineered in a host of useful forms that permit the creation of structures to any useful size. It can be modified materially (alloyed) by the addition of small amounts of many other elements.

Rolled-steel shapes and plate are available in strengths as low as 200 MPa (30 ksi), and readily available to strengths of seven times that. Transmission line structures made of steel are routinely made up of materials with this full range of strength. Where a member does not require a high material strength to perform, as with redundant members in a truss that provide stability to another member or a component that relies on its stiffness to perform, the material of choice is not of the very strong variety. Cables made of stranded steel cables can form very useful parts of structures, and these are available in strengths exceeding 3000 MPa (200 ksi).

An Achilles' heel of steel is its response to being placed in air. It oxidises (rusts). Over time, the material will erode to nothing and be gone back to Mother Earth. We slow the process by several means, such as:

- self-weathering alloys that slow or halt oxidation
- galvanising (hot dip or cold brush on application)
- paint or metallised coatings
- combinations of the above.

Steel has another Achilles' heel, but it is not the metal's fault! We noted above that steel is available in many incarnations. The problem is, very few line designers and engineers understand the differences in these incarnations very well at all. The extent of knowledge of many is limited to having memorised the ASTM – or equivalent – 'label' of our favoured steel alloys (A36, ASTM 572, etc.) while knowing nothing about the performance differences between these various steels. This is akin to shopping in the grocery store and caring little whether you put mustard, potato chips, olive oil or dog biscuits into your basket. It's all food, isn't it?!

We promise you that it matters when you select/specify steel varieties for your work. Do you understand Charpy values and when to apply them? Do you know that all alloys – including some of your favourites – are not routinely capable of passing a Charpy test and will cost you a lot of money if you ask for it? Do you know enough about welding? Have you been told yet that welding practices and standards applicable to buildings and bridges are not appropriate for big steel poles? We bet not, yet there are some serious things going wrong out there because we brush off welding as something to be taken for granted or someone else's concern. Has the scrap metal content of new material risen in recent decades and is today's steel the same as yesterday's? Yes, and it is not. Get curious and stay out of trouble! Be one of the few!

Aluminium

Aluminium – the 'miracle metal' according to a friend in the aluminium foundry business – is a very appealing structural material. Often, its excellent strength-to-weight ratio is well worth the higher 'per unit weight' purchase cost. But, there are limitations to its availability thanks to the world's commitment to commodities rather than choices. It would seem that the best days for aluminium structures are behind us because few manufacturers will commit to their production for business reasons.

It used to be that many of the aluminium structures produced some decades ago were made with unique shapes formed by dies that are no longer to be found – either at all or in working order. Today, the member shapes are not what they used to be and not what we'd like to see. Running aluminium material through manufacturing cutting, punching and drilling machines requires the use of a lower grip pressure on the piece than is used for steel. The machines that are made and used in North America cannot make this pressure adjustment – excluding manufacturers with these machines from providing the products we want.

So, while it is not impossible to find materials and suppliers of aluminium structures, the hunt may be difficult and limited. You may have to chart new territory to make good use of 'miracle metal' in large quantities in structures. It will also require that you get a good education in alloys and so on. Aluminium is smooth, thermally expansive and it flows (creeps) with time under stress. So, the joints are a whole new thing to understand. There is less friction, and grip can loosen with cold and time more than steel joints without special attention paid to them. Have fun!

Concrete

The good thing about concrete poles is their low cost to produce and the fact that they can be made quite large for application on significant lines. Their Achilles' heel is their – no surprise! – weight and poor performance in a freeze–thaw environment. Thus, concrete poles are used for small distribution poles, and are generally found near their places of origin due to shipping costs, and these places of birth are all in regions such as the southern USA where cold weather cannot treat them poorly.

Large concrete poles are spun so that the concrete is forced to the outer ring of the cylinder, leaving a hollow centre. The spinning also forces the water out of the mix, developing very low water : cement ratios and very high compressive stresses. The strength of concrete poles depends on the number and strength of steel reinforcing rods and the concrete mix. This combination of materials and production methods means that the user (you) cannot look a standard concrete pole up in a catalogue as easily as you can a wood or even a steel pole. You tend to need the participation of the manufacturer on your design team to get the poles that you want. How are you at playing with others?

Fibre composites

A few adventurous manufacturers of resin/fibre composites attempt from time to time to develop and offer valued pole and cross-arm product lines made from fibre-reinforced materials for transmission line use. There are some successes, although few have yet permeated the market for extended periods. We should appreciate their efforts for, over time, they will make gains, and the products do fill needs. Without bogging down discussing selected product lines, there are a few things to say in general about resin/fibre composites. These are points often common to the resin/fibre composites used for conductor cores that we discussed above.

First, shapes can be formed either by pultrusion, extrusion or wrapping over a mandrel. The last process creates hollow thin-walled tubes useful for poles. The strength of any such product lies in the choice of resins and, more importantly, the fibres and in the directions in which the fibres are laid. Thus, strength control lies with the manufacturer, as it does with most manufactured products. However, the cost-effective products use reasonably small amounts of material due to material costs, and the materials have moduli of elasticity lower than steel. This leads to poles that can be small diameter and very flexible or large diameter and very stiff. The stiffness comes from the geometry, not the material.

If you have a long history of working with wood poles or steel poles, you may have become lax in your concern for monitoring deflection. This is because the industry has come to accept the natural flexibility of wood poles as within satisfactory margins. Steel poles are also acceptably stiff unless they become quite slender and tall. Composite poles, however, have a very different strength-to-stiffness relationship. You can purchase a very stiff pole or a very flexible pole, and both are very strong.

Consider a fishing rod analogy. The thin end near the tip is very flexible, and can be bent without breaking. The handle end is very stiff – you can neither bend nor break it. Resin/fibre poles can be the same. You can bend the thin ones under load twice the amount that you can bend a wood pole, or you can buy a fatter pole that will hardly bend at all – but you cannot break either one. The point is that you must learn to watch the deflection behaviour of these poles, because deflection will be your deciding factor for selection.

Types

We have two sections of discussion here with labels that appear similar – types and styles. 'Types' refers to structural differences, and 'styles' refers to geometric, aesthetic and functional differences.

Fixed-base structures – tubular

Structures made of tubular, but not necessarily hollow, components have the characteristic that individual members are quite large and probably quite heavy. They can be as simple as a single tubular pole made of a single section, or multiple sections connected end to end and with or without any supplemental tubular sections – arms – for connecting to the conductors (Figures 4.4 and 4.5). They can be as complex as multi-pole arrangements joined by braces and arms or as simple as a wood pole. The pole sections are either slip joints or bolted face-to-face flange joints. Brace and arm connections are pinned or framed. The connection to Mother Earth is direct buried like a typical wood pole or through the use of a large welded baseplate to interface with a concrete foundation via anchor bolts.

Tubular structures are almost always highly populated with frame connections, putting the majority of the structure's main members in bending. In the next chapter, we revisit the following three principles of structure design. These principles are all you need to explain the comparative cost and efficiency of the use of materials. These three principles are:

1. The most efficient structural element in tension is a steel cable.
2. The most efficient structural element in compression is a latticed steel mast.
3. The most expensive use of material is in bending.

Figure 4.4 Unique tubular structures. (Courtesy of Peter Catchpole)

Figure 4.5 A more elegant looking three-legged tower. (Courtesy of Peter Catchpole)

Principle 3 in particular explains our complaint of tubular structures. The other reason for their higher cost is their large foundation, as the structures must also resist bending forces.

The tall red and white tubular tower shown in Figure 4.5 was a proposed three-legged, 145 m tall, long-span design with catenary curved legs and Fibonacci-spaced braces. It was a combined truss-and-frame design. I could not sell the idea at the time. Dang!

Fixed-base structures – latticed

The term 'latticed' refers to the appearance of the structures because they are a 3D truss made up of comparatively small, light members that are pin connected at their joints. Well, almost. The typical latticed structure or major structure component is a four-sided box – square or rectangular and tapered or not – in which the chords along the four edges are designed to resist tension or compression forces. The lacing or bracing between the four chords resists shear forces across the chords' directions and/or provides support to the chords against compression loads. The chords and some of the larger bracing members are often continuous members through the truss joints while the lesser members are closer to pure pin connections.

These structures are not the simple structures that are used as classroom exercises. They are sufficiently complex that they are still studied within the industry for better understanding.

Figure 4.6 A frighteningly tall H frame structure. (Courtesy of Peter Catchpole)

The complexities are derived from joint eccentricities, joint slippage, fabrication tolerances, and even nuances in member arrangements. The very tall braced H frame pictured in Figure 4.6 uses lacing that is welded – not bolted – to the leg chords. How pure is that truss? A full understanding of the behaviour of latticed structures under load is not a trivial undertaking.

Software such as PLS-TOWER makes various assumptions and ignores various realities to set the stage for doable calculations. Make sure that you understand these assumptions and ignored realities before you declare an understanding of the software's analytical results.

Guyed structures

Well, if using trusses to remove the bending forces from the structure's members and foundations to save weight and cost makes sense and yields fruit, is there money to be saved by using cables for tension members? Can structures be reshaped so that members can be tension only and therefore made with cables? The answer is 'yes'. Using guys to support structures provides considerable cost savings compared with self-supported structures. Revisit the three principles above to understand the source of the savings and sensibilities. A very thorough paper (White, 1993) discusses guyed-V structures for high-voltage transmission lines.

Brian White claims to have come up with the guyed-V structure. We can't argue the point. He said that when he saw an internally guyed portal frame in Sweden (Figure 4.7), he

Figure 4.7 A portal frame. (Courtesy of H. Brian White)

envisioned swapping the two sloped legs left for right, including the guys. This put the two masts onto a single foundation point and put the guying outside the tower's space towards the edges of the ROW. It appears to have been a worthy idea because there are guyed-V towers all over the world.

As you analyse the guyed-V design, you find that the bridge of the tower takes on a larger and larger percentage of the tower's overall weight as the voltage rises. The reason is that the number of conductors increases with voltage and the horizontal separation grows faster than the vertical clearance requirement to ground. At 800 kV, the guyed-V design looks quite top heavy. Figure 4.8 is of an early vintage guyed-V tower in a modest 345 kV line.

The final step as fuelled by Brian's seemingly endless imagination was the creation of the cross-rope suspension (CRS) tower in 1974, a tower that was largely inspired by the CRS assembly installed over the avalanche-swept valley in British Columbia and described in some detail in Chapter 6.

In Chapter 6, 'Fun with cables structures', we further described the CRS tower design and its origins. The feature of the tower to notice in Figure 4.9 is that it is basically a guyed-V design with the two masts spread apart and the heavy EHV bridge replaced by a very efficient cable system. The weight saving over the guyed-V design at 400 kV voltages and above is significant for that reason.

Wood structures have used guys to support corners, terminations and highly loaded tangent structures at all voltages since the beginning of electricity. It has been an adventure in the industry to carry this simple concept up to structures of the highest voltages and carrying heavy wire loads. There are a few things to understand about guyed structures in order to avoid big trouble.

Figure 4.8 A first-generation 345 kV guyed-V tower. (Courtesy of H. Brian White)

For any structural member to adopt load (stress), it must adopt strain. In simpler terms, a guy wire must lengthen to carry load. A pole or beam must bend to resist load and so on. If you guy a wood pole, the flexibility of the pole is likely sufficient for it to deflect a visible distance before it adopts much stress in its fibres. An attached guy wire is relatively much stiffer, and is likely to take up the load long before the pole takes up the load. Thus, the pole is supported by the guy almost completely.

Figure 4.9 A 735 kV CRS design. (Courtesy of H. Brian White)

Consider the guying of a much stiffer moment-based steel pole – one that is rigidly bolted to a foundation or embedded in very solid ground. The stiffness of the pole will prevent the guy from adopting such a large percentage of the load. It may be that the pole can overstress before the guy can elongate enough to carry much of the load. You could say that the load sharing is complex and indeterminate. If the pole is stiff enough, adding the guy may look useful on paper but can actually be nearly useless.

If the supporting guys of a structure are long, the strain required to adopt the load requires considerable elongation. If the guy angle is steep, its elongating is increasingly ineffective in supporting the structure's lateral movement. The support that guys provide to structures is based on a complex relationship between the lengths, orientations and elasticities of all members. The historical practice of the design of relatively small wood poles for low-voltage lines has used a linear analysis method in which deflections are ignored, and the results have always been understood as very approximate at best, but acceptable.

For large structures with complex member arrangements and multiple guy attachments, non-linear analysis that accounts for the effects of elongations and deflections is absolutely necessary.

Even so, if the elasticity and particularly the deflection of a member is often not well known, the resulting load sharing between guys and structure members is equally poorly calculable. Think of wood poles with their inherent variability of properties and dimensions. It is something of a joke to think that these are well-understood structures. In our industry, most analysis tools approximate various important characteristics of a structure, and the result must be understood as equally approximate. Much of the complexity is removed when the guyed structure's connections to its foundations and between its main components are pinned connections.

This is the case with the X structure shown in Figure 4.10. This structure is hinged at the base such that it can lie down on the ground along the centreline of the line. It is held upright by the guys ahead and back. This version of the X structure is tubular. In a way, it is a truss structure, and in others it is a frame. Roughly 25% of the weight of the structure is in the knuckle where the four tubes connect to each other. Why? There are large bending moment forces to support right there.

The trick to getting the cost of a guyed structure down as much as possible is to align the guys to attachment points on the structure to reduce or eliminate the bending moments in the structure caused by eccentricities between the guy alignments and the centre of effort of the load developed through the conductor attachment points.

For example, if you look carefully at the guyed 'delta' or 'banjo' tower in Figure 4.12, you see that the apex of the guys is at a level near the bottom conductors, well below the centre of effort of force caused by transverse wind on all phases plus the overhead ground wires. This means that a transverse wind on this tower design puts the mast in bending. If the

Figure 4.10 A longitudinally guyed, tubular X tower. (Courtesy of Peter Catchpole)

guys were oriented to intersect at the apex of the load, the bending would diminish. Then, either the tower could be lighter or the risk of failure would be lowered.

Styles

With tubular structures, you could say that only your imagination is the limiting factor in expressing yourself aesthetically or functionally because we do routinely see architects around the world expressing themselves with tubular steel forms. Not so fast! In the transmission line business, we (1) employ very few architects, (2) have line owners paying for the structures who are rarely trying to express themselves aesthetically and (3) have manufacturers dedicated to the industry but who are very much in the commodity business. These three facts limit the ability to (1) express yourself aesthetically and (2) move very far beyond anything that has already been done. To break this barrier, your best argument is structural performance and cost. But, we present an argument later as to why even cost does not matter. Good luck!

Since latticed structures are assembled from many small stock shapes, it is less costly to develop structure configurations with more diversity than it is with the large custom-made tubular parts. The basic building blocks for latticed structures are stock items, not large custom fabrications. With rare exceptions, the installed cost of latticed structures, with foundation costs included, is less than that of a tubular structure designed for the same purpose. This is true whether the structures are standard configurations or unique. Why? Trusses and their tension-and-compression-only members and foundations employ the wisdom in the three principles stated above.

Pole structures are the structure type of choice at lower voltages. As the span lengths increase and the wires get larger and the number of circuits increases, the pole material of choice moves away from wood simply because large trees are hard to find and expensive. The man-made materials – composites, concrete and steel – become the materials of choice. With steel poles, there seems to be no limit to what is accepted or attempted. In North America, the tubular steel pole industry is served by more than half a dozen capable manufacturers, and they offer products in excess of 10 ft (3 m) in diameter and more than 200 ft (60 m) tall. They support double circuits at 500 kV and up to six circuits up to 230 kV routinely.

Since the installed cost of these large poles and their foundations can rarely compete with the cost of a latticed structure for the same task, one should ask why they are used so often. The most popular answer is aesthetic preference by someone – the owner, the local landowners or related parties. The technically necessary answer is limited ROW space. Placing a line within a usable portion of a road or other type of easement does require a very small footprint, and that forces the foundation into a bending design.

Self-supported (unguyed) latticed towers come into their own at transmission voltages, and can be configured as either quite modest structures or as large as anyone will ever require (Figure 4.11). It is interesting to travel the country and the world to see the variations on a theme that exist. Some seem attractive and some seem quite unattractive. Perhaps the choice lies with your own history. It is usually the other places' structures that look less appealing than the ones that you grew up with. This is true if familiarity is a source of comfort.

Figure 4.11 A 500 kV/230 kV river crossing tower

Figure 4.12 A single-circuit 345 kV guyed 'banjo' tower. (Courtesy of POWER Engineers)

Given the almost consistent cost advantage of latticed structures, it is frustrating that they are so widely viewed as unattractive compared with tubular poles. There is a simple truth about structure aesthetics: latticed towers are complex and unattractive up close yet nearly invisible from afar while tubular poles are simple and more appealing up close but very visible from afar.

Guyed structures almost exclusively support a single circuit (Figure 4.12). Thus, they do seem to fill a niche rather than be a full service option, albeit a significant one. A common complaint is that they have a large footprint, with the guy wires spread out in (usually) four directions. Knowledge of the practices of others might be useful.

There are two purposes for ROW dimension requirements. First, the conductors' electrical issues must be accommodated all along the line. Call this continuous strip the electrical ROW. Second, the structures must be given space to be installed and maintained. This requirement is limited to wherever the structures are placed. Call this the structure easement. The structure easement for self-supported structure types fits easily within the electrical ROW, and all is good. The ROW width is then a single value, serving both structure and wire purposes, and the idea that you should even think about structure easements evaporates from your list of concerns.

Roadside distribution-line ROWs include special add-ons at corners for guy anchors that project out of the road allowance that is used as the electrical ROW. These protrusions are effectively a structure easement. If you take a view similar to that of a distribution engineer, then the ROW width can be tailored to the conductor needs, and add-on patches are a different requirement at the structures. With this view, you can accommodate the large guyed structure footprint without the need for widening the electrical ROW for the entire length of the line. You can end up with a long, narrow ribbon of ROW for the conductors to occupy with periodic and not-so-intrusive wider patches of structure easements at each guyed structure. Each of these categories of easement has different constraints to place on landowners and line owners regarding the use of the location and vegetation clearing. In total, the compiled uses and constraints are less onerous than the present single easement definition imposes. But, old habits die hard, don't they. Keep hammering!

Insulation

We appreciate the viewpoint that a valued purpose of transmission line insulators is structural. A line's structures support the electrified conductors above the ground at a safe distance from things on the ground and from each other. The insulation that separates the conductors from the structure itself, the other conductors and these things on the ground is air. The insulators simply join the conductors to the supporting structures. They pierce that insulating air medium between these two incompatible (electrified and grounded) line components.

Needs and coordination

Electrically, the insulators must hold the conductor sufficiently far from a grounded surface to avoid unwanted flashovers through the air under various high-voltage events, and have sufficiently long surface paths not to allow tracking while contaminated with dust, water or industrial pollutants. There are two sources of electricity to be managed – the stuff in the wires that is being transported and the disruptive stuff from the sky: lightning.

As described in Chapter 2, lightning can strike the line in one of two ways. If it strikes a conductor, it can do considerable burn damage to the conductor metal and considerable electrical damage to the system because the voltage is far beyond high. With this type of strike, the flashover will be from the conductor to the grounded structure. It is preferred that the lightning strikes the structure, and so possibly avoids jumping to the conductor. The purpose of a line's overhead ground wire(s) – aka shieldwire(s) or skywire(s) – is to intercept the lightning strike and hopefully send it to ground through the structures' grounding systems without bothering the electrical duties of the conductors. The higher the voltage of the line, the less likely that the strike will jump to the conductors as the preferred path. Jumping across the air gap created by the insulator becomes more difficult as the insulator and the gap get bigger. As the insulator length increases, the strike will more likely travel to ground through the structures.

Said another way – low-voltage lines with inherently shorter insulators are less likely to provide a large enough air gap to the ground to prevent lightning flashovers from the structure or ground wire to the conductor than are higher-voltage lines with longer insulators. Sometimes, flashovers are controlled by placing shorter insulators on one phase of the line, attracting the flashovers to only that phase.

Structurally, the insulator must have the strength to support the conductor tension and loads that you plan to survive, all with an acceptable factor of safety. Pin-connected insulators have a handy characteristic for the structural engineer. Whether in-line strain insulators or suspension insulators, they always point in the direction of the load that the conductor imposes on the structure. Said conversely, the load on a structure can only be applied through a suspension or stain insulator in the direction that the insulator is pointing. Nice to be able to see load vectors, isn't it?

The structural issue with fixed-base (not pin-connected) insulators such as line posts is that they want to point in the direction of the applied load. When they do not, the insulator is put into bending. Remember bending . . . expensive bending?

Types

As just noted, insulators come in two structural forms – pin connected and fixed base (Figure 4.13). Pin-connected insulators are either a series of short units that are pin connected end to end to build the desired length or a series of longer units, sometimes a single unit of the desired length. Regardless, they are rotationally free at both ends. More accurately, they are meant to be rotationally free. Your selection of hardware at the ends should never hinder this freedom without considerable thought on your part. The common insulators are series of short ball-and-socket units, and long-rod porcelain units and polymer or non-ceramic units. The three types have very different structural properties.

Figure 4.13 Insulator types. (Courtesy of Peter Catchpole)

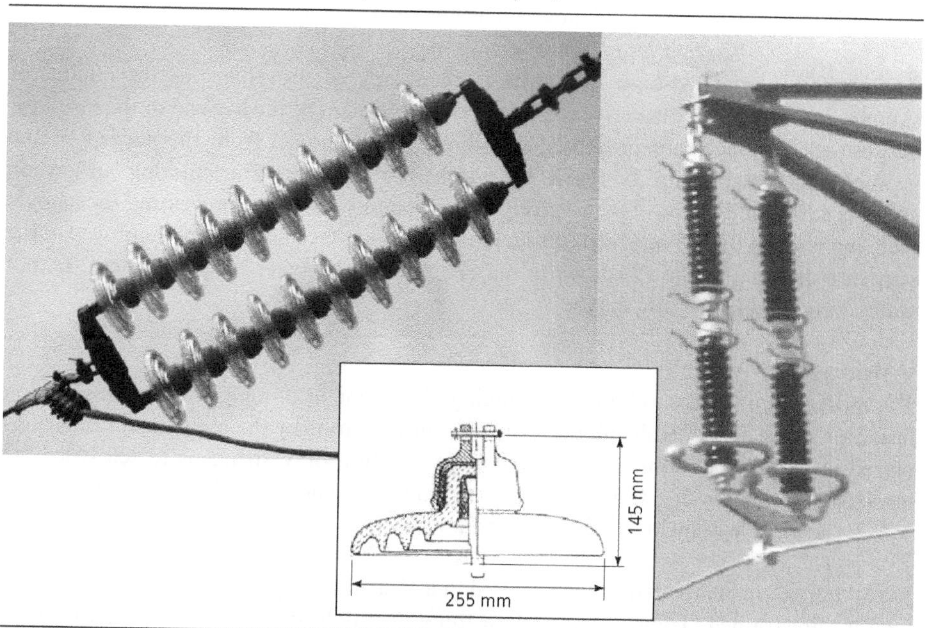

The ball-and-socket string and a diagram of a similar clevis-eye unit are shown in Figure 4.13 alongside a long-rod insulator. The point of interest to the structural engineer is that the tension in the insulator due to the weight of or the tension in the attached conductor stresses the long-rod porcelain with that load but does not stress the porcelain of the ball-and-socket or clevis-eye unit. In the ball-and-socket or clevis-eye unit, the cement between the top cap and the porcelain dish and between the porcelain dish and the pin carries the load, putting the porcelain in some compression and shear. Not that one is better than the other, but they are different beasts – structurally.

The ball-and-socket style of insulator unit is also available in toughened glass instead of porcelain.

Non-ceramic insulators have a long solid rod of fibreglass and resin, with end fittings compressed into the rod and polymer sheds fitted tightly over the rod. The polymer is available in several materials, silicon being the best performer and the most expensive. The non-ceramic materials are more prone to corona damage than the glass or porcelain materials, so they require greater care with corona ring protection and a lower voltage. This has been learned the hard way in many jurisdictions.

Fixed-base insulators are called pin insulators in the low-voltage world and post insulators in the higher-voltage world. They sit vertically or at any other orientation, depending on their function. There is a middle-ground insulator type called the hinged post, which is free in one axis but not in the others. Post insulators are designed for and limited in capacity by the bending forces placed in them. Calculate the cost of a post insulator per unit of force that it can support and compare that with the cost of a suspension or strain insulator (pin connected) per unit of force supported to understand the relative cost of supporting loads by tension versus bending.

The attraction to fixed-base insulators is their lack of movement at the conductor support end. It is often attractive to prevent the movement of the live end of the insulator because the displacement prevented is a dimension removed from the needed spatial geometry of the structure or the ROW width. For example, if a 5 m-long suspension string is calculated to move transversely 1.5 m due to wind, then preventing that movement means that the 1.5 m of separation to the structure face need not be employed in the structure design and the ROW width may be reduced by twice that movement. Is that really a big problem? Only maybe.

V strings

The most common way to limit the swinging displacement of a high-voltage suspension insulator string and thus shrink the structure dimensions is by the use of a pair of strings placed in a V shape. Some say that the main function of V strings is to sell twice the insulators as is necessary. Certainly, the constraints on space should be real and valuable before the money is spent on all those insulators.

Recall that an insulator string such as used in a V string will point in the direction of the load if unrestrained. The most important thing to watch for with V strings is to ensure that a wind force cannot push the windward string of the pair further than the assembly's natural geometry (angle from vertical) and collapse the leeward string, risking the binding of its hardware parts. V strings are set at angles so that all of the loads that are planned to be applied to them cannot collapse the assembly in this way.

It gets tricky sometimes to develop a geometry that contains the load vectors because it is a fairly common practice to include very-high-speed winds in the load case set, making the situation seem unsolvable. We point you to the discussion on wind and wind loads, where you should find comfort. In response to the issue, it is easy to find V strings with a wide range of geometries.

Very deep Vs – with each insulator less than 45° off of vertical – cannot support a large transverse wind or line angle. Very shallow Vs – with insulators much more than 45° off of vertical – can carry much larger transverse wind forces or line angles. The latter geometry causes concern for the tensions in the insulators and for electrical clearances to the supporting structural members above the assembly. The entire design exercise is one of fitting through the small eye of a needle.

Keep in mind that the ease with which an insulator swings or a V string is loaded to its angular limit is a function of the wind/weight span ratio. As the ratio increases (and the weight span decreases), the amount of swinging that occurs for each wind pressure increases. So, in rough terrain where the wind/weight ratio is all over the map, finding V strings with a sufficiently large angle off of vertical becomes more difficult. Managing insulator swing and V string integrity is a good reason to smooth the conductor profile by varying structure heights when in rough terrain.

When used at running corners, the V string is essentially tilted to orient it usefully against the array of load vectors that it must support. As with very shallow Vs, this is where the use of hangers is useful.

Finally, we offer a comment on forces that V strings impose on the support structure. Without perfectly intuitive knowledge of loads, you might think that using a V string in place of an I string changes the loads on the support arm. Consider a V string with each unit at 45° off of vertical – a most common and rational configuration. One end of the assembly is attached at or near the structure interface of an arm and shaft (body) and the other end is attached at the end of an arm. A purely vertical load such as occurs every day or with ice on the wire shares that load equally between the two support points, and the load at the end of the arm is half the load that would be imposed there with an I-string alternative.

Consider the wind component of a load. A lateral force of '1 unit' applied at the apex of the V by the wind-loaded span will impose an equal vertical force of '1 unit' at the end of the arm, as would an I string. So, as lateral load increases, the load on the windward arm approaches that of an I string. Concurrently, the load on the end of a leeward arm diminishes. This is easy to visualise if the wind is sufficient to move the load vector to parallel with the outboard insulator offloading the inboard insulator completely. So ask yourself this: is the structure designed to have an intense vertical load applied to the windward phase while the leeward phase is unloaded completely and the middle phase is loaded as . . .? Even if the answer is 'yes', then ask yourself: is that an efficient structure design?

Horizontal V strings and braced posts

There are two versions of the horizontal V string that we recall were not distinguished by unique labels when first developed. Maybe we weren't listening well! Recall the post insulator. When attached to the side of a structure and oriented near horizontal and perpendicular to the direction of the attached conductor, the post was not very strong against vertical or longitudinal loads. Dang bending! To help this not so great arrangement at higher voltages where the cantilevered post is necessarily long, two things can be done.

First, the end of the post can be supported against vertical load by an insulator in tension placed between the end of the post and up to a point of the structure above, usually at an angle steeper than 45° above horizontal. If the post remains fixed to the structure at its base, this is called a braced post insulator. Left as such, the assembly remains weak against longitudinal unbalanced loads, and can be challenged by them. As an option, release the fixity of the post at its base and allow the braced post to rotate longitudinally on a hinge mechanism due to any longitudinal load. This rotating version of the braced post is called a horizontal-V insulator.

The trouble with braced posts is their limited longitudinal strength. The trouble with horizontal-V insulators is their lack of restraint against longitudinal loads. There have been really annoying and sometimes catastrophic failures caused by the latter. We discuss these in Chapter 7. The feature that must be included in the horizontal-V assembly to provide it with some amount of longitudinal restraint is an axis of rotation that is tilted off of vertical.

If the axis of rotation is vertical, it will have zero restraint against a longitudinal load applied by the conductor at the other end of the post. If the axis of rotation is rotated 90° to the horizontal and now perpendicular to the vertical face of the pole, you have a simple V string as discussed above. Imagine the path that the conductor attachment point at the apex of a V string traces as a longitudinal force is applied there. It is an arc in the vertical plane that is parallel to the conductor direction, and it rises as it leaves the normal position of rest under a support point. As it moves away from the position of rest below the pin points at the pinned support ends of the assembly, the vertical load is coupled with the offset to build a restoring force to counteract the longitudinal force.

For horizontal V strings, the industry seems to have settled on an axis that is tilted 15° off of vertical. The axis passes through the hinge at the base of the post, and the companion hinge that defines this tilt is determined by a rigid connection for the pinned end of the suspension insulator that supports the post that is offset outward from the structure face a suitable distance from above the post's hinge. Now, visualise the arc drawn by the conductor's attachment point to the end of the post as a longitudinal force moves it

Figure 4.14 An unusual insulator bundle from the 1920s. (Courtesy of Tacoma Power Company)

along line. The restoring force relies on the rising of the arc from its low point and the vertical force that would bring it back to the low point. This is a much weaker force than is available to the V string with its horizontal hinge. Thus, we can understand the failures that that arrangement invites when longitudinal loads are applied.

Bundles

When a single insulator cannot support the loads applied to it, a common solution is to double up – or triple or quadruple or more. Bundling insulators into clusters is common with suspension units whether I string or V string and with strain insulators. The challenge is to provide mechanisms that develop equal tension into each parallel unit or string in the bundle. Figure 4.14 from the now dismantled 1926 crossing of the Tacoma Narrows in Washington State, USA, is not the normal solution. Here, 12 parallel strings set between two rigid yoke frames were held to equal tensions with compressed springs. Equality was estimated by spring compression.

The typical equalising mechanism is a triangular yoke plate that has either one back support point and two forward support points all in one plane, or one back support point with three forward support points in a triangular arrangement (Figure 4.15).

It is easy to understand the load sharing that occurs if you imagine that you are building a big, heavy mobile much like over a baby's bed. It is equally useful to understand how the load sharing changes when things start to move towards unbalance.

Consider a triangular yoke plate with dimensions as illustrated in Figure 4.16. If load P_2 exceeds P_1, the plate will rotate about the top hole, making dimension C less than B since B and C are measured as offsets from the centreline of the reaction load vector,

Figure 4.15 A three-bundle suspension string yoke. (Courtesy of Peter Catchpole)

$P_1 + P_2$. With B equal to C, P_1 must equal P_2. This balancing act is caused by the moments $P_1 \times B$ and $P_2 \times C$ that develop due to the fact that $A > 0$.

If $A = 0$, placing the top hole midway between the two outer holes, rotation of the plate will cause no change in dimensions B and C, and the yoke plate is rendered useless as a load equaliser. In fact, the efficiency of the yoke plate as a load equaliser increases as the ratio $A:(B + C)$ increases. However, as that ratio increases, more and more rotation is needed to find balance.

Figure 4.16 Yoke plate principles

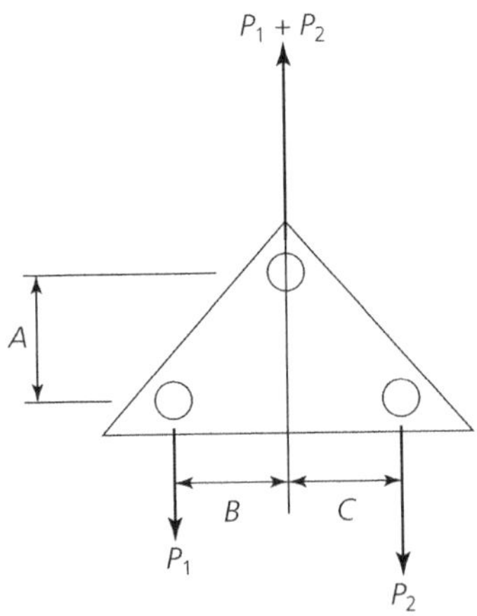

Suppose load P_1 goes to zero. Dimension C then also goes to zero, and dimension B is irrelevant. With dimension A large relative to $B + C$, a very small rotation of the yoke plate is required to align P_2 with the vector $P_1 + P_2$. The rotation is accompanied by a very small movement downwards – as viewed in the figure – of the attachment point of P_2. If dimension A were small relative to $B + C$ and load P_1 went to zero, the movement downwards, as viewed in the figure, would be larger.

Thus, we see that a yoke system – our big, heavy mobile – which may be made of stacked yoke plates to hold a bundle of four or more insulators or subconductors, requires that dimension A be greater than zero on all yoke plate parts, and, if so, the assembly does its job nicely. However, dimension A has a great effect on the load sharing that will occur should one of the attached loads $P_1, P_2, \ldots, P_n$ disappear and rotations are set in motion to find a new equilibrium. If you follow this through, you will find that the loss of a single insulator unit in a string within a yoked bundle of loads will lead to the remaining units being well out of equality of load sharing as a function of the assembly's A dimensions.

In other words, when an insulator string breaks when part of a bundle of insulators is yoked to ensure load sharing equality, the load sharing between the remaining insulator strings will not be equal, and at least one of the strings may be very highly loaded. If you change an intact four-bundle assembly into a three-bundle assembly because one string failed, the load on at least one of the remaining strings will change from $P/4$ to more than $P/3$. Thus, if an insulator bundle breaks when already pushed to a large tensile value, there is little hope for the string to remain intact.

This frustrating condition can be assisted by employing a larger $A:(B + C)$ ratio for the yoke plate dimensions. Such is not our common practice. This bundling arrangement exists by electrical necessity on very-high-voltage lines, and is a compelling argument for never pushing the tensile load usage of insulator strings in bundles to anywhere near their mechanical strength limits because, when something goes wrong in an insulator bundle, there is no good back-up plan. The assembly will tend to tear itself apart with increasing efficiency, especially if the coefficient of variation for the strength of the many units within the assembly is large. The broken state is more likely to find another weak unit.

It is a fact that the risks embedded in arrangements increase with their size and complexity, and the size and complexity increase with the voltage. Frustratingly, we want risk to diminish with voltage. Enjoy the challenge!

Materials

> When I was a kid, three of my uncles worked in the Ohio Brass insulator plant in Niagara Falls, Canada. In the 1980s, they all retired early because the plant got old, fell into poor condition and their product line suffered. Remember the expanding cement problems with porcelain ball-and-socket insulators? That was them. They did it. Probably not their fault.

Porcelain insulators, especially in the bell unit form that connected together to create string assemblies, were the inexpensive and widely used insulator for decades. In the

1970s, the non-ceramic or polymer insulators came onto the market. We think of porcelain as simple and a continuing and singular recipe of clay and ... brown sugar or whatever. We immediately understood polymer insulators as anything but a singular recipe of poly-stuff, and the trouble the non-ceramic insulator manufacturers had was finding a polymer recipe that actually worked well in all necessary ways and was cost-competitive with other formulas that shared technical qualities of value.

Part of the polymer insulator product was the structural component made of glass fibres and resin (fibre-reinforced plastic or FRP) that formed a central rod on which the polymer skin was placed. The FRP rods were designed for tension or bending as needed. The resin recipe and choice of glass fibres also took some time – decades – to perfect. As this material issue was getting sorted out over this long period, units of 'less than desired quality' were sold and installed on lines all over our networks. The ensuing problems that surfaced over the years were expensive to correct, and many people remain quite spooked at the mention of polymer/FRP products for our transmission lines. It would seem that anyone who has followed the progress made over the decades with polymer insulators has been able to find some comfort in the eventual product lines being offered. Still, the spook remains.

All the while, the glass insulators – again in the form of bell units that connect together to form strings – have remained in play. While there are several manufacturers around the world, the best known are the Sediver products out of France. These insulators look simple but are in fact a very high-tech product. The glass is toughened, and the outer skin of the glass is actually in compression. There are considerable stresses intentionally built into the glass. The result is a product that explodes into tiny fragments when mechanical damage is initiated by striking the unit with a rock or bullet.

This is not the case with porcelain because it can be cracked to the point of failing electrically but without appearing to be damaged at all. A primary attraction to glass insulator units is their ability to display in a 'no kidding' fashion their mechanical and, therefore, electrical integrity at a glance from a distance. Besides, they look nice! They tend to be the insulator material of choice for many EHV lines because of their ability to be easily seen as electrically and structurally intact – or not.

Hardware

Line hardware is generally spoken of as falling into two use categories – *conductor* hardware such as insulator assembly parts, including suspension clamps, splices, bundle spacers, deadend fittings, dampers, marker balls, and *structure* hardware, such as guys, guy and anchor assembly components, bolts, braces, links, clevises and ladders.

Recall our definition of the *wire system* and *support system*. It is also useful to consider two hardware categories that fit neatly into these two systems. If you do that, you will recognise that insulator assembly hardware will fall into either system. Strain insulator assemblies and splices are loaded by the attached wire tension, and are part of the wire system. Guy wire systems supporting structures at terminations and corners are part of the wire system. Insulator assemblies loaded by gravity and guy wire assemblies loaded by wind are part of the support system.

The value in looking at hardware in these system categories is useful when assigning safety factors to them based on their comparative importance to holding the line in a serviceable state when things become unhinged. For example, is the structural integrity of a shackle supporting a suspension string of insulators on a support system structure as important to the line's serviceability and security as that of an identical shackle supporting a deadend string of insulators at a wire system structure? The answer is 'no', so seeing this shackle in a role is more important than seeing it as the same shackle.

Types

We make the *system* distinction so that when it is time to assign some form of load and factor of safety to a hardware item, you recognise whether you should be compatible with loads and factors used with the wire system or with the support system. They will vary at times of importance. That said, there are hardware items in both categories that do not participate in the stresses imparted on the systems – for example, marker balls, bundle spacers, vibration dampers and ladders – but they certainly affect the stresses in the systems.

Hardware can have a very predictable strength and be very trustworthy or not, depending on its materials, manufacturing process and its role in the line. For example, bolts are well understood, and perform very predictably, provided they are made with the material and process described by standards. In part, this is because it is tough to misuse a simple bolt. As mentioned, they are well understood – and the material's properties seem to transfer nicely to the item's performance. Other items, such as cast or forged items, can be more complex shapes, but the material's properties bear little resemblance to an item's properties because the item's shape means everything in defining its strength. The best example is cast metal strain clamps used to terminate a wire at a structure. One of our failure stories in Chapter 7 illustrates this point very well.

We will make the point that the structural integrity of the wire system must be higher than the integrity of the structure system because there is much more at stake with regard to the wire system integrity than for the structure system integrity. Coupling the importance of the structural integrity of the wire system components with the varied predictability of hardware items that can be part of that system, and with the truism that the 'devil is in the details', we cannot emphasise too strongly that this is a subject that deserves your full attention.

Materials

Line hardware is almost exclusively made of a metal, although there are sometimes organic materials involved in the form of FRPs and elastomers. Organic materials, especially elastomers, break down over time without very special care. Metals do what metals always do: they corrode for numerous reasons (e.g. oxidise by rusting), they wear on each other and they fracture in fatigue. All of these failings take considerable time – usually, but not always – and are often near impossible to detect without considerable effort (cost). Since these items are found on a line in the hundreds and thousands, it is a daunting task to instil and maintain their integrity.

Many of the very expensive failures that occurred during the ice storm in the Montreal area in 1998 were triggered by the failure of an insulator in a deadend assembly or the failure of a U bolt at such a tower. Items worth a few dollars cost the system many millions to repair. Certainly, any maintenance programme that can successfully get the monitoring job done is worth doing. But the job of the designer must be to play it safe. Hardware integrity, especially within the wire system – is no place to try to save a buck. The structural failure of a $2 hardware piece being the cause of a massively large and expensive line failure is an all-too-common story. Don't let it be yours. Use oversized items where wear can occur. Configure parts to minimise wear. Watch for placing incompatible materials together, be they metals far apart on the galvanic series or metals incompatible with acidic soils and so on. Get vibrations out of the system as much as possible.

Stay on top of the industry's studies of hardware performance with age. In recent years, studies are indicating less than desired performance of compression fittings for conductor splices after years of heated operation. Recent studies are indicating that the polymer of insulators is being destroyed by corona cutting more than anticipated. The use of corona rings on polymer insulators to mitigate the problem is being recommended at lower voltages than expected. Unless you dedicate your career to this subject or include the subject in the bigger scope of line engineering as a whole, staying on top of this subject will be a career-long adventure. Just one more thing to enjoy!

How you would like to select hardware

It is often suggested above that a primary goal of the line designer is the efficient use of material. The employment of probabilistic or reliability-based design methods is aimed at improving our ability to reach this goal. We seem to believe that we can tweak sags and tensions on our wire system to optimise it. We can select from a wide variety of structure types and strengths and spot them along the line to make very efficient use of them.

It follows that we would like to choose hardware that precisely. But wait – your choice is a 10 000 kg unit or a 20 000 kg unit. I need a 18 000 kg unit for 80% of my structures and a 23 000 kg unit for the rest! There goes our ability for precise control over predictable failure sequencing. The hardware industry is not interested in playing along with such a finely tuned design method.

It follows, then, that the hardware that cannot be tuned to the fine detail that we aspire to seek should be stout enough to be left out of the efficiency-seeking probabilistic calculations. That is fine and workable provided that you also understand the effects of hardware on failures well enough to remove them completely from all actual events that may occur on your line anytime in the future. If you have not, you have created a very efficient design that may someday fail and become instantly highly inefficient.

Remember: we design for one set of events, and lines fail for other reasons. More often than not, a piece of hardware of some type or another is in the mix of the failure triggers.

How you have to select hardware

Most important hardware items are presented with load capacities attached to them. Clevises have large strength increments: 20 kip units, 40 kip units, 80 kip units, etc. Insulators, guy wires, anchors and so on are the same. When you put a string of six items together to carry a tension load, you may have up to six strength capacities represented. All are above some minimum, but some will be well above. Simply put – where is the efficiency?

Most utility stores are reluctant to fill their shelves with all of the material options an engineer would like to have. Most line crews would not want to fill their trucks with all of that gear either. Most engineers would not like to see the mess made in the field if the construction crews were asked to make sure that the correct choice of guy wire or insulator – out of six options – got put in all of the right places. That won't happen.

We suggest that the efficiency available through wise hardware selection lies in honouring the issues associated with the stores department, the line crews and construction practices. Any money that you could save via the probability methods way of thinking won't lead to success. In fact, the probability methods recognise this and give up on trying to fit the hardware selection process into the design method.

The strength of an assembly of parts that carries a load is only as strong as the weakest component, and some components are weakened to below manufacturer declarations by the nature of their use within the assembly. In Chapter 7, we describe a hardware item worth less than $10 that will not work should it ever be called upon to 'do its thing' for the simple reason of its orientation within the assembly.

If the devil is in the details, and hardware is the details, then getting the hardware right matters a great deal. Here are some guidelines for hardware selection. Some of these were noted earlier:

- If there are many pieces of the item for the project, be very careful with quality control, limits of use with respect to strength, orientation and compatibility with other items.
- If there are very few pieces of the item for the project, be very conservative. Do not scrimp. You are saving no money by scrimping.
- Watch for wear on parts that move cyclically. Try other arrangements.
- Watch for binding of parts that move when a failure has started. Let things articulate unless that will be a problem.
- Do not use items that are handy but that, with thought, will be seen to force other choices that are not good or necessary.
- Work with hardware with good track records and vet other items thoroughly. Not all new things are good and not all are bad.
- When you hear of a failure, find out if hardware caused it, and if it did, find out why. Learn from failures.

- You can never save enough money on a project by skimping on hardware choices. But you can invite a failure that will break the bank if you try. Lower the project costs by other means, not this.

Foundations

When we speak of foundations, we are lumping two very different materials into a single category. We have the fabricated material (wood log, metal screw and plate, concrete, etc.) and we have the soil in which we place that material. It is necessary to make the distinction between the two.

Our understanding of the strength of steel, wood or concrete is at least a magnitude better than it is of soil, especially untested soil. When it is time to assign a factor of safety to the foundations in the good old strength versus load equation, you are wasting a lot of fabricated material if you assign the typically large (soil) factor to it just because you did not make the distinction noted above. Give the soil – Mother Nature's part of the foundation – a factor of safety of 2, 3 or more; but give the man-made materials a more modest factor of safety of 2 or less.

Types

There are basically two types of foundations – moment-carrying foundations and tension-only/compression-only foundations. You can think of direct buried poles as moment foundations. Otherwise, moment foundations tend to be poured concrete or some piles.

Large moment-carrying concrete foundations in North America seem to be almost exclusively drilled piers. They most commonly support steel poles with an anchor bolt interface. If the soil can be drilled, they seem to be the foundation of choice. No foundation in decent soil could be more expensive except for a concrete foundation formed in an open hole. The reason is best understood by the three structural principles that we try to drive home. The third principle says that the most expensive use of material is in bending. That is why steel poles with drilled piers are expensive and almost always more expensive that any structure with no significant bending forces involved.

The ability to get equipment and material to the structure sites and the right or willingness to tear up the site during installation contribute to the driving forces behind foundation type selection and therefore structure type selection.

It is worth noting that guyed structures and latticed towers tend to remove the need for expensive moment-carrying foundations. Guys use tension-only anchors, and the guyed structure tends to require a compression-only foundation. In good, firm soil, these foundations can be relatively cheap to install. In very soft soils, they can be more expensive but still much less than moment-carrying foundations in the same soil. Be very sure that when you do any structure option studies, you include the installed cost of the associated foundations in the analysis.

Consider another point when designing foundations, including selecting the type of foundation. It is hard to imagine a location where you would accept the failing of a foundation as part of a structural failure. When a foundation fails, a whole category of equipment and materials must be brought to the site to reinstate the foundation compared with dealing with a failure event in which the foundation remains intact and can be reused. The more remote the facility, the more this matters. Try to design foundations that can survive the supported structure's collapse.

Micropiles have come into their own in our industry in the last decade or two. They install quickly and with modest-sized equipment, if you are careful. They can be installed in clusters to carry compression loads and even modest bending, and they are very efficient in tension for guyed structures. The beauty of micropiles is that they can be installed in nearly any material, from soft sands and clay to rock. The method is to drill through whatever is down there to a depth that will provide the bonding length needed to carry the load. You drill until it works. When you drill, nearly nothing comes out of the hole needing disposal. The cost of the installation has more to do with getting to and from the site than it does with how deep you drill when you are there. You can imagine that this is not true of drilled piers with their large volumes of concrete and moved earth.

Select your contractor carefully, as there are expensive drillers and less expensive drillers, and a good product is not always related to that cost. The concept of drilling until it works requires that the driller – the individual with their dirty gloves on the machine's controls – is the person who will make the call. Let them do that but confirm their decisions with a testing programme wherein you test the capacity of the installed micropiles by testing at least one unit at every structure. If that one fails, test another, and take appropriate action thereafter.

The nice thing about micropiles is that the steel rod of the pile is surrounded by a cylinder of grout that protects the steel against soil-induced corrosion. If you trust that the soil will not interact badly with steel and if you have no rock within the depth of foundation that you plan, you can install screw anchors. Some jurisdictions with such favourable conditions have developed the technology to install rather long and large-diameter screw piles with big crawler machines. They go in fast and with little disturbance to the soils – even less than micropiles. But, the conditions need to be right. There is little worse that can happen to a contractor or to the person paying for their services than to find rock where none was expected and the foundation is a design that cannot penetrate. That gets expensive very fast.

Materials

In soil, especially near the ground surface where the water/air interface moves around, remember that wood rots, steel corrodes (even galvanised steel) and, yes, even concrete has been seen to fall apart.

Wood rots fast in sand because the water drains away, giving air its turn at attacking the wood. Wood in clay that is always damp fairs much better.

Galvanised steel can completely corrode away at the ground-line in less than 15 years. It only takes the right kind of soil and electrolytic action to eat it up. If you put in a guyed system, be sure you are protected against this action, and start by understanding the chemistry of the soil.

Concrete is created by a complex interaction of natural materials. Do not think that it is a very simple material to specify and understand just because we have been using it for everything under the sun for centuries. Respect it and get educated in its formulations and uses or, someday, pay a price.

Analysis methods

As with everything else, there are numerous computer programs that address foundation design issues. Most require certain soil parameters, so if you do any soil investigation work, be sure to ask for the appropriate values from the laboratory.

If you get into drilled piers, be sure to use a program that features the ability to analyse piers with large bending moments and small shear and small axial loads. The vast majority of drilled piers are under buildings and bridges, where the bending moment is secondary to the axial and shear loads. Programs aimed at those foundations may not do a good job analysing piers under transmission line poles.

It is necessary to point out to even reputable geotechnical consultants that we in this business have a requirement for structure settlement that is often an order of magnitude more lenient than any other client they encounter – meaning the average owner of a bridge or building. Those folks cannot accept movement of more than a few millimetres or so because more settlement will crack plaster and break glass. But we often don't really mind if some of our structures move 100 mm or more.

We therefore do not share the geotechnical expert's usual level of concern for the nature of soil nor the conservatism that they express with their recommendations. That is to say, we can afford to have them relax their standards, and we need not pay for their usual conservatism. Before you hire a geotechnical investigator, impress upon them the same sense of value that they should have for your project. They may not be aware of it, and you may otherwise feel like you are paying them for too much effort.

The counter-argument to this is that, unlike most of a geotechnical consultant's projects, a line project is spread over a long distance, and you will be asking them to extrapolate their findings further than most of their clients.

The bottom line says that a very good knowledge of soil conditions may be required to bring in a properly priced project. Unexpected soil conditions are the most common sources of contractor requests for more money than budgeted. The combination of lack of knowledge of soil conditions and certain contract language is a recipe for budgetary disaster.

The very best defence against cost overruns with a transmission line's foundation costs is to try to use a foundation type that works no matter what the contractor encounters. That is why we like micropiles and the structure types with which they work.

Summary of useful equations

Here, we collect the various equations presented throughout the chapter for your ease of reference.

$$\text{span}_2 = \sqrt{(\text{sag}_2/\text{sag}_1)} \times \text{span}_1 \tag{4.1}$$

$$\text{Wt}_2 = \text{Wt}_1 \times (T_2/T_1)^k \tag{4.2}$$

$$\text{optimal aluminium area} = \sqrt{(\text{PWL} \times A/k)} \tag{4.3}$$

$$T = 2H \times \sin(\theta/2) \tag{4.4}$$

$$V = T_{\text{back span}} \times \sin(\theta_b) + T_{\text{ahead span}} \times \sin(\theta_a) \tag{4.5}$$

Here:

Wt is the weight of a structure
T in Equation 4.2 is the transverse force on a structure
k in Equation 4.2 is a factor between 0.25 and 0.70
A is an estimate of the needed area of single wire (phase or subconductor in a phase)
PWL is the calculated present worth of losses for amperes/phase/km
k in Equation 4.3 is the cost per extra kcmil of three phases of the conductor plus support costs/km
T in Equation 4.4 is the tension in a wire
H is the horizontal component of the wire tension
θ is the horizontal line angle turned
V is the vertical force in a wire
θ_b is the vertical back-span departure angle
θ_a is the vertical ahead-span departure angle

REFERENCES

Catchpole P (2006) A study of parameters relevant to Aeolian vibration calculations for transmission line conductors. *Proceedings of the ASCE Transmission Line Conference, Birmingham, AL*. ASCE, Reston, VA, USA.

CIGRÉ (2005) *Safe Tension Limits for Single Conductors. CIGRÉ Technical Brochure 273*. CIGRÉ, Paris, France.

White HB (1993) Guyed structures for transmission lines. *Engineering Structures* **15(4)**: 289–302.

Structural Engineering of Transmission Lines
ISBN 978-0-7277-5988-7

http://dx.doi.org/10.1680/setl.59887.153

Chapter 5
Loads and strengths

The basic work of transmission line engineering is making calculations that match structural loads that we think will occur with material strengths to resist that load such that the system will perform to the satisfaction of those who care. We understand the loads that will occur only so much, and we understand the strength of the materials we select and the structures and assemblies that we make from them only so much. We are not sure of either. So, we protect ourselves by including a factor of safety into our work.

If you dig back many decades into design methods and codes, you find that the factors of safety were defined in rather simplistic terms back in those days. As time went by, you can see the definition of these factors of safety becomes more complex. A long-standing formula to express our needs in matching loads with resisting strengths is this:

$$\theta \times R \geq \rho \times L \tag{5.1}$$

where:

θ is a strength factor logically less than 1.0
R is the ability of the material or assembly to resist L
ρ is a load factor, logically greater than 1.0
L is the load applied to the material or assembly with resistance R

Basically, Equation 5.1 tempers our belief in the strength and exaggerates our belief in the loads that the strength must resist. In this very basic relationship, the factor of safety in the strength-to-load relationship is ρ/θ. The simple phrase 'factor of safety' has disappeared from most safety codes and design guides or standards to be replaced by these more descriptive units of strength factor and load factor even to the clarifying point of calling them strength reduction factors and overload factors. The reason for the two-part factoring is to assign the risk of the unknown more appropriately to where it belongs. It is staggering to hear to this day that there are people in positions of power in this business who get the bits and pieces of this relationship all wrong to the detriment of everyone in their sphere of influence.

If we distrust the magnitude of the load more than we distrust the magnitude of the strength of the material or assembly, then ρ will be further from unity than will θ. Call this recognising and managing risk better than we once did. Some codes such as the US National Electrical Safety Code (NESC) and the Canadian Standards Association's

(CSA) C22.3 No. 1 apply strength factors to some things and overload factors to other things, but rarely use both for a single situation. It tends to be one or the other. Wires in spans, including conductors, are limited by a strength factor: 60% of rated wire strength when faced with an ice and wind load that is not factored. But that same ice and wind load is factored upwards by a load factor when applied to the support structures and hardware components. Weird, isn't it?

The International Electrotechnical Commission's IEC 60826 standard has taken the partitioning of strength and load factors to a whole new level, and is the latest incarnation of the subject of matching strengths with loads – and more. The '826' standard was decades in the making. It was championed by an engineer from Hydro Quebec in Canada, and was, from the beginning, fashioned after that utility's approach to the structural design of transmission lines. It is a reliability-based approach that was envisioned many years earlier. When envisioned, it was as if seeds were strewn around the world, and a global conversation began as to how reliability-based design – or, as some called it, probability-based design – should be done.

A most classic method for illustrating the method was to show two bell curves that overlapped slightly on a plot of strength or load on the horizontal axis against the probability of occurring on the vertical axis. The bell curve to the left was the load plot – a likelihood of a load occurring. To its right was the bell curve showing the distribution of the likelihood of the load being successfully resisted by a structural component. If the two curves overlapped, then the overlap area of the plot was where the load exceeded the strength. The overlap area as a fraction of the total area of the strength bell was an expression of the probability of failure of that component of the structure. The objective was to manage the size of the overlap by controlling the position and shape of the strength bell.

It is a great concept, but the detractors claimed that it is impossible to make all of the necessary calculations to ever, ever, ever declare a quantified 'probability of failure'. Indeed. The war of words both written and spoken went on for decades between the two camps. Slowly, the reliability-based method adopted a somewhat concessionary name: relative reliability. Relative reliability declares that you can make one component weaker or stronger than another by adjusting some relevant characteristic of that component compared with its use elsewhere, but you cannot do anything of value by trying to understand its absolute reliability. The absolute reliability of a component and of a system made of many components will correctly remain a mystery due to the forever vagueness of the load and strength input to any and all calculations.

Middle ground seemed to have been found, however strange it was to many that the strength of a system and the odds of its success in surviving for a useful length of time could be expressed in relative terms only – not in absolute terms.

If (when!) you read IEC 60826 with enough concentration required actually to roll up your sleeves and use it as a design guide, you will find that it is a strange mix of overly accurate tabular values for a host of issues and vague, qualitative guidance on other issues. As we have already described in Chapter 4, on the subject of conductor tension limits and Aeolian vibration management, the standard says do something intelligent but it does not offer a clue as to what that might be. To the practised or brave engineer, that is freedom. To others, it will leave them frozen in their tracks.

Load sources

Physical loads on the structural components of a transmission line come from a few sources, all of them important to understand even though some get much more attention than others. The load sources can be categorised as follows:

- from the environment: wind, ice and temperature
- from our actions: construction and maintenance
  discrete events:
  - natural sources: landslides, earthquakes
  - internal sources: failed components.

The amount of attention paid to these sources is generally in the order presented, with the last item getting the least attention. We will cover the subject in this order as well.

Effect of temperature on loads

Figure 3.3 in Chapter 3 shows that the tension in a span is changed as the temperature alters only slightly with long spans but a great deal in very short spans. The proper way to understand this is that a change in temperature will have a great impact on the tension when the change in length in the span due to the thermal coefficient of the wire's materials is of the same order of magnitude as the slack in the span.

Recall that the slack in a span increases with the cube of the span length. The length of wire in a span changes only directly with the temperature change and thermal coefficient. This means that, as the span length increases, the slack to be modified by temperature change quickly outgrows the ability of temperature change to affect it. In short spans, the slack can be measured in millimetres, as can the thermal change. With these spans, the effect of temperature becomes radical. Review the very cold night, short-span line failure described in Chapter 7.

Complementary to the comment above and deserving some attention is the point that the effect temperature has on tension in the wires and thus on the supporting structures is very dependent on the coefficient of thermal expansion of the wire. All-aluminium wires (ASC or ACC conductors and their alloyed versions) change the most, and the high-temperature, low-sag special conductors change the least.

Wind loads

We often think of wind *loads* as being related to wind *speed* by the simple formula

$$P = 0.00256 \times V^2 \times A \qquad \text{(US units)} \qquad (5.2)$$

$$P = 0.613 \times V^2 \times A \qquad \text{(SI units)}$$

where:

P	is pressure (psf or N/m^2)
V	is wind speed (mph or m/s)
A	is area per linear foot of wire (ft^2 or m^2)
0.00256 or 0.613	is the air density factor

This formula looks simple because all of the other factors that are known to affect the answer are rounded off to unity and effectively disappear. Two references that describe a more refined version of the formula are found in the American Society of Civil Engineers' (ASCE) *Manual of Practice 74* (Wong and Miller, 2010) and IEC 60826.

In those references we find something like

$$P = Q \times (Z \times V)^2 \times G \times C_f \times A \tag{5.3}$$

where (in US units):

Q	is the air density factor (varies with altitude and temperature between 0.00317 and 0.00165; the value of 0.00256 applies at 60°F at sea level)
Z	is the terrain roughness factor (0.72–1.42)
G	is the gust response factor (1.4–0.9)
C_f	is the drag coefficient (a function of the Reynolds number)
A	is the projected area of the wire
V	is the wind speed (mph) – which, by the way, increases exponentially with height above the ground

Most of these variables (Z, G, C_f and A) hover around 1.0, and the simplified formula above basically assumes they each equal 1.0. National codes may dictate that they, in effect, equal 1.0 (and the simplified formula applies) but, again, there are wind load cases in which the code has no interest but in which you should have a great deal of interest. So, a refined formula might then be of service to you.

There are some things to know about the 'refined' formula. The relevant building code in the USA, the American National Standards Institute (ANSI) standard for antenna towers, etc., uses a refined formula. Sometimes it looks different but is the same, and sometimes it looks the same and is different. There is nothing particularly truthful (accurate) about the formula, in the sense that some components are basically empirically derived from experimentation, not reason.

Most of the parameters are quite approximate and variable with time and location. You, therefore, need to realise that looking for accuracy within a few percentage points is futile. What the formula does do is let you acknowledge things such as increasing wind speed with height above the ground, decreasing gust impact on longer spans, etc. In certain scenarios, such as long river crossing spans, these are useful points.

You should understand that employment of a refined version of the formula may shift the answer in the right direction when spans are long or short and high above the ground or not or passing through rough terrain or otherwise. But never kid yourself that the answer you get is any more accurate than the accuracy of the poorly understood input values. The ACSE's *Manual of Practice 74* and IEC 60826 are good reading material for the subject, as are a plethora of other technical papers.

Synoptic winds

The formulas above are only useful for synoptic winds. Synoptic wind is a steady state wind that blows evenly across a large area. It is associated with moving masses of air the size of modest countries.

When using the formulas above, understand that different sources use a basic wind speed V of different definitions. Synoptic winds are defined by units such as 'fastest mile', '1 h', '10 min', '1 min', and '3-s gust'. Conversion between these definitions is by factors that are generally accepted as reasonable, but they are most certainly nothing more than general approximations.

For a synoptic wind to load a transmission line in the manner that the equations assume, the wind must strike the wires of the line along a distance of at least three spans and for enough time to move the conductors off their vertical position of 'no wind' rest to a displaced position of 'windblown' rest. For a wind to be that steady for that length of time, it is likely to be high above the ground away from ground turbulence such as a water or deep canyon crossing, and it is likely not to be a wind strong enough to cause structural damage.

We design our facilities for strength against synoptic winds, but it is not likely that synoptic wind will knock a transmission line over. If a line falls over due to a synoptic wind, it is likely that there was something else wrong with the line or there was debris carried in the wind that struck the line and did damage.

Wind rosettes

There is another tool available that can lower the cost of defending your design against wind loads. If you find the appropriate wind data, you can plot the direction, frequency and speed of the maximum wind coming from a particular direction. You get a pinwheel of information with eight or 16 compass points, and each 'spoke' is a length, the longer ones representing wind speeds. The plot is called a wind rosette. Some rosettes are quite balanced, meaning the strength of the wind from nearly all directions can be equally strong. Other rosettes point out the fact that, in many locations, strong winds come only from one or two compass points.

If your line (or a part of it) is perpendicular to the maximum wind, then you have a greater need to strengthen the line than if the line is parallel to that wind. There are locations, such as near the sea or in valleys, where the prevailing maximum wind is quite noticeably from a particular direction. The wind rosette for these locations is quite 'lopsided'. You decide if it's something you can take advantage of.

Wind acceleration

Local land features such as valleys, ridges, hills, etc., can accelerate and decelerate and re-direct wind. When encountering these conditions, it is more practical to modify an existing structure's design application limits than to design a unique structure for the location. The acceleration over ridges can be dramatic, increasing the basic design wind by 30% and the load by much more than that (1.3^2). But remember to reconcile the accelerated wind direction with the line direction. A wind accelerated across a ridge will severely load a line running along that ridge, but will not load the wires of a line also crossing the ridge but roughly in the direction of the accelerated wind. This is pointed out because some experts are enjoying the work of calculating these accelerations and embedding formulas into design guides of this phenomenon. But users of the new guidance need to discern the alignment of wind to the line direction to avoid unnecessary costs being incurred. We have noticed failure to do the discerning on some projects.

Remember that a fast wind blowing along a line such as over a ridge will load the structure on the ridge but not the wires attached to the structure. This type of loading is important, and falls into the category below as a local wind.

Local winds

Local winds are also referred to as high-intensity winds (HIWs) in the industry's technical literature. HIW has been of great interest in the last 10 years or so, in part due to the insistence of Brian White. It became apparent after studying failures that, as noted above, synoptic winds are not causing line failures but HIW events are causing failures. Even in Ontario, Canada – a region not well known for tornado activity – a study some years ago showed that five of six wind-induced failures over a period of time were caused by HIW events.

To the degree that HIWs were not studied and understood, there were failures. This is because the local, high-speed nature of HIW (downburst, microburst, tornado, etc.) loads a tower differently from synoptic winds. A synoptic wind blows across whole spans of wires, and the load on the support structure comes from the wires through their attachment points to the structure. This means that the maximum load on the tower arises at a point on the tower called the 'centre of effort' of all of those wire attachment points. This is a point quite high up in the tower, among the cross-arms.

A gust or local wind event of small dimensions does not load the spans of wires for two reasons. First, it is too narrow in dimension to push on the entire spans unless the spans are quite short. Second, the gust may not last long enough to move the wires from their vertical position to a blown-out position, and the wires can load the tower only in the direction that they hang. Thus, as a worst case scenario, the HIW can only load the entire tower itself with wind blowing directly against the tower, and the centre of effort for this loading is much lower down on the tower than the loading arriving through the conductors.

When the support structure is a tubular pole, this difference is not much of an issue. The bending in the tall, vertical beam is simply lower than with the synoptic wind, and the shear capacity of the tube is very high. For truss towers, such as self-supported latticed towers and guyed towers, this is not the case. Figure 5.1 illustrates.

Figure 5.1 Wind-on-tower loads

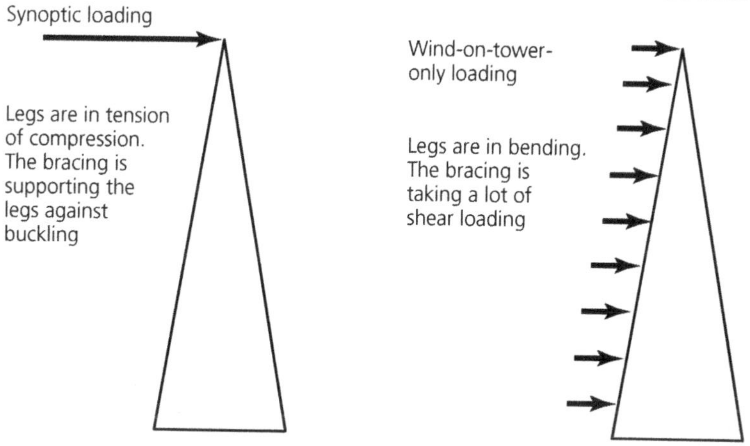

A well-designed latticed tower has its legs aimed at the apex of the centre of synoptic loading. This puts the legs in compression and tension, letting the tower brace with no duties but to provide support for the legs in compression against buckling. When the tower is loaded with wind force up its entire length, the bracing system has to resist shear.

When you see a tower failed by a wind event and the windward face of the tower is pushed right through to the leeward face because all of the bracing buckled, that tower's bracing was not sufficiently strengthened against an HIW event.

> The study that Ontario performed upon realising that five of six wind failures had been due to HIW events was to discover the cost increase required to give their latticed towers enough bracing strength to withstand the forces of an F2 tornado applied to the wind area of the tower only with no load on the conductors, believing that, statistically, 85% of all tornadoes are F2 class or lower. It concluded that the necessary bracing strength increase would increase the cost of towers by about 2%. Well worth doing so, it became Ontario's practice.
>
> Several years later, we were designing a latticed tower for a project, and one load case was for withstanding the F2 tornado force on the tower only. On review of the tower designer's work, we noticed that he had left out that load case for some reason. When we added the load case back into the mix, the weight of our 345 kV tower rose 2%. Imagine that!

Towers can fail if the HIW-on-tower load case is ignored. The cost of inclusion is very small, so do include it, especially in locations where such winds are known to occur.

With guyed tower types such as the guyed-V or cross-rope suspension structures, the synoptic wind loading of the tower through the conductors does not include a comparable magnitude of load on the tower masts. But the HIW loading on the tower puts the masts into big bending. This must be included in the load case list for such towers.

Cost of wind loads

Above, we noted the pretty solid evidence that the cost to accommodate HIW events in a tower's design package of load cases is very modest, at about 2%.

Remember the 'rule of thumb' formula that says the cost of a steel structure C equates to the transverse load as follows, per Equation 4.2 in Chapter 4:

$$C \propto k \times T^k \tag{5.4}$$

where k ranges between 0.3 and 0.7.

This is applicable for synoptic wind loadings on tangent towers and for tension loads on deadend towers. Doubling the life expectancy of the structure against synoptic wind speeds requires a 15% increase in T. The cost increase will be between 1.028 and 1.050 (4% and 10%). Assuming the structures comprise 30% of the project cost, that's a 1.2% to less than 3% increase in the total project cost. This may or may not be considered a trivial amount, so increasing the strength of structures for the purposes of doubling their life expectancy needs to be done with thought, but this is the impact of having the project survive a doubling of the design wind speed return period. Not much if you ask me. Why try to save a bit of money by cutting back on the design wind speed?

Ice loads

The ASCE's *Manual of Practice 74* also has an excellent chapter on ice loads. Your line personnel can tell the designers just what goes on in the night on your system. Do you get icing? Is it solid glaze ice? Is it wet snow or light rime ice? How thick? Is it precipitation icing or in-cloud icing?

If icing is a real issue with your system, you might consider undertaking a programme of collecting ice-loading data. There will be no better data for you. However, it will take a very long time. In the ASCE's *Manual of Practice 74*, there are two ice accumulation maps of the USA – Figures 2.8.1 (p. 41) and I.3-1 (p. 151). The latter is the result of 9 years of data collection (visual mostly) done by the railway companies in the 1920s and 1930s. They observed and noted ice accumulation on their telegraph wires. Around the turn of the 21st century, an ASCE committee rewrote the ice-loading standard for buildings (ASCE 7). The author (Bennett) of the original railway map in the *Manual of Practice 74* was still around, and it was said that he wouldn't talk to the ASCE committee. Odd! Does he think his data stinks? Maybe he knows that none of this matters!

One clue to the absurdity of some ice maps is their apparent conservativeness and poorly thought-out mismatches at jurisdiction boundaries. While Canada and the USA try to coordinate their information, the methods of developing the maps vary, and so the adjoining information shown along the border will flag the point that all is not as simple or as well understood as it seems.

A very easy and absolutely appropriate way to collect ice data is literally to collect a foot or some measured length of ice from wires immediately after an ice storm. Put it into a plastic bag, let it melt if need be, take it to a scale and weigh it! If you want to get fancy, note the diameter of the wire that the sample was collected from and the basic shape of the specimen (radial or dripping). Back-calculate the effective radial thickness based on the weight, length and density of the sample. You are likely to find yourself underwhelmed by the radial values you come up with. There is a lot of drama stirred up when large ice accretions occur. They invariably look bigger than they are from a radial thickness point of view. That is because a radial shape for ice around a cylinder of metal is the most efficient packing of material that can be done. Mother Nature is a slob in this regard by comparison.

From a design point of view, remember that ice is specified in many jurisdictions in terms of radial thickness. The weight therefore varies with wire diameter. Recalling the principle that doubling the weight of the wire doubles the tension, and if the everyday tension of your design is about 20% of the rated tensile strength (RTS), then the limit for the weight increase of the wire is about three times the bare weight, since that will put the tension at about 60% RTS. This is more accurately true for long spans and modest increases in weight than it is for short spans and large increases in weight, under which conditions the tension increase falls short. It is also a conservative estimation to make. It also means that a large and heavy wire can more easily carry a given ice load than a small and lighter wire.

Types of ice load

When big drops of water strike a surface then freeze, the accumulating ice is clear and dense. As the ice loads up on the top surface of the conductor, the conductor tends to roll a bit under the weight, and the annular ring of ice tends to migrate around the wire. When that water does not freeze immediately but runs a bit before freezing, icicles form. When the water is already frozen or nearly so, and comes from much smaller droplets, it can form a less-dense form of ice, and is white in colour. The clear dense ice is called precipitation icing, and the less-dense white ice is called rime ice. Precipitation icing comes from falling rain, and rime ice comes from clouds passing by the wires. Rime is also called in-cloud icing, and tends to occur at an altitude or where wet air rises up a slope and freezes along the way.

Finally, there is simply wet snow. Snow can be sticky if wet by falling at a temperature very close to the freezing point of water. Wet snow can accumulate on a small wire to a surprising thickness. Like the precipitation ice, it tends to roll the wire and form a very clean cylindrical shape centred on the wire. The density of precipitation icing is considered to be 90% of water. Wet snow or rime ice can be from 5% to 20% of the density of water.

If the predominant type of ice accumulation is rime or wet snow – particularly wet snow – the weight of the accumulation may lose importance if the wind picks up, whereupon the diameter of the accumulation matters. Some dramatic photographs of rime ice are found in Chapter 7.

Combining ice and wind loads

Ice, wind and temperatures do not always occur in isolation from each other. They can occur simultaneously. After all, there is always a temperature and, to some degree, there is likely to be a breeze. We seem to take it for granted that the extreme values of each of the three do not occur simultaneously. We prefer to temper the amount of each when the others are also present.

For example, IEC 60826 suggests that, in the absence of combined ice and wind statistical data from which a project-specific answer could be derived, assume that the ice load to be associated with a wind speed should be 40% of the extreme ice load and the temperature at which they are combined should be −5°C. The wind speed is to be 65–85% of the design wind speed, depending on the project's chosen return period.

You can argue with that or not, depending on your fear of venturing into the dark night. Do you believe that you will be safe if you trust a standard such as the IEC's? Do you feel safe if you make your own choice? Who will attack you and on what basis? There is no right answer of course. You can do either, but there will be repercussions either way if, someday, your choice is seen as having failed. Such is the life of the professional.

Standards such as the IEC's offer guidance that presumably has been given extensive thought, but that does not mean that the guidance is applicable to your project at hand. Think!

One shortcoming that does come to mind is the combining of ice and temperature. Often, the temperature required for the ice to form on the wires is the only temperature considered in sag–tension analysis. But, you ought to consider the real possibility of the temperature dropping considerably after accretion occurs, and reviewing the tension increase that results. It may or may not matter to your design, but it is a climatic reality.

It is your responsibility to ask yourself: *what would be the impact on the design of a reasonably likely combination of ice, wind and temperature other than that first considered?* Then, check it out by a quick calculation.

Construction and maintenance loads

The whole point of designing the support structures and the spans of wires themselves for construction and maintenance loads is to keep the people who do that work safe. It is one thing to have a tower fall down in a storm but quite another to have someone on the tower when it falls. It is incumbent on the engineer to learn and understand what the field staff will do, might do and could do while in the field that will cause structural overloads and damage. Then, do what you can to protect those people – from themselves.

Rigging loads

The key to most field activities of interest is rigging. Learn about rigging! Figure 5.2 is a great illustration of the fact that, if you give linemen rigging holes, they will fill them with things, and for good reasons. If you do not give them rigging holes (points), they will get mad at you. The latter is not illustrated in the photo.

Figure 5.2 Rigging holes

Most rigging equipment such as slings, sheaves and shackles has a working strength that is 20% of the ultimate strength, meaning that the factor of safety of the item is meant to be at least 5. Many of these items are carefully stored and tested before they are allowed a long service life. Yet, we attach them to structures that have a factor of safety for the work that may be much less than that.

If a conductor's tension is limited to 25% of the rated strength at an average day's temperature – the kind of day that people climb towers to work, and if some of that structure is stressed to 100% of ultimate member capacities when the tension is at 60% of the conductors' rated strength, then much of that structure may offer the climbers a factor of safety of 60/25 = 2.4. It is easy to understand that a structure may often offer a lower factor of safety.

Suppose you are in a part of the world where there is no ice accumulation and the wind does not blow strongly: the equatorial zone – the tropics. A deadend tower there might be stressed to 100% of its ultimate strength at 30% of the conductor strength because no weather load requires more tension. A suspension tower might never carry anything but bare wire and a modest wind. Under these circumstances, certain activities by personnel on the tower can be the biggest load the tower will ever see. If you have not designed the tower for these activities along with a healthy factor of safety, there is no unintended safety buffer built in, and things may go very bad.

Figure 5.3 Load points on structures

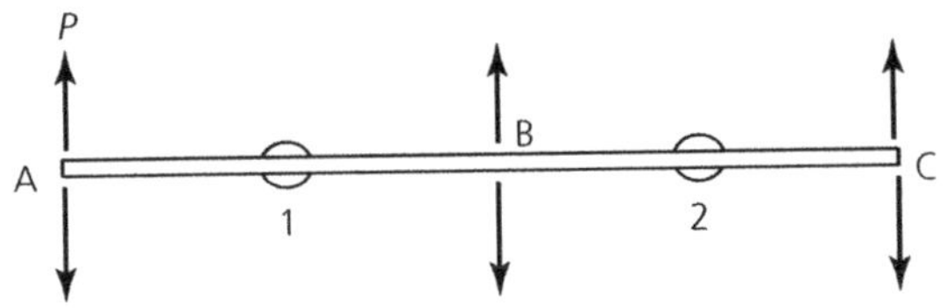

Consider one simple example of structure loading that has been the cause of structural failure. Figure 5.3 represents a deadend in plan view, and could be a two-pole frame or a square-bodied latticed tower as far as the exercise is concerned. The distances A–1, 1–B, B–2 and 2–C are all equal. Call that distance x. When intact, there is a load P applied at all six attachment points, and everything is in balance. A moment calculation about point 1 or 2 shows there is no horizontal reaction, meaning there is no shear load in the tower body bracing.

When the tower carries three phases of load on one side only, the reactions at points 1 and 2 are $1.5P$ each. This is the design strength requirement developed for the bracing by the 'one-side-only deadend' load case. Now, suppose that one of the six loads is taken to zero at point C to mimic the removal of a phase connection at the end of this arm.

A moment calculation about point 1 says that the reaction at point 2 is $3x \times P/2x = 1.5P$. This suggests that in this simple but reasonably representative model of a transmission line structure, the loss of one phase (a broken phase case) requires the same strength of tower body bracing as does the full deadend case. For tower shapes something like this, designing for one broken phase is practically like designing for a deadend. The difference is that a true deadend tower may be designed for the loss of all wires on one side with a large ice load in play whereas the single-phase loss may be with a lesser tension in play. But, read this ...

Suppose you were performing maintenance on this tower, and the tower was not designed for an ice load case and the tension at the time was near the tension that was used to design the tower. Let's remove from one side of the tower phase C and then phase B. Said another way, when putting the phases back into place, let's attach phase A first. Now, the moment about point 1 shows that with phases B and C missing from one side, the shear in the bracing at point 2 is $(1x \times P + 3x \times P)/2x = 2P$. This is a shear load in the bracing significantly larger than the $1.5P$ design load called for by the deadend load case. If you have not designed the tower for this unbalanced event, it will come down. It has happened.

If you remove or attach the three phases in a sequence that attaches the middle phase first or disconnects it last, then the imbalance with phases A and B missing from one side of the tower puts a shear into both bracing faces of $1.5P$, then P – a safe approach. The sequence of work can matter a great deal if the tower strength at the time of construction or maintenance is not protected by an ice load so big that this problem is hidden from you. This is one reason why we beg you to never build a tower with deadend (strain) insulator attachments when the tower itself does not have full deadend strength against any and all load cases, especially this construction and maintenance case as described here. Too many field personnel are not astute enough to see the difference and not do the wrong and dangerous thing.

Working in carts

As a matter of routine, people use a cart or buggy to go out in a span to inspect hardware and splices or to affect a repair. Sometimes, people go out in a span without a cart. They just crawl along the wire. This tends to be done in countries where a cart is a luxury. Either way, it is really useful to understand the effect of a point load on a span's tension.

CIGRÉ Technical Brochure 471 (CIGRÉ, 2011) deals with the subject from a safety point of view, especially with regard to working on wires that are not new. Is it safe? A point load on a span of wire increases the tension in the wire by a surprising amount under some circumstances. The higher the span's catenary constant *C*, the larger the tension increase. The tension increase is reduced or tempered more and more as the adjacent spans have slack to feed into the span due to both their length and the suspension length of the insulators at the span's ends.

The tension increase is minimal when the point load is near the ends of the span and a maximum when applied at the middle or low point of the span. Under rather typical combinations of tension, wire size, span lengths and insulator lengths, a tension increase as much as three to five times the weight of the point load is quite likely. In other words, if a cart with personnel that weighs 200 kg is sitting at mid-span, the tension increase can be as much as 1000 kg. If a wire is small, this is not a trivial tension increase.

Unless there has been a known low-temperature or icing event within the last 2 years or so, it is fair to say that the tension in the span with such a cart load in place is new to the span in its recent history, and being in that cart is a matter of supreme trust if the wire and any hardware in or near the span is old. There certainly are people who go out in carts under such circumstances without thinking but, you, being an engineer, might want to point out the risk in case they are not aware.

It is rare that a conductor has lost a great portion of its strength unless it is very old or in a bad environment. Rare, but not unknown. It is less rare that splices become weak with age and years of heated duty. It is also rare that suspension strings are near their structural limit on an average workable day. Testing a span of wire for point load capacity is difficult to basically impractical except in isolated situations.

Discreet event loads

In the ASCE's *Manual of Practice 74*, these are called accidental events. We will dispense with airplane encounters and airborne livestock in tornadoes as something we can manage. Some things are just going to be left to chance.

Landslides and earthquakes

We try to identify slopes that are susceptible to landslides and then stay away from them. But that is not always possible. We know of one line set on a tropical mountainside that keeps creeping downhill with the hillside material itself. Every few years, the crew goes in and resets the insulator strings that trend out of plumb as one or two towers move downhill and lengthen the attached spans. Then, when that can't be done any more, they go in and relocate the towers back up the hill, back on alignment.

We suggested that they might try setting them uphill of alignment and buy more time. Fortunately, the spans are quite long and they have considerable slack available. That is what gives them the time. The lesson is: use long spans with lots of slack to limit damage should something move.

> One of the frustrations that Brian had with the Kemano–Kitimat line was that the obviously advantageous locations for towers along the valley floors where the rivers ruled the day were not in the centre of the narrow valley away from the very high and steep sides but tucked up along the toe of the mountainside away from the river. In fact, the very best locations were the high ground seen on drawing profiles every so often.
>
> These periodic high points were not recognised by the person(s) spotting the line as talus slopes – the big piles of rubble loaded with new material year after year as mountains send avalanches and rock slides down their many chutes to the valley floor. Every tower placed on these attractive high points was placed right in the gun sights of the avalanche paths. Dang!

Figure 5.4 illustrates one of the many avalanche events that was split by the huge earth deflector placed around the tower that sits in the avalanche path. Defending against avalanches is a brute force against brute force battle if you cannot get completely out of the way. If you try to wage that battle, someday be prepared to lose.

The other natural event that we try to pay attention to is earthquakes. It is generally understood that the nature of transmission line structures and the wires spanning between is that the structures do not need to be designed to withstand earthquake ground accelerations as buildings require. The reason is that their mass is not large enough to require it. Every once in a while, someone in charge brings up the notion that we should design for earthquakes, and a review of the whole issue typically educates the person as to why we don't need to.

It is generally considered to be a good idea to find known fault lines and to place the structures elsewhere rather than on top of them. Again, a long span with plenty of slack between the towers will serve you well should things start moving.

Loads from within

Galloping loads

A very damaging structural load that a system can impose on itself is caused by galloping. Most of the literature that you will find on galloping is concerned with the nature of the

Figure 5.4 Avalanche country

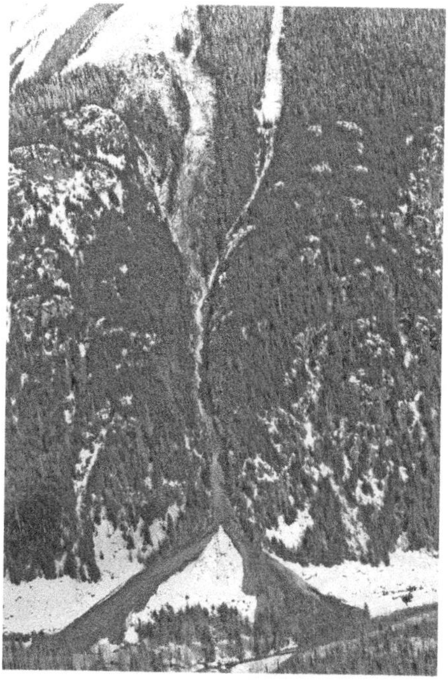

conductor motion as it relates to electrical flashovers. A flashover will not stop the galloping, so once the galloping motion is big enough to flash over phase to phase, you have a persistent structural problem, not a momentary outage such as lightning might cause. But, galloping can also break an insulator assembly, wear out a vang plate or hardware component, rip off an arm, or tear a conductor apart; it can literally tear a tower apart if it goes on long enough and with enough vigour.

A few technical papers offer design loads for dealing with galloping such as the vertical load increase at a suspension point can be 1.5 times the static load, and the tension increase at a deadend can be double the static tension. Be very careful with these factors, because there seems to be no consistency between sources as to the denominator that develops the factor. Some have the factor as that of the bare wire at rest, while others have it as the weight or tension of the wire at rest but with the ice on it that is making it gallop.

It is often noted that galloping can occur and is most likely to occur when there is a very thin skin of ice on the conductor. In other words, when galloping, the static conductor tension is assumed to be very close to the cold, bare wire tension. However, galloping can occur when there is a great deal of ice on a wire. Is the motion slower when fully loaded with ice? Is the magnitude of the motion lessened with a larger ice load on the wire? We have never heard the answer, so it is difficult to know whether the load increase factors should be the same regardless of the amount of ice on the wire.

The good news is that a line to be designed in a location is probably not the first line in that location, and much can be learned from the older lines in the area. Is galloping a problem in the area? Does it occur frequently? How damaging have events been? The answers to these kinds of questions are the best guidance for your new line.

There have been many papers published around the world over the years that deal with the shape and size of galloping wire 'ellipses'. The ellipse defines the boundary within which the conductor is *probably* located while galloping. It does not define the path of the motion. The motion itself is reasonably unpredictable, but it is generally a vertical motion with variable amounts of lateral motion included. Most recently, it is being suggested by the experts who pay the closest attention to the subject that conductors laden with wet snow or rime ice have a greater lateral scope to their galloping motion. The ellipse, they are now suggesting, should be made a circle where the diameter equals the height of the ellipse when wet snow or rime ice is the load on the wire.

Different organisations use various degrees of finesse to describe the shape, size and position of the ellipse and even the required separation or allowable overlap of the adjacent ellipses. Typically, the ellipses are tall and tilted. Most galloping events are discovered not by observation or post-event evidence but by electrical flashover during the event. Successful versions of the formula can only be determined with the hindsight of a flashover record that is interpreted as a satisfactory flashover rate – or otherwise.

It has even been claimed that certain lines don't gallop. In particular, it has been claimed that circuits configured horizontally with phases beside each other and not above or below each other do not gallop. They must have been joking! It is correct to say that, when they gallop, they do not flash over because the elliptical boundaries of the motion are far less likely to conflict than for vertically stacked phases. If you fear galloping, don't stack the conductors vertically.

Most galloping occurs with ice accumulation on the wire – although not all. Galloping requires that the wind across the wire encounters a different surface shape on the top of the wire than on the bottom. This can occur with oblique wind across a stranded texture. The lay of the strands causes a different shape to be encountered by an oblique wind. The most famous bare-wire galloping story is of the Severn River Crossing in the UK: see Davis *et al.* (1963) to learn a lot.

Watch for load effects, particularly at running angle structures where the large and heavily loaded motion of the near-horizontal insulator string can do unpleasant things to undermine your intentions.

It is fine to spend your time trying to determine the phase spacing required to avoid galloping-induced outages. It is equally important to remember to design for doubled vertical and tensile loads. A good many line failures have been initiated by these large loads bashing away at and breaking wires.

One reason for conductors breaking and towers cascading in the big icing event in Montreal in 1998 was galloping with a large amount of ice on the conductors. The tension was very high and the sag was very large because the spans were quite long. The suspension clamps on the line had virtually no vertical radius in the support clamp's saddle. This meant that the conductors turned the departure angle over the nearly sharp edge of the flat clamp in a hard bend right at the clamp lips. The bend angle was large due to the large iced sag, and the tension was high for the same reason. The conductor broke at several towers right at the lip of the suspension clamp. It actually looked like it was ripped off at that point. If you expect big ice and galloping, be very selective with the suspension clamp shape. Give the conductor a complete seat to rest in as it turns the fullest departure angle. That is when it matters.

Finally, *CIGRÉ Technical Brochure 322* provides a new set of calculations for establishing phase separations to manage galloping flashovers (CIGRÉ, 2007). It has gained some respect in the industry simply because it comes from CIGRÉ. Whether it offers a better solution than any of the many other criteria out there will take a long time to discover. Still, it is worthy of consideration. Some very smart people wrote it.

Loads from failures

It is a widely held understanding that, someday, something on your transmission line will fail, and the line will go out of service. That view is widespread but not universally held. Either way, a prudent engineer will plan for the event by doing something to reduce the effect of that component failure. It is like saying: *something will fail – now what can I do to limit the damage*? The classic title for this subject is 'failure containment'.

The essence of a component failure on a line is that the failure changes the slack somewhere. Most of the time, the failure injects a large amount of slack into the system somewhere, but sometimes it removes slack. Either way, the sedate nature of an intact and properly functioning transmission line depends on the slack being where it is supposed to be – by your design work. When something upsets this balanced situation, redistribution of slack gets underway in a hurry.

Support system failures

Support system failures are dropped suspension insulators or failed tangent structures. In either case, breaking wires are not part of the triggering mechanism, and it is quite probable that the wires do not break as such a failure propagates.

Consider the dropping of an insulator at a tangent structure. On either side of the structure, the intact spans have some value of slack within. If the spans are level, the slack is as per Equation 3.9 in Chapter 3. Assume the two spans are 300 m and the catenary constant is 1200 m. The slack in each span is 0.78125 m. If the insulator drops the conductor, the span is now the sum of the two spans, and the slack is the sum of the two slacks. Back-calculate the catenary constant, to get 2400 m. The catenary constant has doubled, meaning that the dropping of the conductor has doubled the tension in the new long span.

A span normally requires $2^3 = 8$ times the slack of a span half its length. The dropping of the wire only offered double the slack – the sum of the two original intact values. The shortfall of 4 times required the tension to double ($\sqrt{4}$). Elasticity of the conductor will temper the doubling. Movement of the insulators as permitted by their length and the slack available in the spans beyond will temper the doubling, and the tension will relax considerably if the conductor finds and rests on the ground. Still, the tension will increase, and the two adjacent structures must deal with that.

Suppose this happened when the conductors are laden with ice.

Suppose the ground is not level and the structure that drops the conductor is higher than the adjacent structures. More exactly, suppose the attachment point of the conductor that fails is higher than the adjacent attachment points. First, understand that as a span becomes inclined, Equation 3.9 increasingly overstates the slack in the span. Second, the more detailed slack calculation useful to inclined spans is based on Equation 3.7. Calculate the slack on either side of the span's low point based on the two unique values for the distance x; sum them and subtract the inclined distance between the span's attachment points to get the slack. It will be less than Equation 3.9 suggests.

When the conductor is dropped from a high point, the new long span is shorter than the sum of the two inclined spans, and much more slack is inserted into the new, long span than simply the sum of the two original spans. In this case, the insertion of slack can be so large that the tension drops, especially if the conductor also finds the ground. So, in the level-ground scenario, the dropping of a conductor from a structure can increase the tension a great deal. But, if the line profile is not flat, the tension can be reduced, and the whole effect on the adjacent structures is the opposite.

When an entire support system tangent structure fails, what will be the repercussions? It depends! When an entire structure falls, the nature of the repercussions depends on whether the falling consumed slack – had to pull wires from adjacent spans – or whether it fed slack into the line that had to be sent into the adjacent spans and onward.

Wire system failures

Wire system failures are broken deadend assemblies, failed splices or a broken component of an angle structure. When a deadend tower or assembly or a splice fails, an infinite amount of slack is fed into the phase or line. Infinite (very large) slack is synonymous with no tension in the wire. The next component of the line that is connected to the failed conductor(s) will be required to support the tension once held by the failed component.

When a corner structure or a phase attachment to a corner structure fails, we have the same scenario as painted above with the dropping of a conductor from a suspension tower set high above its adjacent towers, except that the story is horizontal, not vertical. The failing of a corner feeds a large amount of slack into the new span defined by the location of the structures adjacent to the failed angle structure.

As with the failing of a support system component, the repercussions depend on the amount of slack being inserted into the system. But, with wire system failures, that amount of inserted slack is much more likely to be large, and the repercussions are likely to be much more serious or difficult to manage. Therefore, pay attention to the wire system. Give it integrity or pay a big price.

Summary comments on loads

There are two subjects that conclude the conversation on loads. Both of them point out the complexity of the real world compared with our usual means for modelling it.

No strain = no stress

All of the materials that we would ever use for the physical components of a transmission line have a describable stress–strain relationship. This means that every line component from the wires to the foundations and the soils cannot be stressed (i.e. cannot carry a force) without being strained (i.e. shortened or lengthened). What this means is that structural analysis of stretchable, bendable materials by methods that ignore stretching and bending produce inaccurate results. Accurate results are entirely dependent on the relative elasticity or flexibility and malleability of attached components.

The simple formulas in Chapter 3 are useful to convey basic concepts, but most ignore wire elasticity, so be wary of getting accurate answers from their use. Conversely, there are times to be concerned for the second-order accuracies these formulas ignore and there are times not to be concerned.

The guying of structures is interesting. Basically, guy wires exist to hold a structure either reasonably upright or up at all. If a guy wire attached to the top of a pole is very steep, the pole top must move laterally a long distance to stretch the guy wire enough for the wire to take on significant load. If guy wires attached to a pole that turns a small line angle are placed in line with the two conductor directions and not on the bisector or even towards the bisector, the pole top has to move a long way into the angle on the bisector to load the guys. To be efficient, guy leads should be long, and the guys on shallow angles should be on the bisector when possible, or at least away from in line with the conductors and towards the bisector to gain efficiency.

Small guy wires attached to a very stiff pole will never take on a share of the load such as large wires will take on when attached to a flexible pole. Load sharing depends on relative flexibility and elasticity – and geometry.

All of these examples should tell you that the static or linear analysis methods that were in use in our industry before computers existed to make more complex calculations easy should be discarded from your tool box. They can be very wrong all because stress requires strain. Make non-linear analysis your standard practice except for the most basic of design concept explorations.

Dynamic loads

The other reality of our existence is Newton's second law of motion: $F = m \times a$, where m is mass and a is acceleration. For things to move, they require force to accelerate or decelerate.

> When we interview young engineers out of school and new to the industry, I tell them that the reason I find this business interesting is that transmission lines are 'large-deflection' structures.
>
> The structural engineering that we all learn in schools requires that the deflections of the structures be very small for the formulas they teach to be valid. And, if you accept that the wires spanning between support structures of a transmission line are part of one grand structure, as I have been describing, then you must understand that a transmission line is a large-deflection structure because the wires can and do move all over the place with wind, ice and temperature. Even some tall pole structures or those that hinge at the base are beyond the limits of the schools' formulas. Our engineering is outside the box of what is taught in school, and that makes it interesting.

Practically all of the analyses that we exercise in this business ignore the $F = m \times a$ issue and the resulting dynamic loads. Mostly, we can get away with this because ignoring the point leads to conservative results more often than not. When a gust of wind blows on a span of wire, it takes time for the span to move into the blown-out location that our calculation says it will go to.

If the gust does not last long, then the blown-out displacement and the lateral force resulting from the displacement will not manifest. Until all of the blown-out displacement is achieved, most of the energy – and initially all of the energy that the wind puts into the wire – is translated into motion energy, not a lateral load on the support structures. This is why only synoptic winds fully load structures through the conductors.

Some folks entertain themselves by making dynamic models of ice shedding to see what that could tell us. But, the efforts are elementary and worth taking in with a very large block of salt, as the saying goes.

Nevertheless, the very fact that large deflections, dynamic loads and structural members with considerable strain capabilities are our reality but are often poorly addressed by our design tools and criteria means that there is a lot of uncharted territory for the energetic but practically minded of you to dig into. You will enjoy your career much more if you find this stuff interesting. The conclusion to my interview statement with young recruits is this:

> The last thing you want to be is a substation engineer. The first thing your company will do is put a fence around a patch of ground and say to you, 'Don't leave!' Whereas we drive on dirt roads in trucks, fly around in helicopters, get stuck in mud, look at storm damage and make outside-of-the-box calculations. Think of a substation as just a tumour on a transmission line.

Strengths of materials

At the start of this chapter, we describe the basic relationship between loads to be resisted and the strength of materials and structures that we develop to resist those loads. We describe the use of factors that we use to deal with our levels of distrust in the magnitude of the loads and in the strength of the materials and structures. The approach we take has evolved into having the ability to exaggerate the loads just in case we might have underestimated them and to downplay the strength of the materials and structures assembled with them just in case we might have overestimated those.

Here we discuss the basic strength of certain important materials and we will discuss the importance of using these various factors not only wisely but properly.

Working strength and ultimate strength and

The extra 'and' in the heading is not a mistake.

As noted above, materials exhibit a certain relationship between stress and strain. A stress–strain relationship exhibits an elastic zone at low stress and strain values. The relationship transitions to a plastic zone until, eventually, the material ruptures at some high values of stress and strain. The elastic zone is the range of stress and strain within which, upon relaxation of stress, the material returns to its original shape, meaning no change was permanent – it is elastic.

In the plastic zone, any stretching of the material with increasing stress causes some amount of permanent shape change. The material does not fully recover back to its original length or shape when all stress is removed. All materials have elastic and plastic zones of different magnitudes. For example, glass and carbon fibres have no plastic zone. The material goes straight from elastic to rupture. Bang! Annealed aluminium has a very small elastic zone and a huge plastic zone, as measured by the range of strain within each zone. Steel has a respectable elastic zone and an even larger plastic zone. Wood has a very small plastic zone, and so on.

When you design a building or most other normal structures, you have a choice between using a working stress design method or an ultimate strength method. As you know, both methods place factors of safety somewhere in the 'load–strength' equation for the purpose of ensuring that the load that the structure is able to support is sufficiently greater than the load that the structure is expected to see.

The working stress method factors down the strength of the material to a safe level within the elastic zone and may factor the loads upwards a bit. By comparison, ultimate strength methods apply significant factors to the loads and run the unfactored, unrestrained material stresses towards yield or rupture values. As we have noted here, we try to place the factors where they belong based on what the issue is: uncertainty with the strength of materials or uncertainty with the magnitude of the loads. But, there is more to it.

The ultimate strength method argues that it is able to make use of much of the rather large plastic zone of steel whereas the working strength method stays away from that zone. That is fine when you have a material such as steel that has a large plastic zone worth accessing. If your material has a different balance between elastic and plastic ranges, the merits of the methods will change.

When you design a transmission line, you will likely use some amounts of both methods. The Canadian standard CSA C22.3 No. 1 and the US NESC, for example, declare an ice thickness and wind pressure to be applied to the wires for their safety load case. When applying this load to the wires for sag–tension calculations, they have you factor the strength of the material significantly to 60% of its rated strength, as we have already described. That looks like a working stress method. Why? Sag is a deflection calculation, and that calculation forbids factored loads. You cannot exaggerate the load with a load factor and expect to get the correct deformation (sag).

When you apply that same ice and wind load to the structure via the wire attachment points, they have you factor the load significantly and not the material strength, but you can now run the materials of the structure to yield. That looks like an ultimate strength method. Remember that all of the deflections that a calculation offers up when the loads are exaggerated will also be exaggerated, and if there are significant deflections in the structure, the answers are, well ... wrong.

Every load case that you might use outside of any required by a national standard such as CSA C22.3 No. 1 or the NESC is typically handled as an ultimate load applied to a material that is run to 100% of capability (yield or rupture). It's a strange world.

Strength of wood

The equivalent standard in the USA for Canada's CAN3-O15, 'Wood utility poles', is ANSI O5.1. Both have appendices that are not part of the mandatory standard but which shed light on the definition of wood pole strength. Let's deal with the US standard.

Western red cedar, for example, is given a maximum fibre stress of 6000 psi (41 MPa). A bunch of small, clear samples of the wood were broken to get this rupture value. Then, for a chosen lateral load applied 2 ft (0.6 m) from the top of a pole, a minimum circumference at the ground line is calculated, assuming that the pole would break at the ground line based on that stress limit being correctly applicable to the pole as well as it was to the small, clear sample. A minimum circumference is assumed for the top of the pole, and the calculated ground line circumference is then transferred to 6 ft (1.8 m) from the butt, based on the defined taper.

A set of loads is established from small to large for application 2 ft (0.6 m) from the pole top. A set of top circumferences was established, and this allowed a whole table of circumferences to be calculated as described for a range of pole lengths. Each load defines a class of pole. Each species of wood has different fibre strength, and therefore has a different set of circumferences that varies with the species for a given pole length. The objective of the process is to know that a class of pole has a specified strength regardless of the species of the pole.

There are some things to know about wood poles.

First, the strength of wood is published as a mean strength with an accompanying coefficient of variance. Paragraph A2 of Appendix A of CSA CAN3-O15-M83 states:

> that for poles with a (published) modulus of rupture (MOR) of 8,000 psi. and a standard deviation (of strength) of 1,000 psi., normal probability theory says that (to paraphrase) '170 poles per 1,000 will have an MOR below 7,000 psi., 25 poles per 1,000 will have an MOR below 6,000 psi., 1.5 poles per 1,000 will have an MOR below 5,000 psi'.

Similarly, equal percentages of the poles will be that much stronger than the mean. No other line product is represented this way.

This is why the load factors used with wood poles are so much larger than they are for other materials. The large strength reduction factor applied to wood is there to shift the usable strength of the wood down to a low value that captures many more than half of the poles. It is meant to redefine the strength effectively as an exclusion limit, not a mean value. But, the factor used does not shift the usable stress limit down to a really low exclusion limit – meaning one comparable to steel's exclusion limit of about 5% or less. It does not go that far, so wood poles tend to be used at loadings that, on the face of it, should see a troubling number of pole failures. There are reasons why this does not happen, one of them being that the loads we design for tend to show up rarely in most locales. Other reasons are described here.

Poles are tapered, and the tall ones won't break at the ground line, as assumed in the calculation described above. In fact, they are not as strong as the calculation assumes. A tapered, homogeneous material used as a beam or column in bending will show maximum stress at a point where the diameter of the tapered column is exactly 1.5 times the diameter where the load was applied. If the load is applied 0.6 m from the top of a pole where the diameter is, for example, 30 cm, the maximum stress will occur and the pole will break where the diameter is 45 cm. You can check the mathematics. It is exact. A wood pole is not a purely homogeneous material, given tree branch knots and other features, so the pole will really only break in the vicinity of that location.

So, when you see poles broken by a load applied near their top, notice that the tall ones break well above the ground line and the shorter ones break at the ground line.

Also, strength may lessen with height, etc., so you can see that the entire process is created to develop pole classing more than to define real stress. The CSA standard warns that the fibre stress of 6000 psi (40 MPa) is usable only with the minimum diameters matched to it for the standard. You cannot take that value and design wood structures with it. At what stress do poles actually break? The appendices get into that.

Appendix A offers a formula for reducing fibre stress limits with pole height, due largely to increased knots. Appendix C is based on work done by EDM of Fort Collins, Colorado, USA. Its project broke large poles and calculated the stress at the ground line according to both ANSI minimum and actual pole dimensions – but, again, regardless of and ignoring where the pole actually broke.

So, EDM got closer (with its actual) numbers than those the CSA standard uses, but it is still 'faking it', and its fibre stress and modulus of elasticity values are not exactly useful to the pure engineering of wood.

One point is clear. In all cases, as noted, the wood pole is the only product that uses a mean value. If you don't factor the load–strength equation somewhere, somehow, and you use all of the poles to their design limits, theoretically half of them will fail. The factor is large to move from the mean value to an extreme limit.

When you understand the strength of wood poles in this way, you should begin to sense the absurdity of trusting the calculation's results as representing reality with any accuracy at all.

Strength of steel

When you buy a product made of steel – a bolt, a length of wire or a cast hardware item – it comes with a published strength (44 ksi, 10 000 kg, 300 MPa, etc.). That value is described somewhere, perhaps obscurely, as being a working or an ultimate strength value. Either way, you know that it is not a mean strength value in the sense that the product has undergone a rigorous testing procedure that ensures that 99% or something like that of the items are guaranteed to be that strong or stronger.

If you dig deep enough into materials standards such as those of the ASTM or the CSA, you can find the strength assurance programme and published strength of any manufactured item.

You will also find that the coefficients of variation of man-made materials are much smaller than those of natural materials such as wood.

For these two reasons, the strength reduction factor applied to man-made materials and products is much less than that associated with wood and yet achieves a comparable level of security.

It is useful to understand as well that a material that is formed into a product by hammering, moulding, bending and so on adopts a strength more defined by the shaping than by the raw material. It also follows that the strength of that product depends on using the product in the manner assumed or intended by the manufacturer. The last failure story in Chapter 7 illustrates this point: a product was used outside of the presumed but unwritten limits of use, and its strength was drastically reduced because of it.

Strength of fibre-reinforced plastic

We have alluded to the point being made here when we described the FRP core of ACCC conductors. Since some sensible FRP pole products have come onto the market in recent years, it is worth recognising how they are materially different from wood or steel poles. To understand them, we revisit the analogy of a fishing rod.

Like a fishing rod that is also fabricated from glass fibre and resin, FRP poles tend to be purchased in sections that push together to make poles that can be quite long – up to 43 m in four or five sections, as far as we have seen to date. A slightly simplified description of an FRP product line is that ten sections of 10 m in length are available, and you can select a 30 m pole of different diameters and strength by selection of three modules in the series at the large-diameter end of the range, the middle range or the small-diameter end of the range.

In other words, you can select a pole that is large diameter and strong or small diameter and not as strong. But – and here is where the fishing rod analogy is useful – if a fishing rod has five sections and you select the two fat pieces from the handle end, you have a rod (think pole) that will neither break nor bend. If you select the two thinnest sections of the fishing rod, you have a rod (pole) that is very, very flexible but which still cannot break. Such is the nature of FRP poles compared with wood or steel poles.

It has always been the case that you need to check a structure design for strength and for deflection. But, in our industry, we have often become quite secure in the idea that a wood pole and sometimes a steel pole will be limited by strength and not by a problematic deflection. Compared with FRP materials, wood and steel are weaker and less elastic materials. So, many of us have forgotten that we should check deflection of our structures as part of our routine as a responsible designer. Let the use of FRP poles bring you back to your full senses.

Unless you chose the large-diameter, very stiff poles, you are going to find that FRP poles are likely to be limited by their deflection limits that you set long before they risk rupture. Compared with wood, FRP poles are fully capable of providing you with a line that flops around in the wind if you are not careful. The fact that loads exaggerated by load factors cause exaggerated deflections is an exacerbated problem with selecting FRP poles properly. That is to say, FRP poles are not tailored well for the ultimate strength methods. They have no meaningful plastic zone in their stress–strain relationship.

Strength of stranded wires

This has been discussed at length, so here we make comments in line with the theme of elastic zone and plastic zone, noting that different types of conductors are quite different in this regard. Remember the comment that the strength of a product has more to do with its assemblage than with the materials from which it is made. This is certainly true of conductors made of aluminium strands with, perhaps, a core of steel strands or another material altogether.

If you are comfortable with your understanding of ACSR conductors, this does not mean that you understand ACSS or ACCC conductors very well at all. Annealed aluminium has a radically different stress–strain relationship compared with the hard aluminium (of ACSR) from which it derives. The FRP core of ACCC has a completely different stress–strain relationship from a stranded steel core.

In addition to the stress–strain relationship, with bi-material conductors we also have the metallic flow or creep of the aluminium with tension and time, and we have the strand settlement with initial application of tension complicating the wire's behaviour. These have been described. But, when you strip the creep and settlement away, most conductors have a rather limited elastic range and a considerable plastic range. We tend to accept that we understand the plastic behaviour up to about 60–70% of the breaking strength, but do we really know it well?

A recent presentation by a large utility showed that because of testing it had done, it uses stress–strain polynomial values for its sag–tension calculations that are unique to each of five manufacturers of the same conductor. To use the right values, the supplier of the conductors for a project had to be known before the design work was done. They claimed this to be their normal process, and saw nothing odd about it . . . OK!

But, this points to the notion that every supplier and maybe many reels of a conductor from any one supplier are unique to some degree in this regard and to the degree that it matters to these engineers. We suggest that you have a choice to make: you can split these hairs for the duration of your career or you can adopt an approach to the work that will serve you very well while you rest satisfied with the degree of precision that the nature of the subject admits. Where have you read that before? We prefer the latter approach, as you have undoubtedly determined by now.

Blending loads with strengths

Having discussed the strength of materials and products and then the nature of loads and how we think of them, it should be well understood that you cannot make any rational sense of strengths and loads unless they are paired up properly.

If the strength of a structure is made very secure by the use of strength reduction factors against a load that is extremely rare, then you have a structure that is not only very secure but also fairly expensive. If you do not strengthen the structure very much against a load that has a high probability of occurring, the cost of the structure is probably quite low, but you are more likely to find yourself replacing it someday before you expected. The business of spending your money wisely so that no one claims you wasted it and so that your structure lasts a respectable length of time, barring an unusual disaster, is kind of an art. Standards such as IEC 60826 try to offer you guidance as that artist.

Safety codes – care they enough?

Remember that Canada's CSA C22.3 No. 1 and the US NESC are *safety* codes. They are not comprehensive *design* guides or standards. CSA C22.3 No. 60826 is a design standard, as is the basic IEC 60826 that many nations use. A standard may need to be adhered to within your jurisdiction for the purpose of safety, but doing so doesn't mean that your line has been well designed. Be very concerned for designers who claim that 'designing to code' is all that they need to do.

It may be true that you can get away with designing only to code under certain conditions. Your lines may never show poor performance if they are placed where the ice and wind conditions are modest compared with code requirements, but that doesn't mean that you did a good design job. It means that the conditions let you get away with something. The bad news is for the designers. Designers can get lulled into thinking that a code is a pretty good design tool. As a rule, they are not –because that is not their objective.

Some codes – and the aforementioned Canadian and US codes are two – are slowly morphing into more comprehensive design guides by the inclusion of more sensible formulas and more topics of concern. The reason for this is that codes tend to be authored by engineers, not safety experts. But this is a frustration. As long as a standard is declared to be a safety code, then loading it with incomplete subject matter that may be good engineering guidance for system integrity and performance requirements but has little to do with safety will ensure these documents remain a source of trouble.

Because some codes are authored and updated by committees of engineers and remain completely or partially committed to safety by title and do not cover all of the subject matter necessary to also make the facility economical and a good performer, then we should understand that codes alone are not a sufficient guide for your work. You must go above and beyond if you want to produce an efficient facility of value.

Design methods – deterministic versus probabilistic

Basically, a deterministic design method assigns specific strength, load and safety factor values to a situation. The point is that strength and load limits are predetermined and rarely questioned, and these values are sometimes locked in as required by law. While the mix of fixed values may have its basis in a probabilistic discussion or calculation, that basis for the values employed is not known to most designers. Nor are they organised in a way to allow adjustment that could account for variable situations. Without doing research to (re)discover their basis, the values are used blindly. This opens the door to blindly employing them in situations of varied nature. Doing so effectively alters the impact or value of the result. For years, this has been our method.

As a simple example, if you apply a factor of safety of 2 to prevent failure against a wind load, and you place such a design in two places where the probability of that wind load's occurrence varies, then the probability of failure due to that wind load varies as well. Money spent with variable results is considered by many to be money not well spent.

There is full knowledge within the minds of good engineers that the efficiency of designs resulting from the deterministic method is variable, generally unknown and therefore not as good as we can achieve. The attraction to a probabilistic design method stems from the desire to produce an efficient design – one with known and managed risk commensurate with its role in the network and its location in the system – by making use of tools that can adjust strength, load, and factor of safety values in acknowledgement of local condition changes.

Our good friend Brian White told the story that back in the 1970s he and some of his line engineering peers decided that a 'probability' approach to transmission line engineering would be useful, as it had become so in the fields of bridge and building design, for example. The seed for the 'probability-based design method' for transmission line design was planted, and it flourished – like a weed.

> I first encountered the principle in 1977 when I entered into my career as a transmission line engineer. I never liked the concept as proposed. It felt unworkable. During my 5 years at Ontario Hydro in Canada (1982–1987), where the proponents of a 'probability' design method were numerous and hard at work, I felt like a leper arguing against the idea. During the ensuing years, I read

> the various papers by many folks on the topic, and found I kept agreeing with this White fellow. In 1985, I introduced myself to Brian at a world conference on the topic in Toronto, and thanked him for his voice because I agreed with him that the probabilistic method for line design, as presented, was unworkable. We have enjoyed each other's company and work experiences ever since.

Now, more than 30 years after its germination, the probability method for transmission lines is being seen as unworkable in its pure form. But, it has been modified to a more workable form, and goes by the name of 'relative reliability'. The difficulty with the original approach was the inability to determine and manage the 'absolute reliability' of so many things important to a probability calculation.

The probabilistic design method does acknowledge the reality of the natural world by properly having us express a *probable* strength against a *probable* load. If you acknowledge the variability and unpredictability of reality, your answer to the question 'Am I safe?' would be 'You are *probably* safe.' One major problem is that people have difficulty dealing with that answer. It seems evasive or insufficient to those who are used to having quantified answers.

Probability mathematics gets onerous and is foreign to most of us. The understanding of the maths and its results is *very* foreign to us – to the point of being unknown. How are we going to convince our peers, managers, landowners and lawmakers of the safeness and competence of our designs if we produce them from a method that is so difficult to explain? One of the barriers to the employment of a probabilistic design method is our inability to understand and explain it.

This is separate from the commonly expressed point that we will never identify reasonable absolute probability functions for the essential load–strength relationships of a transmission line. Most every paper, report or article that attempts to explain or provide formulas for probabilistic methods effectively states somewhere that 'in lieu of better information, assume . . .'. Within the realm of absolute probabilistic design methods, we will have considerable difficulty satisfying ourselves or others as to the efficiency of our results.

Remember that the attraction to probabilistic methods is the ability to define the natural loads and forces more properly by accounting for their unknown aspects with *probable* values and thereby achieve our goal of more efficient designs. Given this tool and this goal, we will move in that direction. The probabilistic method sets us up to manage the margin of safety – the gap between the load and the ability to resist the load (strength) – and we will do so to meet our goal of efficiency.

The relative reliability method basically says that since you cannot express the reliability or probability of some component in absolute terms, you can look at that component being used in an existing facility in a comparable situation and make adjustments to its use so that its performance is likely to be improved or even put at higher risk should you think that is appropriate.

However you look at the subject or feel about it, the future of transmission line engineering lies with the probability-based method. If you have not done so already, get a copy of the IEC 20826 document and absorb it. Don't just read it, understand it. It may give you a headache, but you will be better for it when done.

Factors of safety

One final thought on factors of safety. Design criteria impose factors of different magnitudes on the design loads as a method of increasing the strength of the structure against the actual (design) loads. The thinking is – if the structure can support the exaggerated loads, then it will support the actual loads with a level of risk reduction commensurate with the magnitude of the factor(s) used. Well ... be careful! Be very careful.

The first thing to understand is this. Members that we use to assemble our structures have variable capabilities to support tension, compression and bending forces. For example, a cable member is very strong in tension but supports no bending or compression. Long, slender members can be very strong in tension and have comparably low capacities in bending and compression.

If you load a structure design made up of these many types of members, the analysis and design process will lead you to orienting and sizing the structure members such that they will transmit the forces through various paths to the foundation of the structure. If you take the time, you can choose the members to make them quite efficient: not too small but not too big so as to be wasting material. If you do this ... well done!

If the loads on your well-designed structure were to be cut in half – all of them – the forces in the members would be cut in half as well. If all of the design loads were increased by some amount, the forces in all of the members would also all increase by that amount, and the loads could all be increased this way until one member breaks or buckles. As soon as one member is taken out of action, the forces in the structure find other paths to ground – see ... structural forces are the same as electricity and water in this regard. Hmmm! And you will have to start all over again to design a viable structure with that member omitted!

Consider the following scenario that we got from a Brian White failure story some years ago. Figure 5.5 shows the left side of the bridge of a guyed-V tower. Member A is of interest to us. In Figure 5.5(b), we have the design wind and conductor weight acting on the end of the arm of the tower. These are referred to as the real loads, and they come from the conductor suspended from that point on the tower. The instruction to the tower designers was to apply a factor of safety to the tower by increasing all applied loads by a factor of 2. So, we see the factored design loads applied to the tower on the left of the figure. These vectors are twice the size of the real load vectors that were intended be supported by the tower.

You can make the weight of the conductors anything you want on such a tower, and the guys supporting the tower will not be loaded. All purely vertical loads go into the legs of the tower, to the foundation. The load in the guy wires comes from the lateral wind load.

Figure 5.5 The effect of load factors: (a) factored loads; (b) real loads

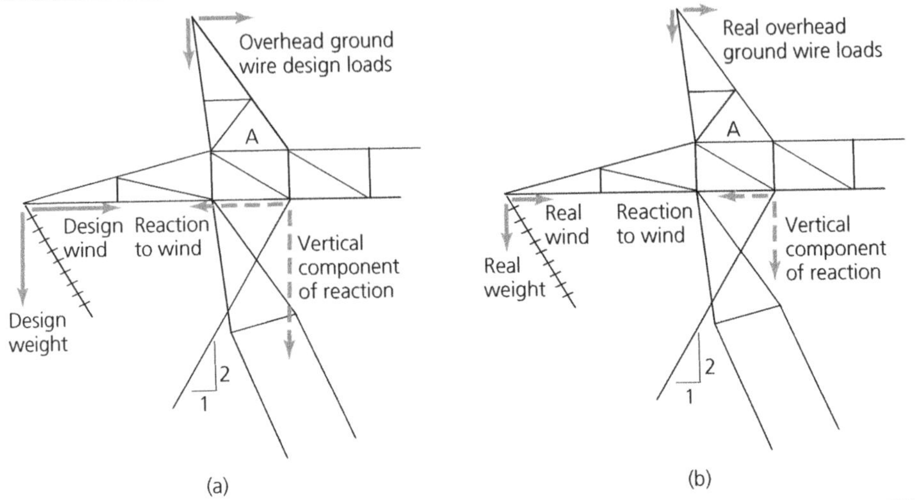

As such, the horizontal reaction to the wind that loads the guy wires must be equal to the wind load. The guy reaction is not affected by the vertical conductor load. The vertical component of load in the guy wires is a function of the reaction to the wind and the slope of the guy wires.

So, we have the bridge of the tower supported by the leg that is attached to the bridge between the vertical loads of the conductors and the reaction load in the guys. These two downward forces with the upward reaction between them are trying to bend the bridge. The bending is resisted by the top chord members of the bridge – such as member A – being in tension and the bottom chord members being in compression.

As noted above, if the loads are all doubled, the tension force in member A will double, and vice versa. For the tower that failed in Brian's story, this wind load case was the primary load case to design the tower. Thus, member A was designed and sized for tension in accordance with Figure 5.5(a). It was not very large and quite incapable of carrying much compression load, if ever required to do so.

Then, one day, along came Mother Nature, and she applied a wind load to the conductors and shieldwire that was less than the unfactored loads – less than the load that the tower was designed to support – and the tower came down.

The tower came down because member A collapsed in compression and the arm rose up and folded over onto the top of the bridge (Figure 5.6). How weird is that to fail upward!? Here is why…

When the wind blew that day, the weight of the conductors and shieldwires did not change. This tower, like most towers, supported a span shorter than the design span used

Figure 5.6 Wind failure

to design the tower. This meant that the vertical load vectors of the conductors and shieldwires were less than planned. The vertical reaction in the guy wires was related to the wind, and the tedious balance between these vertical loads on either side of the mast of the tower supporting the bridge went out of the balance envisioned by the design process, and member A went into compression. End of tower!

What are you to learn from this? The headline of the lesson is: 'Load factors are dangerous!' The details are this.

- All load vectors used to design a structure can be increased in unison (equally), thus predictably increasing the forces in all of the structure's members until a member fails. Then stop!
- If the load vectors that are actually applied to a structure are anything other than some globally increased or decreased version of the design vectors (i.e. if the resultant vector at any load points are reoriented in any way), the forces in the structure's members will be distributed differently from the design loads, and the strength of the structure against the actual applied loads may not be anything like you intended.

In the simplified case described above, the actual strength of the tower was less than half of the intended strength and below the strength required to support the actual wind expected.

You should feel great angst when faced with the unequal overload factors presented to you as required by various safety codes, because the load vectors they create will never be real. The case above used the factors that were equal in all directions and, even so, the problem remained.

Load factors can do this. Load factors exaggerate deflection, and this will falsify the stresses in flexible structures. Load vectors that are other than planned in the design calculations change the resultant vector, and the forces will be other than calculated. Load factors are dangerous, so be very careful. Your best protection is to run the load cases required by the safety codes and design codes, and then develop load cases that will represent reality.

We watched a safety organisation – a branch of government that took responsibility for the subject of safety – argue that a factor of safety of 4 applied to a calculation of wood pole strength against wind load meant that the basic wind pressure in the load case could be multiplied by 4 and that the pole should be expected to withstand that wind pressure. The same standard being debated said that the factor of safety could be 2 if the pole were steel or concrete. Thus, the argument being presented by the safety organisation would have to mean that poles made of other materials need only stand against winds of half the pressure (70% of the wind speed) that the wood pole must withstand. I guess that means that if you expect really strong winds, you are better off using wood poles!

The depth of understanding of the most basic components of line design methods can be staggeringly shallow.

In our office, we very often, but perhaps not often enough, take load factors and invert them, so they become presented as strength reduction factors for the computer. We run the calculations that way. Loads become 'real'; deflections and deflected stresses become 'real'; all tension or compression forces are calculated correctly, and all of the protection in the calculation is represented by a stress reduction on materials and members, even though the source of the value came from a reasonable consideration of the probability of load occurrence.

But, you can take that method only so far, because it tends to presume that the structural members are all elastic. It has trouble dealing with plastic behaviour. The better solution is to increase the actual or unfactored design loads to more extreme values and keep all of the load factors as unity and the strength reduction factors to values near unity – values that require minimal adjustment to the real stresses. Only then can you say with any accuracy: if the structure can support these extreme but realistic loads, then it will support the actual loads of lesser magnitude with a level of risk reduction commensurate with the magnitude of the load exaggeration used.

The subject of loads and strengths is complex, with quite a number of traps set for you. We close this chapter with the same advice as we do most of the others. Think!

REFERENCES

CIGRÉ (2007) *State of the Art of Conductor Galloping. CIGRÉ Technical Brochure 322.* CIGRÉ, Paris, France.

CIGRÉ (2011) *Working Safely while Supported on Aged Overhead Conductors. CIGRÉ Technical Brochure 471.* CIGRÉ, Paris, France.

Davis DA, Richards DJW and Scriven RA (1963) Investigation of conductor oscillation on the 275 kV crossing over the Rivers Severn and Wye. *Proceedings of The Institution of Electrical Engineers* **110(1)**: 205–219.

Wong CJ and Miller MD (2010) *Manual of Practice 74. Guidelines for Electrical Transmission Line Structural Loading*, 3rd edn. ASCE, Reston, VA, USA.

Structural Engineering of Transmission Lines
ISBN 978-0-7277-5988-7

http://dx.doi.org/10.1680/setl.59887.187

Chapter 6
Fun with cable structures

In the early stages of this book's development, the discussion on cable structures – support structures composed almost entirely of cables – was embedded in the discussions on all of the other forms of support structure. But, some new opportunities to get involved in the engineering of cables took place, and the very unique nature of the work and the importance of these types of structures were reasons enough to move the subject to its own chapter.

Understanding the engineering of cable structures is important because cable structures used in the appropriate situations are not only a most cost-effective and structurally secure solution but they can allow a facility to be constructed where no other solution is possible in practical terms.

> When I look back on my career, even long before it was getting lengthy, I became well aware that the cables – conductors, guy wires and so on – were the most intriguing components of a transmission line for me. I had reasoned that their behaviour being well outside the confines of the usual structural engineering box – that of being large-deflection structural members that do not comply with the small-deflection constraint that allows the application of the formulas normally taught to engineering students – was why I was intrigued.

> It does seem that interest in work attracts the work. Going back to 1993, I had the opportunity and real pleasure to give thought to cable structures. Over the years since, I can count five cable design projects that I undertook, another that I came to understand well and yet another – the grand-daddy of them all – that we developed conceptually and would dearly love to undertake someday. The execution of each of these cable projects laid the foundation for knowing better how to tackle the next.

The things that we can convey to you, teach you about cable structures, are embedded in the stories of these projects that came to us. Because the cable projects are so clearly linked to each other, as you will see by the stories of them, this chapter will tell you the story of these projects as they occurred. Much of this chapter, then, is presented in the first person.

Cable structures are worth understanding because they have great structural integrity and low cost. Being willing and able to undertake the engineering of one yourself when the opportunity presents itself could prove rewarding.

Remember the structural principles:

- the most efficient structural element in tension is a steel cable
- the most efficient structural element in compression is a latticed steel mast
- the most expensive use of material is in bending.

These are the reasons why a cable structure is so efficient. The location and circumstances for their installations are the reasons they are interesting.

The transmission line catenaries

At this time, we are aware of five catenary structures on transmission lines somewhere in the world. These are structures that support electrical conductors and are composed almost entirely of cables or a single cable from which the conductors are suspended. Since a suspended cable takes the shape called a catenary, these types of support structures have been given that name – at least among the folks that we encounter.

We suggest that the first such catenary was designed and installed in 1955 on the Kemano–Kitimat transmission line in British Columbia, Canada. The engineer was Brian White. Thereafter, we know of a light-duty catenary supporting a 69 kV line on the Island of Oahu, Hawaii, where the line passes over that island's sharp mountain spine; one in Russia squeezing an EHV circuit through a tight space in a mountain pass; a second catenary on the Kemano–Kitimat transmission line designed by this book's two authors and installed one span away from the first in 2008–2009; and finally a rather elegant and unique catenary recently installed in South Africa.

The two conceptual designs noted above plus one temporary catenary that was installed, all designed by one or both of this book's authors, were all in the British Columbia Coast Mountains as well.

The 1955 Brian White catenary

Being in this business in the 1970s and onward in Canada, it was inevitable that I would encounter Brian White's work and eventually, by decision, Brian himself. Brian's very first foray into transmission line engineering was by being tasked by his employer to take the lead role in the design of the Kemano–Kitimat double-circuit 287 kV transmission line in the British Columbia Coast Mountains. This was about 1950, and Brian was 26–27 years old.

Brian much later said that if this had not been his first transmission line project it would never have been one he would tackle. The line is in extraordinary country, and a knowledgeable person would have said 'no thanks' to the staggering challenge. Thankfully for his own career, he was not yet so knowledgeable, but the project sure educated him fast and continuously through his entire 50-year career.

When the line was in construction in 1952–1954, a 20-year-old kid from Manitoba arrived on site to work on the line. This kid became a crew boss with eight men under his charge that first summer. Why him? He said because he was the only one who did not get stinking drunk every night. The second season, this now 21-year-old was given a much larger crew to direct. This hard-working and very smart and self-sufficient young man was Adam Charneski.

Adam tells me that now, at the age of about 80, he is compiling his story of what turned into a 55-year relationship with that transmission line. I dearly hope he gets his book completed, so I will not relay many of his stories. They are his to tell, and he sure has the stories. I have met several men who were there in the 1950s for this transmission line's construction. That event defined their lives as much as a man's life was defined by being in the Second World War a few years earlier.

I will tell you this about Adam's career. He went on to be the maintenance superintendent for that transmission line for 40 years. It became his job to make improvements over the years that the original designers did not recognise as needed. The line is 50 miles (80 km) long, running from sea level up one river valley, over a mile-high pass above the tree line and glaciers, down into a second river valley, over another modest mountain and across a major river delta to its destination at sea level. Although there were at least five serious avalanche events that destroyed structures in the high country over the years and there were other rock falls and deep-snow problems, the seasonally raging rivers that scoured new channels every year on the valley floors were Adam's biggest problem.

Much has been written about the installation of the catenary in 1955. See the Brian White references 1a, 1b and 63 in Chapter 10 for much of the story in Brian's words and some intriguing photos. Here, I offer a synopsis. The transmission line went into service in late 1954. On 25 January 1955 – when less than a few months old – both circuits had structures wiped out by an avalanche where crossing the floor of a mountain cirque. This is a place called Glacier Bowl. It had been argued by Brian that the line should never have been placed on that valley floor. But, there it was placed, and there it was summarily wiped out early in its first winter.

Brian's proudest achievement was to come up with the catenary solution and oversee its installation such that the circuits were both back in service by 15 September of that year – a less than 8-month timeframe. At that time, the equipment and access roads to the site of the catenary's installation were still in place and fully serviceable due to the recent line construction. People and all equipment and materials could basically drive to the site from the sea-level town of Kemano some 20 km away and 1000 m below.

Over time, through simple interest and eventually necessity, I came to understand quite a bit about that catenary's design and construction. The design criteria and features of the mountain portion of this transmission line are unique, and its description requires big numbers. In 1950 when the design criteria were in development, there was not a lot of data available to lean on. Brian *et al.* tapped into Bonneville Power Administration in Washington State for its high mountain experience. They tapped into the Swiss for the same. They came up with a conservative design ice load case of 40 lb/ft (about 60 kg/m) ice on the conductor. This large ice on the conductor – a large conductor called 'Emu' at 3364 kcmil (1704 mm^2), 2.29 in. (58 mm) in diameter, weighing 4.76 lb/ft (7.2 kg/m) itself and with a breaking strength of 135 400 lb (602 kN) – set the everyday tension in the conductor to a loose 15% or so. This conservative value frustrated Brian in later years because it forced spans shorter than he would otherwise love to have used, although some of the spans are respectably long at 600 m and more. That 40 lb/ft of ice load was allowed to stress the conductors and all components of the line to 100% – an unusual combination of load and strength limits but one fully compliant with the approach we suggest at the end of the last chapter.

This combination of large ice load and big spans required to span to the catenary above the floor of the Glacier Bowl called for catenary cables made of 3 in. (76 mm)-diameter galvanised steel rope. The catenary is made up of two such cables spanning across the transmission line path attached to anchors composed of a cluster of 25 ft (7.6 m)-long rock anchors high up on the valley's mountain sides. The two anchors on each mountainside are a few hundred feet (40–60 m) apart, causing the entire rig to have a very elongated X shape in plan view.

The locations of the anchors are such that the catenary cables are not exactly perpendicular to the direction of the six conductors hung from them. It was surmised and shown by scale model testing that a large movement of the conductors from something like ice shedding would initiate a transverse component of force in the catenary structure. Movement in a conductor that translates to movement of the catenary would trigger all sorts of unattractive movement in the other conductors, Thus, the two cables with spread anchorages were used to provide longitudinal stiffness to the rig.

Each of the two cables across the valley is actually composed of three sections connected end to end by large poured zinc sockets. One 1000 ft (305 m) length of each cable reaches from an east anchor to a yoke plate that marked the start of a catwalk section of catenary above the suspended conductors. Here, the two cables are 1 m apart. The catwalk at 400 ft (122 m) in length uses the second lengths of cables as handrails from which the catwalk and the conductor insulator assemblies are suspended. The third and final length of each cable reaches from the west side of the catwalk section to each cable's west anchorage. The total length of each of the two cables is 3800 ft (1158 m).

The spans from the nearest structures of each circuit to the north were 2500 ft (762 m), and to the south 4000 ft (1220 m). The avalanche that caused the line's renovation wiped out three conventional structures, and the catenary allowed the removal of three more – six structures in total from the valley floor. I have been to the site in the winter quite often, as you will read, and the flowing of deep snow across valley floor where these towers had been is a winter-long event every year. It truly was no place for a transmission line. Since the catenary's installation in 1955, the conductors have been held well above the valley floor out of harm's way, and the catenary is considered some 50 years later to be the line's most secure structure.

Because access to the site was easy and equipment was already on site, the catenary's installation went quite well. During the installation, one circuit was kept in service by – get this – a very long temporary span of 4/0 ACSR conductor operating all summer

Figure 6.1 Brian White, aged 32, standing at his catenary site while its cable is pulled to the anchorage. The Kemano River valley is in the background, 1000 m below. (Courtesy of H. Brian White)

at 300 kV. I am told the 4/0 danced in the air, buzzing and pushed around physically by extraordinary corona. The big cables were laid out on the valley floor below their eventual aerial positions and the ends were pulled to the anchorages by large block and tackle for pinning. Figure 6.1 shows Brian, when he was 32 years old, standing at an east-side anchorage site with the cable socket pulled nearly to the pinning position. The troublesome Kemano River along which the transmission line runs can be seen in the background 1000 m below. The catenary cables were lifted into place in this way, passing through the operating circuit with a detailed step-by-step process. Then, the repaired Emu conductors were laid out on the valley floor with required lengths calculated, and lifted one by one up to the catenary for suspended connection. Literally the day after the work was completed on 15 September, winter blanketed and closed the site in its first serious snow.

The original plan for access to the catenary's catwalk over the conductors – a feature set in place for inspection and maintenance of the conductors' insulator assemblies – was to use carts that rode on one of the cables and that were stored at a cable anchorage. It did not take long to discover that these were not safe places for storing carts, as the winter snow was deep and it crept relentlessly downhill, crushing anything that did not yield like a willow tree branch. It was not many years before access to the catwalk was converted to helicopter access via a small pad set on the cables over the catwalk midway between the two circuits. That small 3 m × 3 m helipad is 135 m above the ground below, and the

Figure 6.2 A helicopter landing on a catenary

maintenance crew became very adept at stepping out of the helicopter for catenary maintenance or inspection work (Figure 6.2).

The final features of the catenary to describe that lay the foundation for future endeavours were its vibration dampers and the yoke plates that bracketed the catwalk section (Figure 6.3). The two yoke plates transmitted the tension in the cables through themselves as each cable connected to a corner of the basically rectangular plate that is about 1 m wide and 0.5 m long.

The presumed Aeolian vibration of the big cables was managed by a series of lead weights laying in a large plastic tube 'sleeve' about 6 m long set over the cable at each anchorage. The pipe sleeve was sloped with the slope of the cable, of course, so the lead weights were held in place by a stainless steel wire tied back to the high anchorage end. These locations were buried in deep snow every winter, and the tubes allowed the weights to lift up and down, freely rattling any vibration movement in the big cable to a calmer state.

Over time, this damper system proved to be not very good as every so often the stainless wire would break and the weights were to be found in the spring laying in the grass and shrubbery down the slope having shot out the end of the tube like lazy missiles. Several years ago, plan B was put in place, designed by a good vibration expert with no particular knowledge of the site. These were a series of standard Stockbridge dampers set on the cables near each anchorage. When I flew by them a few years later to see how they were doing, nearly every unit had a bent cable between the two weights. They were buried in

Figure 6.3 The bundled cable concept for the Hanging Valley catenary

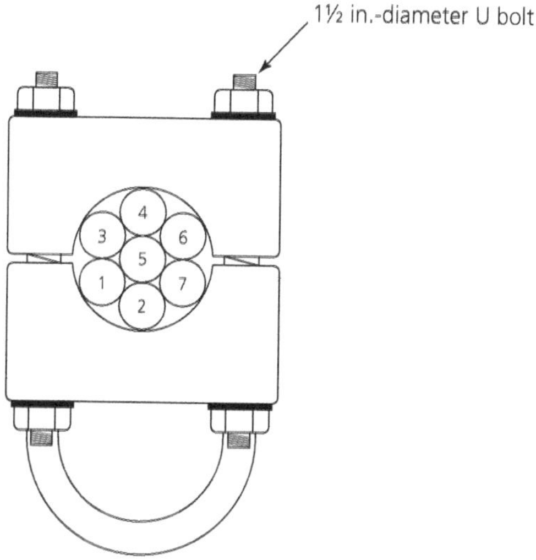

winter snow each year and the snow creep down the slopes had bent the units until they were quite useless.

I had nothing to do with the 1955 catenary design and installation. After all, I was 6 years old that summer! But everything I have just told you was guidance for things to come.

The KitiKat

In late 1992, an avalanche brought tower T113R on this transmission line to its knees. Brian, still being the go-to engineer for the line, was called to help get the line back in service. Constant scrambling by many people over a mere 3-week period had the line back in service in January 1993. His solution was the little hinged tower that is described in detail in the next chapter. It is described in the next chapter – a chapter on structural failures – because although a brilliant solution for the moment, it had insufficient staying power for the reasons to be described.

Brian was very excited and proud of his fast return to service solution and had agreed with the line's owners that there were other locations on the transmission line that deserved some proactive attention. There had been other structure losses to avalanches in 1975 and 1986, so the owner understood that attention to the line on an ongoing basis was useful to the line's integrity. Since Brian worked alone from his home, he wanted some engineering production assistance. As it turned out, I was the lucky guy who agreed to go to the site and help him out.

In September of 1993, I stood with Brian and few hosts from the line owner's staff on Kildala Pass looking southward over that mountain cirque area and to the Kemano

River beyond. Before us was his hinged tower, and beyond that his pride and joy – the catenary. He said to me, 'More interesting transmission line engineering to learn from has occurred that we can see from where we now stand than anywhere in the world.' I could not disagree, but neither of us had any idea of what would happen in this same place over the coming months and next 15 years.

Brian's hinged tower came down on a blustery night in November 1993 – 10 months after being installed and 2 months after he and I stood looking at it. It is worth noting that the linemen who looked after this line were very sceptical of the hinged tower's capabilities. They were right. While these trades people have no formal engineering knowledge on which to base their views, they have other reasons for understanding things. When they speak, do consider that they may be right.

One of the two very important circuits was out again in a pile of mangled steel and shredded conductors. The whole story of that event and the reasons for it are provided in the next chapter, because the lessons to learn are many. But, beyond simply getting the line back into service again, the owner asked that we take some proactive actions as well on the assumption that Mother Nature was not done with us.

We undertook two exercises. We designed a large catenary structure for another location on the line where an avalanche had destroyed a structure in 1986. This was in the Hanging Valley. It took several months to get the circuit back in service after the loss of the hinged tower, so we also developed a plan to expedite that exercise should it require repeating. This was the installation of a temporary catenary at the fallen tower's site.

The avalanche event that destroyed the tower (T113R) that Brian replaced with the hinged tower was a high-speed powder event that crushed the tower with dense air wind pressure. The tower fell uphill as directed by the conductors, and the conductors, notably the centre phase, were trapped in the tower steel wreckage. When Brian's hinged tower fell, it too went uphill, and the conductors were again trapped and damaged in the wreckage, and it took nearly 4 months to unravel the mess and get back in service. The line was put back into service with a tower identical to the original. The rationale was that the original tower had lasted 38 years, so this one should do well also. It took nearly 4 months to get back in service because avalanches have the nasty habit of occurring in the winter when access to the site and work there is very slow and hard.

We devised a plan to place a cable at the site such that it could be quickly tensioned into place above that sort of wreckage, have the conductors cut out of the wreckage, repaired and lifted to the cable's suspension insulator attachment points, put the circuit back in service within weeks (not months) and clean up the rest of the mess below at our leisure after the winter passed. This cable was labelled as a temporary catenary so that no one would come to believe that it had the integrity to be permanent. This temporary status also allowed lowering its design loads a bit.

This cable was a single 1.75 in. (44.5 mm)-diameter galvanised bridge strand cable 2200 ft (670 m) long. It was pinned high up on the mountainside to the east and to the rock at the base of the 'at-risk' tower, T113R. It was placed downhill of the tower in the expectation that the tower would fall uphill as it had twice before. If the tower fell, the conductors would thump into the snow over the temporary cable, trapping it there. The cable was made in several sections, and a final section was left on a reel and stored on site in a storage shed. If needed, that cable would be attached at the joint, laid over the fallen conductors and tensioned to pin its west end to a point above the tower to the west. This would put the cable about 20 m above the wreckage, and the repaired conductors could be lifted to it easily.

After a long ordeal, this temporary cable was installed and pinned to the rock next to the base of the T113R tower. The fact that this rock anchor broke in the first winter is a short tale of admitted screw up on my part in the next chapter. But, with the able help of my friend Adam Charneski, who happened to be installing towers on a reroute of another part of the line 40 km away, the temporary catenary was reset in place and sat there like an insurance policy waiting for a real reason to exist. I have to admit that the very idea of the temporary catenary was Adam's by way of a black marker scratching on a terrible fax copy of an old photograph of the site. He said, 'We can do this . . .' So, we did.

One never wished to see an insurance policy have to be put into action, but 10 years later the temporary cable was put into action for one winter. It worked as planned. By that time, we had so many catenaries in our minds, this one was humorously dubbed the KitiKat in honour of the town Kitimat and its small size compared with the others. The name stuck, and I believe it remains draped down the mountainside still attached to the high end anchor, having done its job.

The Hanging Valley catenary

The Hanging Valley is the name given to the place where the Kemano–Kitimat transmission line drops off the north side of the Kildala Pass and heads north-west towards Kitimat. The valley is 700–800 m above sea level and a few kilometres long, with mountainsides rising steeply on each side for more than another kilometre upwards. At the lower end of the valley, the ground drops steeply to the lower Kildala River valley at near sea level. Entrance to the Hanging Valley from below is by a poorly maintained switchback road built years ago for the line's original construction.

I expect the valley's name is due to this geographic nature in which the entire valley seems to be hanging up there like a room isolated from the rest of the valleys around by its steep entrance. It is a stunningly beautiful place to be sure. On the high mountainside at that hanging entrance, a pristine but shrinking glacier clings precariously to the steep slope, as if ready to slide down into the valley below at any minute.

In 1986, a large avalanche sped down the south side of the Hanging Valley and blew away a very stout structure. We were asked if a catenary structure could be placed here, and if so whether it would reduce the risk presented to the many structures along the valley floor. All these structures were measured to be at some risk of avalanche damage, even though many were protected by very large earth berm deflectors. After some study of the terrain, our answer was 'yes'.

We came up with a catenary that spanned more than 1500 m across the valley, with anchors 800 m above the valley floor, which supported the two circuits with a weight span of about 2150 m with attachments more than 300 m above the valley floor. This catenary allowed the removal of nine conventional towers on the valley floor that were deemed to be at varied risk of eventual avalanche damage. The constraints to its construction caused its design details to vary considerably from Brian's 1955 catenary.

We assumed two cables with anchors separated on each mountainside by a few hundred metres because the catenary orientation was not anywhere nearly perpendicular to the circuits' direction, more so than with the 1955 catenary. We assumed a catwalk system over the conductors, and we assumed that power could not be shut off from both circuits simultaneously since the town of Kitimat and the viability of the company's smelter operation would not tolerate that. In addition, the anchor sites had very little to no place for setting large installation equipment. Getting large equipment into the Hanging Valley itself was a daunting challenge since the road was long, winding, steep and without bridges across the sometimes unruly Kildala River.

I developed a design and installation plan that allowed the catenary to be built from the valley floor with modest pieces. It amounted to this: each of the big cables would be made up 'in place' by bundling seven 1.125 in. (28.6 mm)-diameter galvanised bridge strand cables together (six fitting around a central seventh) to form a cable 3.375 in. (85.7 mm) in diameter (Figure 6.4). This allowed the transporting of many small cables

Figure 6.4 Catenary 1 yoke plate at the catwalk ends

to the site and forming the necessary large cable in place, one element at a time. The rationale was reminiscent of the construction of a large suspension bridge cable from many wires, one at a time – although much simpler than that.

The idea also came from recognising that Brian's 3 in.-diameter galvanised steel rope was actually constructed from seven cables of 1 in. diameter: six of them factory wound around the central seventh. The seven parts of this final, large cable would not be wound around each other but parallel like wires in a suspension bridge cable, and would have to be connected together with some sort of clamps at intervals.

The two circuits suspended below the catenary were 123 m apart, so each circuit had its own shorter length of catwalk, avoiding the cost of a very long system with limited value. It was the plan to get the first small-diameter cable into the air and then raise the others up from the valley floor, using it as a trolley cable to guide each new cable towards the anchors.

The design and construction method was carried through in considerable detail ready for pricing and construction, and those details are not worth describing here. I am sure that with added thought should the design have gone to construction, a few edits to the details would have ensued. I never expected this catenary would get installed, and it is not a surprise that the design report languishes still on my shelf at home – waiting!

In fact, the catenary that we did design and install 10 years later took these basic concepts and put them to good use, but we did simplify many of the details. That later effort was Cat 2.

Cable system modelling

When Brian's catenary was being engineered in 1955, a scale model of the cable and conductor system was made to try to understand the dynamics of ice shedding and so on. We did the same for the Hanging Valley design because the skewed alignment of the catenary cables to the conductors was of concern (Figure 6.5). This meant that the model had to be scaled both physically and elastically. Here is how you can construct a model so scaled.

Linear scale:

$$L_{\mathrm{m}} = L_{\mathrm{a}} \times R \tag{6.1}$$

where:

- L is the length
- R is the size ratio (scale)
- m denotes the model
- a denotes actual

Elastic scale:

ratio of load = ratio of strain

Figure 6.5 Anchorage of the Hanging Valley catenary model

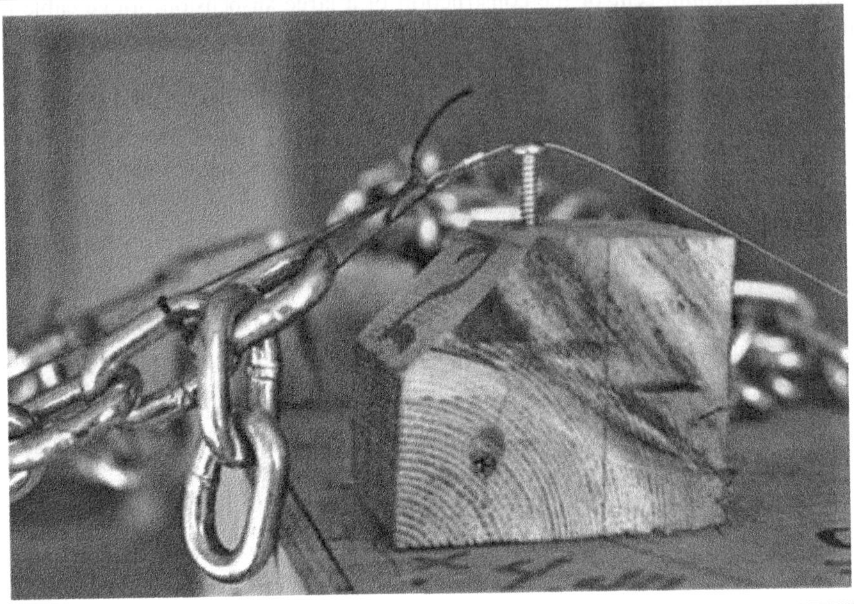

that is,

$$H_a/H_m = (A \times E_a)/(A \times E_m) \tag{6.2}$$

where:

H is the horizontal tension
A is the cross-section area of the cable
E is the modulus of elasticity of the cable

Since $H = w \times L^2/(8 \times \text{sag})$, and since the shape of the model is the actual shape, that is,

$$L_a/\text{sag}_a = L_m/\text{sag}_m \tag{6.3}$$

then

$$W_m = W_a \times R \times (A \times E_m)/(A \times E_a) \tag{6.4}$$

With this formula, you can select the weight and area of the model's wires. You will find that you will use small piano wire weighted with chains that carry no tension.

It is of interest to note that the formula for the swinging period of a pendulum should also let you scale time when you watch the model's motion.

Cat 2

When you look back on your career path, you may recognise times when a tiny circumstance sends your path in a direction that would have otherwise been very different. The best example of that in my career occurred in late March 2007. At that time, I had not been to the Kemano–Kitimat transmission line for 10 years and had sadly relegated my exciting times there to my history book. Then, the phone rang.

Tower T113R – the one we stood up in 1993 after the demise of the hinged tower was down again in an avalanche. Could I help? Absolutely!

All of the people in charge at that site had changed in the intervening decade, and nobody there knew me. They were asking among themselves who they should call. It just happened that the man for whom I had worked 10+ years earlier was in the room, even though he no longer worked there on a full-time basis and his presence that day was simply a rare circumstance. He gave them my name.

This time, the avalanche was different. It was not a repeat of the 1992 high-speed dry powder event. If you are to learn any one thing about Mother Nature, it is that she has a bottomless bag of tricks from which to throw things at you. This avalanche came from the same place high up on the mountainside, but it was a wet, flowing event. This time, the tower did not get crushed by a wind blast and fall uphill. Rather, it was scraped downhill off the face of the Earth by a mountain of moving wet snow. Think of that snow as having the power of flowing concrete. There was nothing left of the tower except its anchor bolts in the rock.

All three phases of the large Emu conductor broke, and the centre phase was trapped in the tangle of steel that was sliding down the valley floor of the Glacier Bowl. The tower – or most of it – travelled 3 miles (5 km) to the river a mile (1.5 km) below.

To keep the story short, we sat in a meeting to plan the recovery, and the question was asked, 'Can a catenary fit here?' I said, 'I think so, and I will check.' Decent topographic information for the area did exist. Within the same day of reporting that we could fit a catenary into this location, it was decided to pursue it. That was simple! The primary reason they chose to replace the conventional towers in this high-risk location was that any conventional towers placed in harm's way had to have some non-zero risk of avalanche destruction assigned to them, and the catenary did not.

I should also mention that, after saying that, yes, I could help, I called Adam Charneski. I really wanted to – and did – benefit from his four-decade passion for and knowledge of that transmission line (Figure 6.6). So, we embarked on the design and installation plan of a catenary one span uphill from Brian White's 1955 catenary. What do we call them now, since Brian's was to this date known only as 'the catenary'? Can they be the White Catenary and the Black Catenary? We settled on Catenary 1 and 2: Cat 1 and Cat 2.

Recall the conditions at the site that defined the design and installation plan of the 1955 project. It was clear that nearly all of the controlling criteria had changed. Access to our

Figure 6.6 Adam, happy to be engaged again on his beloved transmission line

site – now 1800 m above sea level – would be by helicopter only. The access roads and bridge across the Kemano River had been gone for decades. The two operating circuits would remain in service throughout construction, excepting for short single-circuit outages to transfer the conductors from existing support towers to the catenary. There were no accommodations at the site and, excepting the possible emergency use of an old camp building, there would not be. Housing and material staging would be from 20 km away at sea level in Kemano. The useful construction season was 15 May to 15 September. Outside of this time period, the snow would occupy everyone's time.

The design and installation planning work began immediately, in parallel with the actual 'back-in-service' work of repairing the damaged conductors, which included hanging them from the KitiKat as envisioned 13 years earlier (Figure 6.7). The objective was to complete the new catenary that summer, but it became very obvious in early July that we were not going to make it. We hardened the anchorages of the KitiKat, and asked it to support the circuit through the 2007–2008 winter season. It did. Then, we started very detailed installation planning for the 2008 season through that winter.

On 17 May, a Chinook Boeing 234 helicopter lifted a snow cat and excavator onto the relatively large area of flat ground at the top of the mountain on the west side of the project site. The snow was 4 m deep. The area was prepared, and on 27 May the Chinook took over

Figure 6.7 The KitiKat supporting about 25 tonnes of conductors

100 tonnes of materials and equipment onto the mountain in 21–40 km round trips – all in $6\frac{1}{2}$ h. It was a fun day!

The main equipment lifts were rope pullers of various sizes, a tensioner and 28 reels of $\frac{3}{4}$ in. (19 mm) galvanised bridge strand cables each weighing about 6000 lb (2000 kg).

This catenary was a bit bigger than Cat 1 in terms of its cable lengths, averaging 4000 ft (1220 m), and the weight span of the supported conductors at about 2500 ft (760 m). The big cables were again planned to be a pair, with anchorages separated by 100 m or so on each mountainside and brought together over the circuits to allow load sharing and the installation of a catwalk for inspection and maintenance. We tapped into the 10-year-old idea developed for the Hanging Valley of fabricating large cables from smaller cables (Figure 6.8).

It was necessary to limit the weight of cables on reels and of all equipment to the capacity of the Chinook (25 000 lb or 11 300 kg) at sea level. If we built a big cable from only seven smaller cables, each small cable was quite large in itself. So, we decide to make each cable from 14 cables, but organised as two pairs of seven cables set 42 in. (1.067 m) apart. The weight of each of the $\frac{3}{4}$ in. cables on a reel was not the limiting factor. Rather, it was the size of tensioner and puller equipment that could install each to the tensions we sought.

The west side of the project site was a large and relatively flat area whereas the anchorage sites on the east side offered no meaningful level areas for equipment at all. The puller

Figure 6.8 Cable geometries of Cat 1, Hanging Valley and Cat 2

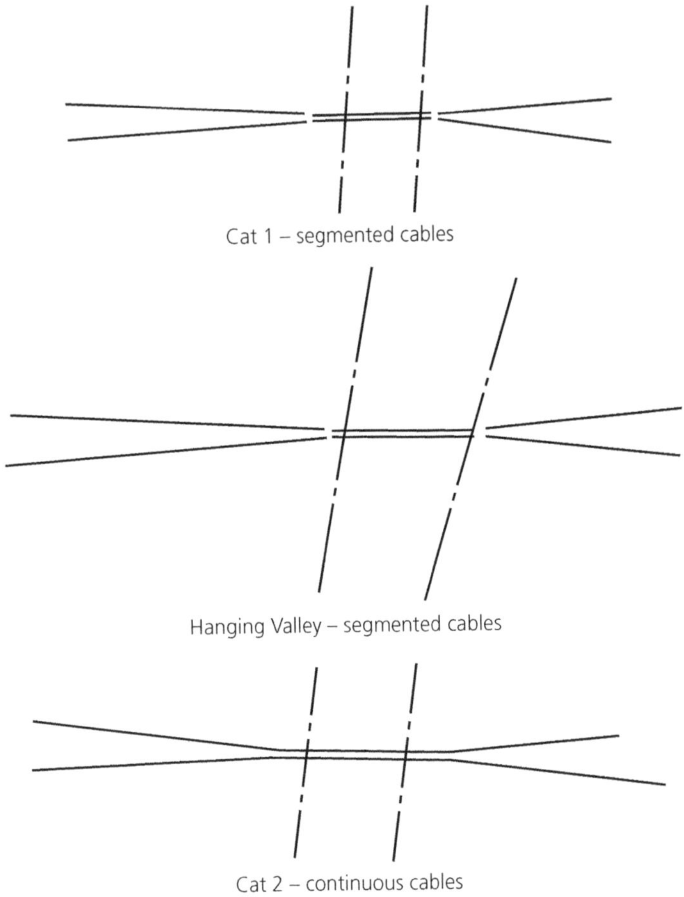

and the tensioner were both set up on the west side, and all of the 28 cables were pulled across the valley from west to east over the operating circuits about 300 m below.

First, the pulling rope was flown across with a helicopter and placed in a turning block then back to the puller. The first of 14 cables between two anchorages were pulled across and pinned. The remaining 13 cables were pulled across using a trolley riding on the first. This trolley arrangement reduced the risk of dropping something onto the circuits below – a fatal event should it occur.

All cables were continuous, anchor to anchor. One set of 14 crossed over the other set in a very elongated X arrangement. All cables were pre-stressed in the factory to 50% of the rated strength and cut accurately to a calculated length. This eliminated creep and factory slack elongation issues from the length calculations. It was quite nerve wracking to watch each cable get pinned to its anchors and discover that its length matched the length of its mates – or not. Credit to the supplier – they were all right. Their claimed accuracy for measuring was about 6 mm per 330 m.

Several ideas floated around as to how we could tie the seven cables of each of the four sets of seven into a bundle worthy of being called a 2.25 in. (57 mm) cable – albeit one in which the six outer cables were not helically wound around the core cable but were parallel. We settled on custom-sized preformed aluminium helical rods at 50 ft (17 m) intervals along every cable.

One of the assumed benefits of this loosely connected collection of cables was that, despite the design tension being at about 22% of the cables' breaking strength, the collection's innate or natural self-damping capability would likely be quite good.

Figure 6.9 shows the four sets of seven cables where they join at a yoke plate at the end of the catwalk, and shows the helical preformed rods, vibration dampers and marker balls as the cables head upward to the two anchorage sites on the west side. The vibration dampers were many – 88 in total – because saving a few dollars by reducing their number to some presumed correct figure was not a rational concept for a single and important installation. The dampers were placed at the catwalk yoke plates far from the anchorage ends of the cables where snow had continually ruined the units on Cat 1.

The work of the aluminium plate yoke plates was to hold the cables together laterally at both catwalk ends. Unlike the Cat 1 design, the continuous cables did not transfer their tensions through the plate. This allowed the plate to be much lighter.

Figure 6.9 Catenary 2 jewellery

Figure 6.10 Dropping in a catwalk section by helicopter

The cables were pulled vertically and laterally into place for joining at the two big yoke plates by working on the cables themselves from carts and a special platform. The 466 ft (142 m)-long catwalk was flown into place one 16 ft (4.8 m) section at a time, and constructed in place (Figure 6.10). It has 1 m gaps at every conductor suspension point.

Unlike Cat 1, which was assembled on the ground and winched up into place, Cat 2 was assembled in place, working from the anchorage sites high on the mountainsides. Both designs used pre-calculated cable lengths.

When T113R was carried down the valley in 2007 in a mountain of wet snow, it had one 2500 ft (760 m) length of conductor anchored to it and spanning back to Cat 1. After passing under Cat 1, that length of cables tightened and pulled on Cat 1 as it headed for the river, like an archer pulling the string of a bow to release an arrow. The 135 400 lb (600 kN) rated strength conductor broke in tension, but not before twisting the catwalk of Cat 1 (Figure 6.11). The reason for the catwalk gaps in Cat 2 is to allow the conductor suspension assemblies to move longitudinally if needed without doing such damage (Figure 6.12).

The everyday bare-wire weight on the Cat 2 suspension assemblies is 10 tons (9 tonnes) per phase. The design load driving parts to 100% of strength is 75 tons (68 tonnes). The insulator assemblies themselves weighed 2900 lb (12.9 kN) each (Figure 6.13).

Figure 6.11 Cat 1 damaged by the broken phase conductor

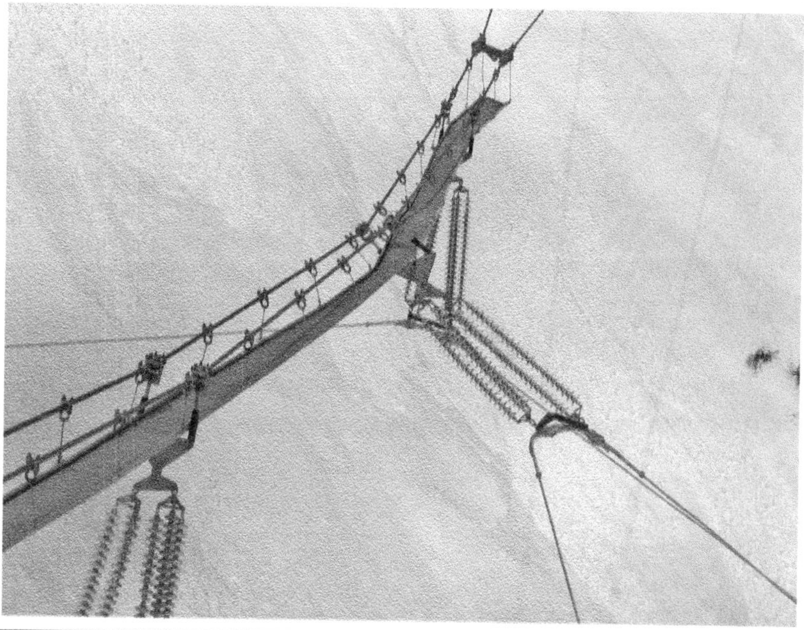

Figure 6.12 Cat 2 from 150 m below

Figure 6.13 Hanging the first phase from Cat 2. (Courtesy of Erik Ruggeri)

The anchorages are designed to the breaking capacity of the 14 attached cables – 476 tons (432 tonnes). The anchorages had two designs, the second design being developed for lack of natural geometry at one anchorage to apply the first design. With hindsight, we preferred the second design, as it would have made the work easier if applied at all sites. Live and learn!

The anchorages had an elaborate and heavy yoke plate system to ensure load equalisation between the seven cables and six rock bolts that supported them (Figure 6.14).

The alternate anchorage design was developed because one anchor site had no vertical wall of rock to work with. This time, we laid a pair of large flat plates on the rock and angled the rock anchors through the plates (Figure 6.15). The design was much simpler. About 3–5 m in front of each anchorage we set cable saddles to direct the cables over the rock where the slope broke from near level to steep. The cables angled down 20–25° from horizontal at each anchorage.

One other purpose of the saddles was to isolate any vibration action in the cables from the anchorages.

The last step undertaken was to construct snow sheds over each anchorage. These steel frames were topped with a thick timber roof that was sloped to deflect any avalanche material coming from above with minimal impact loads. One anchor was located where

Figure 6.14 The author with the yoke plates of the vertical anchorage design

Figure 6.15 Cable anchorage yoke plates and saddles

Figure 6.16 Cat 2, a transmission line structure in the sky. (Courtesy of Alex Brown)

up to 12 m of snow accumulates on its top every year. The design load was 1200 lb/ft^2 (5900 kg/m^2).

One of the great satisfactions that we got from our engineering effort of Cat 2 came when we surveyed its position in place once completed (Figure 6.16). The rig comprising all of those cables assembled in place was within 6 in. (15 cm) of the coordinates in the computer model. As a final comment, we went into the engineering of Cat 2 and the Hanging Valley catenaries thinking that these specious mountain valleys allowed us to believe we had some freedom with cable tensions, anchor site locations and where the conductors attached to the catenary. In both cases, the work was a lot more like fitting something large through the eye of a needle. We actually had to be quite careful with these design features.

While we were in the middle of constructing Cat 2 that summer, visitors from another company arrived for a tour, having heard of our project. They said they had a project elsewhere in this coastal mountain range where they thought catenaries might be useful. Would we go take a look to see if we agreed? You know what we said, don't you!

The mother of all catenary projects

If you think the mountains around the Kemano–Kitimat line are spectacular, they hold no candle to where we went next. Over the next few months, I, with the intelligent help from my co-author Buck Fife, provided conceptual line layouts along about four routes out of the back valleys of the mountains to attractive connection points on the local grid.

In doing so, we 'spotted' and dimensioned about 40 catenaries – one here and a few there. It was all pretty interesting. Then, one day as we were taking another trip into that back country, the helicopter pilot asked if we wanted to take a shortcut. Sure.

He flew us up a valley that on the map was truly a shortcut to our destination, and we wondered why it was never considered as a route. As we went up the valley, the mountainsides became steeper and higher. Buck was sitting in the front seat of the helicopter, and he turned around to me in the back with his eyes wide open. His silent message was clear. 'This (place) is perfect (for catenaries).' We said as much to our hosts, and they asked us to look into it.

By the time we were done with this preferred route of about 80 km of line, we had spotted 18 catenaries – four in a series from sea level up a steep, deep V-shaped valley with no road access, to over a pass at 2000 m elevation, two more later on to clear an avalanche-ridden area and later a series of 12 catenaries with only four standard structures at two safe locations within the series. This series of (almost) nothing but catenaries covered a distance of 18 km. For 18 km, we touched the valley floor in only two locations.

Every one of the catenaries was larger than Cat 2, one spanning 2.3 km across the valley with its conductor attachments at its central low point of sag being 300 m above the valley floor below. Having just completed Cat 2, we had very good cost data and some new design ideas. The ratio of labour and equipment cost to facility material cost for Cat 2 was 12 : 1. In other words, to lower the cost of a catenary, it is imperative to lower the cost of the labour and equipment to install it. It is not of much value to lower the cost of material unless in doing so the labour cost is greatly reduced. Thus, the proposed design of these many catenaries changed quite a bit from the design of Cat 2, yet hindsight would not change the design of Cat 2.

The sag of Cat 2 from the anchorages to the catwalk elevation is about 215 m. If you consider that the catenary is like a very big swing weighing about 75 tonnes and that 60 tonnes of conductors are hanging from its low point, you can recognise that the speed at which it could swing back and forth longitudinally is very slow. If the spans of conductors hanging from the catenary are uneven, then they are going to dampen any such longitudinal swinging. This rationale suggests that we need not make each catenary from a pair of cables in an X configuration but can make each from a single cable.

I was quite satisfied with this arrangement. Knowing that the labour is cut in half if there is half of the anchorages and cables to erect, we see that halving of materials made a lot of sense. So, we designed each catenary as a single cable, and we assumed Cat 2's second choice of the flat plate anchorage design as the standard design. In fact, I wondered if the rock anchors themselves couldn't be single large-diameter galvanised steel cables inserted and grouted into the rock in deep holes rather than using high-strength, very strong steel rods. We tested that idea 4 years later. Read on.

Some of these catenaries were to be installed at relatively low elevations, and others much higher on the mountains. This meant that some catenaries could do lighter design duty and that careful planning could allow us to work at low elevations when higher elevations were closed off by weather. The summer of 2008 when we built Cat 2 delivered terrible weather all summer, and we achieved 50% site access all summer. The ability to apply principles of mass production would improve productivity considerably.

We dispensed with the multitude of catwalks and helicopter pads, and reasoned that a portable system would be sufficient, given the frequency of use.

All of these design changes and the opportunity for mass production had us estimate the per unit cost of these catenaries at about 40% of the cost of Cat 2, situating the 'per km' cost of the line in this location in the range comparable to that of a conventional structure placed in harm's way on the valley floor but with no risk of avalanche damage. Our experience with Cat 2 told us that this was technically a very constructible design. We wait for the call!

Comments on the evolution of catenaries

In this chapter so far, we have described a series of transmission line cable structure design and construction events. The main point of the string of stories is not so much that you can now run off and design a catenary structure. Perhaps, with the help of some of the reference documents noted in this book, you could do that. The main point of the stories is to illustrate how a concept adapts to the specific conditions of a location and time and, most importantly, how we learn by doing.

There are people who might expect that you, as an engineer, can step up to the plate to solve a problem armed with all of the best information and experience to be found and just solve the problem to perfection. Those people would be wrong. We learn by doing. Brian used to say even in his early 80s that he was still learning. So he was, and so he should. As you work your way through the various projects that become your career, you will bring the best that you have to each one. Over time, your best keeps getting better.

The cross-rope suspension structure

Here is a brief story of a monumental 'learn by doing' moment. One of the other names for a catenary is 'cross-rope'. It is called this because Cat 1 is literally a steel 'rope' across the path of the transmission line. Brian and I called it 'the cross-rope', but when I arrived on site in 2007 after a 10-year absence to deal with the avalanche damage, the owners were calling it 'the catenary'.

I adjusted.

The creative mind of Brian White saw upon completion of his 1955 catenary that, if he did not have two large and high mountainsides to which he could anchor his cables, he could tie the cross-rope to two guyed masts and apply the design idea in flatter terrain. Thus, the cross-rope suspension (CRS) tower design was born from Brian's fertile mind. Having done one thing, he saw that he could do another.

There are CRS towers in Quebec in Canada (735 kV), Argentina (500 kV), the USA (500 kV) and – many – in South Africa (400 kV). In every one of these cases, Brian was a consultant to the utility involved, and the CRS exists on their networks by his direct influence. Eskom in South Africa has been perfecting the design of the CRS over the years ever since, and has now about a half dozen incarnations of the design.

Since the CRS design adheres to the three structural design principles listed at the start of this chapter with more perfection than any other transmission line structure design in existence, you will find that the design offers the most cost-effective, failure-resistant support system for a transmission line of all the choices possible.

The hurdle that most often stops its adoption is an unwillingness to learn maintenance methods working from its cable system and to deal with the large guying footprint. The maintenance methods are well developed by those who do the work of learning, and the way around the latter issue is described in Chapter 4 on structures, and it takes only the willingness to change to find these issues are very solvable.

King of the highwire

Just when you think that you have had a good time having exercised your design expertise outside the box on a respectable number of times, you should hope for another phone call to kick it all up another notch. I got that call, and two further adventures ensued.

Nik Wallenda is a seventh-generation daredevil performer. In this age of internet information, it takes no time to learn who he is and who his family has been. In March of 2012, Nik called, referred by a friend who thought that someone in the transmission line engineering business might be able to help him with a project. He wanted to know if I was interested in helping him and his usual people design and plan the installation of a big steel cable across the Niagara River over Horseshoe Falls so that he could perform a tightrope walk over the Falls. Other tightrope walkers had walked across the Niagara River Gorge on a cable, but nobody had ever walked right over the big waterfalls itself. I said ‘yes’.

Niagara Falls

I said ‘yes’ because it just seemed like a cool challenge. In addition, my parents, most aunts and uncles and every cousin I have were born and raised in Niagara Falls, Ontario. I grew up only 15 km from the city, and all of my grandparents lived there. I knew the location for Nik’s event very well. I thought, ‘If anyone is going to engineer the installation of a cable across Horseshoe Falls, it will be me.’ I would be quite depressed if anyone else got to do it.

Nik had worked for nearly 2 years to get Ontario and New York State laws changed so that he could lawfully perform his stunt. The laws against such a stunt had stood for 100 years but were repealed for him for the summer of 2012. The event was to be – and was – televised live around the world.

Nik’s usual team for typical tightrope walking events consisted of a group of friends helping with the installation under the experienced guidance of his father, a retired performer and his Uncle Mike (UM) a retired US Navy mechanical engineer. The usual cable is a $\frac{5}{8}$ in. (16 mm)-diameter steel cable pulled to about 6000 lb (2700 kg) of tension and spanning several hundred metres. The cable was invariably tied off laterally to the ground with sway guys – light ropes or small cables at 8–12 m intervals along the cable, to give it stability.

At Niagara Falls, the deep gorge at the Horseshoe Falls was 300 m wide – cliff side to cliff side – and full of energetic water. There was no place to put these sway guys, meaning the cable had to be a free or unrestrained span. So, before, we were asked to be involved, UM had already decided that the cable would be a much larger 51 mm steel cable and pulled to a tension of 60 000 lb (27 tonnes). This much heavier cable would offer its own inertia to unwanted motions. I had no reason to suggest that the cable be otherwise.

In my first conversation with Nik, I suggested – and he agreed – that my job was to give him a cable that had no surprises when he was walking on it – no unexpected bumps in the night. Since Nik had never walked on such a long, heavy, unrestrained cable before, he organised a rehearsal site elsewhere in Niagara Falls, where he could practise and get the feel of such a cable under his feet but in a safer, lower-to-the-ground environment.

A 3700 ft (1128 m) length of the 2 in. cable was purchased and cut into three lengths: 2350, 1250 and 100 ft (716, 381 and 30.5 m). Each piece had a large female poured-zinc socket attached to one end, and the other end was left untouched. The long piece was for the river crossing. The middle length was the rehearsal cable, and the short piece became known as the tail piece. The tail piece was to be overlapped with either of the longer cables, and attached by a series of Crosby clamps to form a cable with a socket on both ends and of the length desired based on the installation geometry sought.

The rehearsal site was an installation of about 1200 ft (365 m) in a parking lot. The cable would be about 1–2 m above the ground at one end and held higher (15–20 m) at the other end, and would mimic the tension of the longer river crossing installation. We learned a few things when installing the rehearsal cable.

Micropile anchors were drilled through the parking lot, and the distance between was measured accurately by survey. The concept was to lay out the cable, splice it to the calculated length, set the low end over a steel drum about 2 m above the ground and to lift the higher end with a large mobile crane to a height of about 17–18 m, by which time the cable would be at the desired position above the ground and at the desired tension of 30 tons (27 tonnes).

Of course, all of the hardware pieces at the cable's ends were accommodated in the calculation, and we were aware that any minor adjustments could be made by lifting or lowering the hook of the big crane. One of the pieces of equipment was a digital dynamometer set at one end of the cable. So, all of this work was done and the cable was lifted upwards by the big crane. We watched the dynamometer and we watched the cable begin to lift off the ground.

To our muted horror, the crane hook reached the calculated height and then some, but the dynamometer reading was depressingly low, and the cable remained lying on the ground. We had really messed up that calculation! My colleague and I poured over the calculations and could not find an error. Then, one of the linemen who was a very hard-working, smart and practical man said, 'Let's get the hook down, undo the splice and remove 10 ft [3 m] of cable from the overall length by increasing the splice overlap.' I had no better idea.

That work was done, and the cable was raised again with the crane hook, and the installation looked very much like we wanted. What had we missed? It then dawned on us we had forgotten one characteristic of the big cable. It had not been factory pre-stressed. On our first application of tension to the cable of up to 15% of its strength, the looseness in the cable as created during its factory manufacture came out. That 1200 ft (365 m) length of cable lengthened about 10 ft (3 m) – about 0.8%.

Remember that we had pre-tensioned all of the cables used to build Cat 2. Good thing! Remember the factory slack component in the Canadian Electricity Association (CEA)'s sag–tension program STESS?

That evening, we learned a second thing about this big, free-span cable. In his enthusiasm to get on the cable, Nik and his entourage went out to the parking lot after dinner and, in the dark, he stepped off the roof of a small truck onto the cable. It rolled out from under his foot. I describe it as like trying to walk on a log that is floating in water. He said, 'If this cable has to do this, I can't walk on it.' The order of the following day was to solve this rolling problem. A very long cable has very little torsional stiffness. How do we stiffen it?

While it was not our first thought, we came up with pendulums set at intervals as the solution. With pendulums in place, the cable cannot rotate without rotating the pendulums, and that is difficult to do. At 250 ft (75 m) intervals, the cable was still a bit wobbly. At 200 ft (60 m) intervals, Nik was happier, and he practised on that cable for nearly 2 weeks with a daily crowd cheering him on. With these lessons learned, we developed the detailed design for the crossing of the river.

Where there is a waterfall, there is rock. We designed micropile anchors, believing that we would find rock into which we would grout the high strength steel rods. Here is a small story about our small world. We had obtained a price from a local driller to install the anchors but we were a bit uneasy with the offer. I had a question to answer, so I made a phone call to a driller in Vancouver, Canada – nearly 3000 km away – to ask my question. This was the driller who had installed the big rock anchors for Cat 2 over 4 years earlier.

'By the way' I asked, 'do you ever work far from home?'

'Sure. All the time', he said. 'We are going to Ontario in about 3 weeks.'

'Really?' This was promising, I thought. 'Where are you going?'

'To the Niagara River to install rock fall protection along a section of the river for the power company.'

'No kidding!' I told him of my requirements about 10 km away upstream on the same river. 'Will you take a few days and install my anchors?'

He did our work. In fact, the man on the drill was the same man who did our Cat 2 anchors. It is a small world. We found rock only 2.5 m deep on the Canadian side of the river, and we drilled 4.5 m into the rock and tested each of the four anchors in the cluster to 15 tons (14 tonnes). On the US side, we found rock at 20 m depth, and drilled into it a very small distance. The long bond with the earth was sufficient to hold the 15 ton test tensions.

The distance between the anchors set in the two nations was surveyed. The profile of the crossing was reasonably well understood, and we calculated the cable length and the placement of a big crane on both sides of the crossing. This length of cable had yet to be tensioned, and was about 1800 ft (550 m) long. We had to estimate how much it would lengthen on application of first tension. We had to get it right because, unlike at the rehearsal site, there would be no opportunity to lay it back onto the ground (into the river!) to make a correction. We assumed 13 ft (4.0 m) of lengthening. After the cable went up, one of the linemen said he had added his own 6 in. (0.15 m). I just smiled.

You have read elsewhere in this book that we transmission line engineers get a bit concerned about Aeolian vibration in a conductor (cable) when the catenary constant (the ratio of tension in the cable to the unit weight of the cable) exceeds about 5000 ft (1500 m). At tensions that cause the catenary constant value to be greater, we expect to need vibration dampers on the cable to avoid damaging vibration. The catenary constant of this installation would be 8000 ft (2440 m) – significantly above our standard threshold. I asked for a damper recommendation for the big cable from one of the best in the business – Chuck Rawlins, retired from Alcoa.

Since we had only one span to provision, there was again minimal concern for optimising the quantity of dampers needed. Chuck said, 'It might require as many as ten dampers.' We bought 20 units. In the meantime, Nik asked that we also reduce the spacing of the pendulums from 200 ft (60 m) to 150 ft (45 m). We had the pendulums in two lengths – 10 and 20 ft (3.1 and 6.2 m). We staggered their installation: 10–20–10–20 ... The swinging period or frequency of a pendulum is a function of its length. Longer pendulums swing slower than shorter pendulums, and we did not want adjacent pendulums to get into a harmonic dance with each other. The weight on the bottom of each pendulum was a stack of steel plates weighing 40 lb (18 kg). The span between the crane hooks required nine pendulums.

Perhaps the most interesting aspect to this project was the installation constraints. All the work of moving ropes and cables across the river had to be done between 7 pm and 9 pm – overnight because the river was captive to the *Maid of the Mist* tour boat operations that would not be shut down so we could do our work. We laid out our equipment and the cable on the ground and waited for 7 pm Tuesday evening. The live TV event was Friday – 3 days hence.

The plan was to pull the lead end of the cable off its steel reel and through a big tensioner on the US side of the river by a double-purchase rigging system anchored and wrapped on a truck-mounted puller on the Canadian side. Once the lead end of the cable reached the Canadian side of the river and could be grabbed, the double-purchase rigging system would be changed to a six-purchase rigging system, and pulled from about 10 tons (9 tonnes) of tension to 25 tons (23 tonnes) of tension, and pinned to the Canadian anchor.

As the cable paid out from the reel and through the tensioner, we were looking for a mark put on the cable by the manufacture at a prescribed distance from the lead end. This mark would let us know where to attach the tail cable and achieve the desired overall length with the tail cable attached to the US anchor. As the cable was paid out, this lap joint was made. All of this pulling and tensioning was done with the two big cranes holding the cable just high enough above the ground to keep it off the ground and out of the river.

With both ends pinned to the anchors and the dynamometer reading 25 tons (23 tonnes), the crane hooks were raised to the desired heights (Figure 6.17). Having had the

Figure 6.17 Pinning the cable to the yoke plate. (Courtesy of Steve Behal)

experience we had at the rehearsal site with this manoeuvre, the very slow and cautious raising of the hooks seemed to take forever. But, on some days, things go well. We nailed it! Perfect geometry and tension match.

It is interesting to realise that when the tension is this high as denoted by the catenary constant being 8000 ft (2440 m), it takes a very small change in a crane hook's height to make a tension change. The target tension was 60 000 lb (27 215 kg). On the afternoon of the event after the cable had been loaded with its pendulums weighing a total of about 500 lb (225 kg) and after the cable had been at tension for nearly 3 days, one of the very eager TV reporters asked what we were going to do next. I said that we were going to raise the tension by a few thousand pounds but that she would not notice. 'It will be like watching grass grow (not good TV).' We did raise the tension several thousand pounds by raising the Canadian side crane hook about 50 mm.

Finally, we decided to not attach any vibration dampers to the cable, because the summer weather was quite warm and calm and we felt no vibration in that cable at all despite the high catenary constant value. We have much to learn about this vibration business, it seems. If you were to see a photograph of the Canadian side of the installation, you would see two vibration dampers on the cable. I asked the contractor why and he said, 'For show.' We both smiled.

On a final note, the linemen who went out on the cable at night to install the pendulums had more fun than a person should ever expect. Nobody – not even Nik yet – had ever been out there 200 ft (60 m) above the broiling white water of the river at the apex of the amphitheatre of the very big waterfalls – a curtain of white water 2000 ft (600 m) in length curving around them. They were in the midst of the relentless mist, soaked to the bone despite their rain gear, and they could hear nothing against the thundering noise of the falling water even when shouting in each other's ear. 'Niagara' means thunder.

Nik walked across the cable in about 26 min through the same heavy mist and thundering noise under the powerful floodlights of the TV programme (Figure 6.18). It was quite a show, with a live audience attendance of about 120 000 people.

The spectacular show ended at 11 pm, and by 1 am the TV platform was out of the way. The contractor began to uninstall the cable, and by 7 am it was all cleaned up as if it we had never been there. Six months later, we began the work of planning the design and installation of Nik's next objective – the Grand Canyon. I have told you all of the facts of the Niagara event so that we can contrast it to the Grand Canyon event. Just when you think you are going to have a simple repeat fun and interesting adventure, you instead get handed the opportunity to learn something new – again!

The Grand Canyon

As with the Niagara Falls event, the Grand Canyon was something that Nik had had on his bucket list as a tightrope walker for a very long time. The location chosen for the event offered a dramatic 'hole in the ground'. The span across this location was about 350 m rim to rim – a bit more than at Niagara. But the hole was 450 m deep with rock walls

Figure 6.18 Nik Wallenda walking over Horseshoe Falls, 2012. (Courtesy of Steve Behal)

so sheer that you have to lean forwards at the edge with heart pounding and toes gripping the corner of the rock to even see the dry riverbed below.

But the nature of the site forced a different installation plan. At Niagara, you needed a few dollars, a passport and no criminal record to walk or drive to either end of the cable's installation across the bridge joining the two countries. At the canyon, you would need a helicopter to get to the north end of the cable's alignment. At this location on the Little Colorado River, the river makes a dramatic S bend, and the land within the interior portions of the S has vertical walls more than 300 m high – like an island without water. The situation was reminiscent of the Cat 2 site in that all major equipment needed to be on one side of the canyon – not for lack of flat ground but for lack of a budget to fly it to the island.

The steel reel with the same 2 in. (50 mm) cable as used at Niagara was positioned on the south side behind the tensioner and beside the puller. This time, we flew the pulling rope

Figure 6.19 Four anchors and cables yoked to the walking cable

across the canyon, with the helicopter paying out from the rope puller, placed it in a turning block on the island, and flew the lead end back to the cable reel and attached it to the cable's big socket. The cable was pulled across the canyon in a single-purchase pull to a tension of about 5.5 tons (5.0 tonnes). When the lead end reached the island anchorage, it was pinned. At this tension, the cable's sag was about 150 ft (45 m). We did not have the depth at Niagara to allow such a big sag without risking getting the cable wet in the lower river.

On the south side, in line with the cable on the tensioner, the 100 ft (30 m) tail cable – remember that piece? – was attached to that anchorage and laid out on the ground parallel to the main cable. The parts in the cable's end assembly were 20 ft (6 m)-long cables to each of the four rock anchors, the big load-equalising yoke plate and a dynamometer. You see all of this in Figure 6.19.

The same assembly was at the end of the cable at the island end, including another dynamometer. With the big cable across the canyon at 5.5 tons (5.0 tonnes) of tension, it was tied off on the south side and a six-purchase block-and-tackle system was positioned between the anchorage and the cable. The puller then pulled the cable out of the canyon, raising the tension and reducing the sag.

A note on the cable's anchorages at the canyon is useful. Recall that, for the conceptual design of the many catenaries in the British Columbia mountains, I pondered the notion that using a piece of flexible cable grouted deep into the rock might be a simple and viable idea compared with the use of high-strength steel rods? I tried the idea at the canyon. Four anchors with at least 10 tons (9 tonnes) of working load capacity were needed. The big yoke plates configured like a large mobile were designed for Niagara and reused at the canyon to share the load from the cable equally with all four anchors.

At Niagara, the rock was at a depth, but at the canyon it was at the surface. We drilled down 18 ft (5.5 m) at a 45° angle with a 6 ft (2 m) lateral separation, and sent a 28.5 mm, 12 ton (11 tonne) working-capacity steel cable into each hole, and grouted the bottom 3 m into the rock. The top end of the cable had a sling eye to which shackles were attached. The cable was guided out of the rock by a small saddle made of a half pipe on a 12 in. (30 cm) radius. This avoided sharply bending the cable on the rock's edges when under tension. The idea worked just fine, although a different cable would be needed if the installation had to last for 50 years rather than 3 days.

Based on the Niagara experience, Nik said he wanted a higher tension in the cable, believing that it would give a more stable surface for his walk. As with Niagara, he took the Niagara rehearsal cable to a site near his home in Florida and practised again for about 2 weeks with the local fans cheering him on. When we began the design process for the canyon walk, Nik asked for a stayed-cable system. This idea did not last too long when it became apparent that this would mean purchasing and paying to install about 15 km of stay wire or rope due to the depth of the canyon. In addition, this amount of stay wire would add considerable tension to the main cable.

When the plan reverted to an un-stayed cable as at Niagara, Nik wanted the rehearsal event. Thus, the Florida installation was planned. At the Florida rehearsal site, the fixed-length cable was set between two carefully positioned anchors and lifted to a few metres above the sand beach by a big crane near each end of the cable. As at Niagara, these two cranes could provide any tension in the cable that Nik desired. The tension was read off a dynamometer installed at one end of the cable. He tried 35 tons (32 tonnes) of tension, but the cable vibrated too quickly, driven by his walking motion. He tried 32.5 tons (29.5 tonnes), and found it to his liking. This was about 5% tighter than at Niagara.

So, at the canyon, our tension target was 32.5 tons (29.5 tonnes) or 65 000 lb (289 kN). I have described the installation plan as pulling the cable to tension with the six-purchase block and tackle, and when at tension pulling the tail cable taut alongside the big cable, removing any slack from its assembly, and then clamping the two cables together with a series of 16 large Crosby clamps (U bolts). In doing so, we would end up with a cable of a fixed and correct length provided the tension held. We had no big crane in the mix with which to adjust the system by lifting or lowering its hook.

The design process was to set the cable at this target tension of 65 000 lb (289 kN) with all desired pendulums hanging from it and contributing to that tension, assuming a cable temperature and then calculating what that unstressed length of cable would be in the span. Another way to understand what the 'unstressed length' of cable means is that it defines the 'amount of metal' in the span. We then placed that length of cable in the same span in a computer model without the pendulums attached. This mimicked the installation condition. The result was a tabular set of tensions at a range of temperatures without pendulums.

As the block-and-tackle system sucked up the cable from out of the canyon, we watched the dynamometer on the island end of the cable, and stopped the pull when we reached the tension we had calculated for the estimated temperature of the cable. This was a bit tricky because the calculations showed that the tension varied about 600 lb (270 kg) per 10°F (12°C) and the digital readout of the dynamometer was not steady but bouncing around by 400–600 lb (180–270 kg) either side of a fast-moving average. Still, we stopped the pull when we decided the dynamometer read 62 500 lb (28 300 kg).

We tried a quick experiment. We marked the cable at the south rim saddle and then pulled 1 in. (25 mm) more wire out of the span. The tension rose 2500 lb (1133 kg). We released that inch of cable back into the span, left the tension at 62 500 lb (28 300 kg), and UM declared the installation to his liking.

When the tension in the six-purchase block-and-tackle system was released by backing the rope out of the puller, the dynamometer on the south end of the cable engaged. To our amazement, it read about 55 000 lb (25 000 kg) – 7000 lb (3200 kg) or more than 10% less than the dynamometer on the island. There was no good reason for this discrepancy, and we had to make a decision. Which dyno was right? Was either one right? UM said he was happy to go with the island reading because the cable appeared good to him, and to decide otherwise would require a half day of redoing the entire tensioning exercise. He saw no need. We moved on.

The following day – the day before Nik's live TV walk, the linemen went out on carts and added the pendulums. These were essentially the same as used at Niagara except that we used the 35 lb (16 kg) vibration dampers purchased for Niagara as the weights at the bottom of the pendulums. The pendulum lengths were the same: staggered 10 ft, 20 ft, 10 ft (3.1 m, 6.2 m, 3.1 m), etc. Initially, the target spacing was 75 ft (23 m), half of the spacing used at Niagara. Nik was seeking more stiffness in the cable. The number increased when he decided that since we had 20 dampers, we should install 20 pendulums. The pendulum spacing became 58 ft (17.5 m). The pendulums weighed about 1200 lb (544 kg) in total, and it was calculated that this would increase the cable tension almost 5000 lb (2270 kg) by their addition. As they were attached, the tension increased, but not as much as expected.

It was hot and windy the entire week that this work was going on, and the weather forecast was for the wind to decrease significantly before the day of the walk. It did not. Nik had been practising his cable walking in pretty high wind, so this was not a major issue in his mind. The walk was scheduled for 6 pm, and the day was hot and quite breezy. Wind gusts greater than 50 km/h were recorded. It took Nik about 22 min to do the walk, and most people who watched seemed to think he looked quite spooked or nervous compared with his Niagara walk a year earlier (Figure 6.20).

Nevertheless, a lifetime of wire walking, deep respect for the dangerous work that wire walking is and no fear of heights allowed Nik to put another first among his peers into his résumé.

Figure 6.20 Nik Wallenda walking over the Grand Canyon, 2013

But, he was unhappy with the tension, stating that he wanted 65 000 lb (29 500 kg) and got 55 000 lb (25 000 kg), having read the south dyno, and the wire was too loose, thus contributing to making the walk quite tedious.

Several weeks after the event, we got a good photograph much like the one above, and we made a calculation. By drawing a line on the photograph between the anchor points and using the length of the 10th and 12th pendulums near mid-span to scale the vertical distance, we estimated the sag in the wire and therefore the tension. It does appear as though the island dyno was wrong and the south dyno was near correct. We estimated the tension to be actually near 57 000 lb (25 800 kg) – about 12% lower than sought.

If you look carefully at the pendulums in Figure 6.20, you see they are not hanging vertically. In fact, they were all swinging slowly back and forth up to 0.5 m in the 30–50 km/h crosswind.

Lessons learned seem to be with hindsight, by definition. That is frustrating to be sure, but perhaps unavoidable. I suppose it is worse not to learn at all. Here is what the Grand Canyon project taught us.

For years, I have had a rule never to sag a wire by measuring tension. Past project experience taught me not to trust dynamometers. We did well at Niagara, so I simply forgot my rule. Always install a cable by sighting the sag if at all possible. Seeing is believing! We did not even bring sagging equipment with us to the site. I had left my rule far behind. Never again!

If we had determined that the tension was low, our equipment set and cable arrangement had no mechanism such as the cranes that we had at all three previous installations to adjust the tension at the simple pull of a lever. The importance of tension to Nik was paramount, and we had no back-up plan to deliver it this time. Never again!

A pendulum of half the length would have halved the wind area, but a swinging speed increased by only 40%. The spacing of 150 ft (46 m) at Niagara was not very problematic, so a spacing of 125 ft (38 m) and pendulums of half the length would have reduced the pulling forces on the cable due to the brisk wind to 25% of what we had. The pendulums were too long and too numerous for a windy site. Never again!

Read into these lessons what you will.

A little summary of cable projects

This series of cable project stories was laid out to show you the evolution of ideas. It was told to provide you with some insight into the nature of cables so that you can venture there someday. The reference documents will add considerably to that point: see Catchpole and Ruggeri (2009) and Catchpole and Rilkoff (2009).

And finally, the stories lead to the fact that I consider the Grand Canyon work to have been a form of failure. It was not a catastrophic failure, but things did not go as planned, and analysis of that shortfall in planned outcome led to interesting lessons learned. What an awesome segway to Chapter 7!

REFERENCES

Catchpole PG and Rilkoff J (2009) Big cats in the mountains. *T&D World Magazine*, October.

Catchpole PG and Ruggeri E (2009) Large catenary structures for high voltage transmission lines. In *Electrical Transmission and Substation Structures 2009. Technology for the Next Generation* (Vogt MW (ed.)). ASCE, Reston, VA, USA.

Structural Engineering of Transmission Lines
ISBN 978-0-7277-5988-7

http://dx.doi.org/10.1680/setl.59887.223

Chapter 7
Lessons from failures

> We learn very little that is new from our successes, for when a line remains up for many years it may be that it is over-designed or that the climate has been cooperative or the design is right on. So be it, and we are content.
>
> However, when some part fails, and the damage expands, there is often a chance to learn something new, but you usually have to dig for it.
>
> H. Brian White

The failure – the structural demise – of a power line is a fascinating source of invaluable information. You should never miss the opportunity to study each and every failure that you can get access to. You will learn things about transmission lines that you can't imagine. Here are some truisms about transmission line failures:

- The devil is in the details.
- We design against certain 'events' but lines fall down for other reasons.
- If you cannot explain the sequence of the failure event in minute detail and have that explanation be in full compliance with the evidence, then you cannot say that you got it right.
- The conductors (all the wires between the support structures) have an enormous say in how the failure will play out.
- Most failure events go unstudied entirely or are misunderstood for lack of realising the value in knowing what really happened.
- You will learn something with every study.
- If you never study failures, you will remain quite ignorant.
- It is all about the *slack*!
- Anomalies start failures, and anomalies stop failures.

Line engineers are often chastised by the maintenance and construction people for not getting out and seeing power lines built and renovated. If you don't go out during construction, you soon lose touch with all sense of scale and the weight and forces within the things that comprise a transmission line.

Similarly, if you look at the aftermath of a failure, be it one bolt or insulator or 100 km of cascaded line, you will be reminded graphically of these matters of scale, weight and tension, and of Mother Nature's powers, the entity against which you are usually trying to design in your cosy, 20°C office. A failure is the best teacher of perspective and mechanisms.

That's the good news. The bad news is that someone will, rightly or wrongly, believe that the reason for the failure, as you find it, could/should have been addressed beforehand, and you are pointing your finger at the person(s) that did not do so. Don't enter into a failure investigation for the purposes of conducting a 'witch hunt'. Don't let management send you there for that reason unless you are prepared for the courtroom.

A failure is a source of design knowledge. Try to work with the management and operations personnel so that they allow you the opportunity to get on the scene fast, before evidence is picked up, melts, drains away, blows away or is stolen. Otherwise, this opportunity is largely missed.

In an investigation, you want to know why the failure started, how it continued and why it stopped. The ASCE's *Manual of Practice 74* has an appendix devoted to failure analysis. The text of the 1991 version of the manual was authored by Brian White. The appendix to the 2010 edition (Wong and Miller, 2010) was authored primarily by a colleague, Ron Carrington. Both men have conducted various failure investigations, and drew from the experiences.

Important understandings

Before diving into the description and study of some failure events, we discuss four things that are foundational to understanding the mechanics of failures. These are: triggers – the singular event that causes the failure; stored energy – a form of measuring the strength of the failure event; anomalies – the features of a line necessary for a failure to start or stop; and, finally, slack – the property of a set of spans of wires that defines the ease of propagation of a failure between support structures. These four things are discussed in detail or in passing in each of the failure stories that are presented below. Underlying these understandings is a basic fact of physics regarding line failures: the structures almost invariably will fall where the attached wires direct them.

Triggers and victims

The most important step in a failure investigation is to find the trigger – the precise item or component whose failure can lead to the subsequent failure of dozens of components and structures that might otherwise happily have remained in service to the delivery of electrical energy.

Transmission lines are long structural systems with thousands of components tied together by quite strong conductors and overhead shieldwires. A simple failure of a single wire or item such as an insulator may pose little problem, but when that failure expands into some form of cascade, the problem can turn into a complex situation where it is often not easy to trace the path of the event backwards, via investigation to the trigger event.

Searching for and finding the trigger implies that the initial single-entity event expanded into the failure of a whole structure or a cascade of many structures as a domino effect. It is important to realise that, in a failure, many components or many structures were not destroyed by the attacking event, rather the attacking event caused the breakage of one component and all other breakages were caused by the breakage of that first component. In other words, if the first component had not broken, neither would the others have broken. That first component's breakage is the trigger event. The breakages of more components due to the breakage of the first are victim events.

Often, the trigger event and the reason for the first component's breakage is all that is important to learn. Sometimes, the reason for the continuation of the event is important, depending on your assessment as to whether preventing the trigger event would have prevented all damage and whether any subsequent breakage should have also stopped even further damage.

A large measure of curiosity and persistence is required to find the trigger event. Finding it requires thinking about what is before your eyes, and creating a detailed scenario of what happened, and then ensuring that *everything you see* makes sense and fits that scenario. If the detailed steps do not make sense, then you have probably misidentified the trigger, and you need to devise another scenario. When everything in your devised scenario of step-by-step events of the failure makes sense – does not defy logic or physics – then you have a candidate for the triggering event.

When you are satisfied that you have sifted through the dramatic and distracting wreckage of a large failure and found the trigger that caused it all, then you know what to fix to avoid a repeat event. Not to do so is to remain exposed to a repeat event. Finding that singular weakness and putting that gem of information into your knowledge base time and again is what can make you a better line design engineer.

Potential and kinetic energy

We have suggested repeatedly that you recognise the wires as structural members, operating as long, flexible cables joining other structural members. With this view, you will understand an entire series of support structures and their connecting wires as a single and spatially large structure. A failure anywhere in this *structure* can lead to effects anywhere else in that structure.

The deceiving feature of such a spatially large transmission line structure is that the most powerful members within it are so often the sedate-looking wires basking in the breeze against the beautiful blue sky. What you cannot see is that these wires are holding more tension within themselves on an average sunny day than it takes to break the stout-looking structures that are holding them up. Think of the tension in the wires and the weight of the wires as stored (i.e. potential) energy. When something upsets the normal condition of equilibrium and releases that potential energy, converting it to kinetic energy, well . . . get out of the way!

Consider two simple and common transmission line arrangements.

The first is a typical single-circuit 138 kV wood pole H frame line with three (477 kcmil) 250 mm^2) Hawk ACSR conductors on 600 ft (200 m) spans hanging from a cross-arm attached to 60 ft (20 m) class-3 poles. If the wires are tensioned to 15% of the rated tensile strength (RTS) on that average sunny day, the tension in the wires is about 2200 lb (1000 kg). If there are two overhead ground wires (OHGWs), their tension may be about 60% of the conductor tension, at 1300 lb (600 kg) each.

If an adjacent structure is lost and all of that tension on one side of our subject structure drops to zero, the total load on each of the two poles is half of $3 \times 1000 + 2 \times 600 =$ 2100 kg applied a bit above the cross-arm level on the pole. Include in our energy measure the weight of the now falling wires at roughly 3×0.9 kg/m $\times$ 200 m = 540+ kg. The average breaking strength of a class-3 wood pole is 1360 lb (615 kg) applied 2 ft (0.6 m) below the pole top. You can ignore the nuances of moving insulators that reduce tensions and all that – or not – and you can see that, in simplistic terms, there is enough energy stored in the wires of this sunny afternoon example to break the poles should that stored energy be released.

The second arrangement is a typical single-circuit 345 kV wood pole K frame line with three two-bundle phases of 400 mm^2 (795 kcmil) Drake ACSR conductors on 300 m spans hanging from a cross-arm attached to 23 m class-1 poles. At 15% RTS on that same sunny day, the conductor tension is about 2100 kg. If this tension energy is released, along with that of its two OHGWs, the total load on the poles is about 4400 kg, whereas the average strength of class-1 poles is about half of that value at 2040 kg. The weight of the falling wires and even the weight of the falling head frame of a high-voltage structure is significant.

The point to note is that as lines get bigger as defined by conductor size and tension and by voltage, the kinetic energy stored in their spans of wires increases dramatically whereas the strength of the typical tangent structures to resist that energy release does necessarily increase with it at the same pace. Big extra-high-voltage (EHV) lines should be understood as having large amounts of mechanical energy stored in the wire system, and, when released, that energy is very destructive. You need only study one EHV line failure to see this fact on display. The carnage can be amazing.

Anomalies

An anomaly is anything that is different from the common 'everything else'. Failures start and end at 'anomalies'. An anomaly is required to initiate a failure, and another is required to stop it. Basically, a perfectly flat, straight, evenly spanned line of identical structures contains no anomalies, excepting its end points. Without an anomaly, such a line will not fail. If an anomaly is introduced and a failure starts, it will not be stopped in such a line. Such lines do not exist, of course, but some lines are more heavily populated with anomalies than others.

Inherent or 'built-in' anomalies are things such as long spans, short spans, grade changes, corners, unique structures and unique hardware. Inherent anomalies are sometimes unavoidable, and others are selected by the engineer. An anomaly can be injected into a line by nature or human action – a structure removed or altered; a wire cut or dropped by lightning, a falling tree or human action; broken or faulty hardware or unevenly applied or shedding ice load; and so on. Throughout the stories in this chapter, we will point out the anomalies that played roles in the failures so that you can learn to think in these terms.

Slack

We defined and discussed slack in Chapter 3. To refresh, slack is the difference in length of a wire in a span compared with the straight line distance between the wire's end support points. Slack describes the amount of wire that can be pulled out of a span as you attempt to make the length of the wire equal to the distance between its attachment points. As you pull the wire tighter in this attempt, the tension rises. To get the two things equal and the slack to zero requires infinite tension so . . . you will never get there! Tension increases exponentially as slack is removed. Review the unit curve conversation.

If you have a lot of slack in the span – many metres – as is the case with long spans or very loose spans, then pulling a relatively small amount of that slack – say 0.2 m – will generate only a modest increase in tension. But, if the span has only 0.3 m of slack, as will be the case with short spans or tight spans, then the same attempt will bring the wire to its breaking point. The discussion in Chapter 3 was meant to convey that understanding slack is to understand the ability to transfer load along the line when things change – say, when a structure moves or a wire breaks.

We offer this simple reminder. We have long said that 'we design lines for one set of imposed loads and they fall down for other reasons'. The next subject is the classic case in point. Only in the last decade or two has attention been paid to a load set called 'high-intensity winds' (HIWs).

Wind events and cascade types

By far the most frequent instigator or trigger event of transmission line failures is a HIW that is the outpouring of warm air turbulence in the form of micro-bursts, downbursts, tornadoes or, at times, simply what are termed 'gusts'. To distinguish, we classically design for synoptic winds or for wind values extrapolated from synoptic wind data. A good paper by Alain Peyrot (2009) tries to describe the folly of this extrapolation.

When we examine failures caused by HIW events, we are frequently looking at the extended cascading of many structures in what are simply called 'cascades', and then we find that the extent of these cascades is greatly influenced by the slack in the wire system.

Cascades come in two forms – or, rather, fall under two labels: longitudinal cascades and transverse cascades. For a definition of a cascade, try this: a cascade is the falling of structures one after the other, each caused by the falling of the adjacent structure. This is to say that, when you see a long string of structures lying on the ground in a similar manner, it is not because they all fell independently due to a common cause imposed on each – they fell because the failure of the first, caused by some event, triggered the failures of the others for a reason not shared by the first tower's cause of failure.

The distinction between the longitudinal and transverse cascades is rooted only in the way that the structures fall: transverse to the line direction or along the line direction. When a structure falls transversely to the line direction for some reason, the subsequently falling structures each fall transversely but they trend towards falling more and more along the line, depending on the slack in the conductors. By 'more and more along the line', we mean the top of each structure tends to lie further and further forwards of its base in the direction away from the source of the failure.

Transverse failures are a lesser recognised form of cascade, but worth noting because they are less likely to be readily identified as a cascade. The classic cascade, in the mind of most, begins with the failure of the initial structure, either vertically or along line minus the tug to one side of the subsequently falling structures, as is the signature of the transverse cascade. The image is of dominoes falling away from the source of the failure and towards the structure that will be next to find it challenging to remain standing. Another thread woven through most of these initial stories is the development of an understanding of line behaviour and how their designs have been affected.

A classic longitudinal cascade

In July of 1993, 100 km of 345 kV line supported on wood pole K frames came down in a summer storm. You should imagine that 100 km of line down is ample opportunity for more than one set of anomalies to be in play, and they were. This was one of our early investigation efforts, and initiated the realisation that each investigation will teach us something unexpected. The cover shot of this book was taken during this investigation.

A notably fast-moving and rather intense summer storm front moved west to east across Nebraska, and unfortunately travelled right along the alignment of a very straight, west-to-east transmission line. There were many instances of the tops only of some trees within small groves having been shredded by very local intense winds, probably small tornadoes embedded in the front. Grain bins had collapsed or were missing altogether, while others a few metres away seemed unharmed. Such was the nature of the wind.

Travelling the line revealed structure after structure collapsed eastward – their poles splintered, almost exploded. Each was a variation on a theme, and their nature pointed to a failure trigger back to the west. A few structures were amazingly still partly standing, held at a dangerous angle of about 45° off vertical by shieldwires that were snagged on the fallen structure to the west. To the west, the nature of the carnage began to look different and, eventually, there was a point where the structures fell to the west rather than to the east.

An interesting feature of failure investigations is that you are on the scene primarily to collect data with an open mind, not to presume things and solve the problem right away. Solving the problem can come later, as paying attention to it in the field interferes with a focus on data gathering. The cover photograph of this book was unwittingly taken of the span just two spans west of where the trigger event occurred. At the time, this was not known. The location was only 1.5 km or so from the west end of the carnage caused by the storm that came from that direction.

We will explain the trigger location and nature in a moment. First, we note that the cascade, once initiated, played out like the classic domino event. It travelled about 5 km, where it met an in-line deadend structure. This very straight line initially had *no* deadend or strain structures for all of the 100 km that encompassed this failure, with one notable exception – a big substation 50 km into the cascade. Each small angle in the line was made with a three-pole structure that was guyed only on the bisector. This lone deadend structure, 5 km into the cascade, had been installed about 15 years earlier, well after the line's initial construction. The reason was that the line had then been crushed by an ice storm event, and this deadend was installed at the end-point of that event's carnage to facilitate reconstruction.

That lone deadend structure's very presence in the line is, by our definition, an anomaly. It was not like anything else in the line in that area. The cascade sailed through the deadend as if it was not there. That was disturbing and warranted attention. We will get to that. The cascade continued for 45 km until it encountered the next anomaly – the large substation with strong, steel termination structures. This anomaly ended the cascade. However, as bad luck would have it, the failure – or, rather, another failure – started at the first wood structure on the east side of the substation. Structures fell like dominoes for another 50 km.

The reason that the failure was restarted after the substation was that the wind storm was still moving down the line, pushing at every tender transmission structure on the line. The speed at which the line failed equalled the 60 km/h speed of the storm front. Where the cascade finally stopped, the anomaly in the line was subtle indeed. The line turned slightly to the south – only a few degrees. The storm did not turn, and the two slowly parted company, lessening the force of the wind on the structures. The cascading actually stopped at a typical structure just before a structure that was about 6 m taller than the others before it because a railway was being crossed. This constituted another anomaly whose effect made sense.

As the wire tensions were lost on the west side of each structure, the tensions to the east pulled on the structures. Being wood pole structures, they exhibited considerable flexibility, bending easily to the east. The bending moved slack into the eastside span, reducing its tension, albeit never enough to prevent the poles' breakage. This was until the poles on the structure to the east were so tall that they did move enough slack to reduce the tension to the east to a value less than the neighbouring poles could support.

We know luck was not with the line's owner, because the cascade had restarted on the east side of the substation. The first structure to fail was the second one beyond the station deadend. It was a standard suspension structure. The first structure outside the substation was a five-pole deadend structure. That second standard-suspension structure was only metres away from a barn and a collection of grain bins. The barn and most grain bins were unharmed but one grain bin was caved in on the west side and another was ripped from its foundation and was sitting in a crumpled mess about three spans down the line. In other words, the localised, severe nature of the wind in this location was easily on display.

Along the entire length of the cascade, every small angle running corner that was guyed only on the bisector of the line angle folded over to the east as if on hinges near the ground line. As for the ineffective deadend 5 km into the cascade, it was noted that every helical wire grip set in the anchor eyes on the west side of the structure was separated from the anchor eye and the five big strands of each grip were all broken in the jagged signature of a fatigue failure. In other words, all of these grips were fully broken long before the day the cascade failure arrived on the scene.

The structure carried no line angle so the tension in the guys could be developed only by pre-tensioning during installation and could be maintained only if the structure never settled or the anchors never rose with load. But, this was Nebraska. Nebraska's soil is all dirt with no rocks. It was clear that even a couple of centimetres of settlement of the structure shortly after being placed 15 years ago would loosen the guys, removing the manufacturer's recommended 10% of breaking strength pre-tension.

For all of those years the structure wavered back and forth in the gentle breezes that passed over that corn field. Yes, there was Nebraska corn everywhere! The wavering action tugged on each guy, and the relentless load cycling through zero stress fatigued every strand of the helical grip within the 15 years prior to the cascade. Pre-tensioning guy wires matters, but maintaining that tension over time is the real and necessary goal to prevent their end fittings' fatigue failure caused by subtle cyclical loads.

In the failure investigation report, it was declared that any guy on a line that is not held in permanent tension by conductor or shieldwire tensions or by an 'unsettleable' structure *must* be routinely inspected for fatigue failure at its end fittings. Without tension sustained by the attached conductors' tension or without a rock-solid foundation that prevents settlement and guy tension loss, the guys will loosen and cycle through zero stress. They will eventually fail in fatigue, and be quite useless when a loading arrives.

The triggering location for the entire event was ripe with anomalies. First, the wind smacked the face of a structure and, unlike all of the other structures being smacked in turn, one of its poles snapped. On inspection, that pole – being a typical fir pole – had the misfortune of having a collection of branch knots all in a plane and exactly at the ground line. This unfortunate anomaly made this pole weaker than usual. Figure 7.1 shows at least seven such branch knots at the ground line plane, and they caused the rare flat plane breakage of the pole. Most poles broke in long splinters. Unusual! The shock of that breakage at the pole butt sent a wave of energy up the pole, and its top 3 m or so above the cross-arm framing snapped off and fell towards the ground, head first. This should not have mattered too much but for the presence of two more anomalies.

Only a few metres east of this broken structure, a distribution line passed under the line. When the shieldwire attached to the top of the falling pole top struck the distribution line, the line did not shut off immediately. Distribution lines don't have the instantaneous shut-off capability of the higher-voltage lines. The shieldwire was burned into two pieces (broken). Now, in the midst of tonnes of falling wood and lost tension in one of the shieldwires, things are getting a bit hairy. Enter the final anomaly to play its role in setting off the cascade.

Figure 7.1 Radial knots at the ground line of the first pole to fail

In the span that is three spans west of the one just described, and is in the book's cover photograph, we found that the other shieldwire was also broken, but in an odd fashion. Four strands were necked down in classic tension failure, but the other three were broken in the brittle fracture signature of a classic fatigue failure. In addition, the strands broken in fatigue had very small burn marks on either side of the fracture. We did not pursue an understanding of this odd breakage until a year later. See 'Nebraska, part II', below.

Regardless, the loss of this second shieldwire meant that the pole tops in the vicinity were now fully unsupported, and if broken by the unbalanced tension would send the unbalance to the next and the next. And so it went for many kilometres. It is impossible to know whether the absence of the ring of knots, the distribution line crossing and the oddity of the second shieldwire would have meant that the cascade would not have been triggered elsewhere by the force of the wind. But, we are confident that their absence would have made it more difficult and less likely to have occurred.

Are there lessons here? You cannot manage well the placement of knot rings at the ground line. You cannot avoid distribution crossings. But, you can inspect or even avoid unsuitable guy installations, and the fatigue-failed shieldwire can be addressed to a point once you understand it. See below. And you can certainly install beneficial anti-cascade structures at intervals of less than 50 km.

Icing events

Many of the world's line engineers need to know little of the problems of combating ice and snow loads on transmission lines, but those in northern climates or in regions with bits of terrain that protrude above the plains have need to learn of a sometimes intriguing subject.

In this section, we will spend most of our time on the particular subject of in-cloud icing that usually produces the most surprises, then some on precipitation icing or freezing rain – a somewhat prosaic or dull topic. Slack will rear its head and intrude, as it does when studying so many aspects of line behaviour.

In cloud, icing occurs, as the name suggests, when vapour-laden clouds rise up the slopes, become chilled and deposit ice of various densities on objects such as trees, blades of grass, transmission line wires and structures. The densities can range from high-density almost-clear ice, if the water droplets are large and flow on contact before freezing, to very-low-density rimes caused by small droplets that impinge and freeze on contact. The latter looks much like the inside walls of most refrigerator freezer boxes – mine for sure!

The images in Figure 7.2 show that build-ups can be devastating, sometimes so large and heavy as to effectively prohibit the sensible building of an overhead line.

While precipitation icing or freezing rain simply falls to the ground, in-cloud icing is very much dependent on exposure to the winds that bring in the water droplets, and it is this

Figure 7.2 Ice-laden structures. (Courtesy of H. Brian White (top left/right) and Dean Gatien (bottom))

dependence on exposure and the presence of one or more long spans with excessive slack that produces most problems and sometimes dramatic failures.

The bottom photograph in Figure 7.2 shows modest rime ice – something that can occur easily. The two top photographs invite the reaction of 'wow!'

Odd ... how things play out

> On 31 October 1991 I was sitting in an office in Idaho undergoing a day-long job interview. My wife and I had flown there from Toronto, Canada, the day before. The weather was not the best along the way, and we arrived to see the entire place covered in an early winter blanket of snow. It looked lovely, and may be a good part of the reason that we decided to accept the job and move to the small mountain town. It was only about 6 months later, after I had begun working at the firm, that I was asked to investigate a line failure in Minnesota. That line had fallen on my interview day due to the same winter weather that attracted us to Idaho but had manifested as an ice storm by the time it got to Minnesota from the west.

The investigation of this failure took place only via photographs taken from the air, as it had occurred 6 months beforehand. Someone had flown down the line, taking good photographs from above of nearly all fallen structures. The failure was about 30 km in length along a very straight alignment, and took all day to occur: 8 am to 5 pm as the ice built up. How frustrating must that have been to watch? It was reported that about 20 mm of ice accumulated on the conductors – not a conductor-breaking amount. In fact, the conductors were not broken, but they did define the event.

This was a single-circuit, 345 kV line with two-bundle Drake 795 kcmil (400 mm^2) ACSR conductors and two shieldwires, all supported on wood pole K frames. Figure 7.3 shows the typical K frame configuration. As the industry had become prone to do, the structures were renovated to improve ground clearances so the line could have an increased capacity. Basically, every second structure was renovated to raise the two lower outer phases about 1.5 m at the revised structures and 0.75 m in all mid-spans adjacent to the change. This was done by cutting off and removing the timber arms outside of the two poles and replacing them with hinge-based braced posts – or what was later known as horizontal V insulator assemblies. The new assemblies positioned the two outer conductors at basically the same elevation as the middle conductor. This improvement to the ground clearance was the point of the change. This arrangement was, at the time, illustrated in the insulator vendor's catalogue and promoted as a line-capacity-increasing method that happened to sell insulators.

Study of the photographs taken from the air – once past the drama of all that busted stuff – revealed, by the direction that various structures had been pulled down by the conductors, that this was not a single cascade (after all, it took over 8 h to ruin only 30 km of line) but a whole set of independent events, each initiated by the crushing of a renovated structure. No original, unchanged K frames triggered any collapses. In addition, all

Figure 7.3 The 345 kV K frame structure

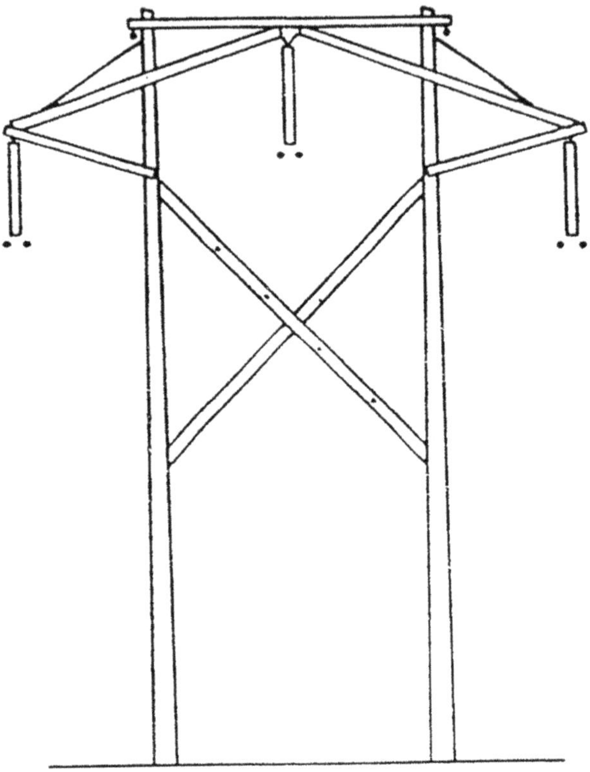

triggering structures were of the renovated structures and had their poles broken well above grade. Why?

H frame or K frame structures are flat panels with notoriously little longitudinal strength relative to the tension in the conductors. If the design criteria for a line involves longitudinal load cases, such as broken wire cases, then such structure types cannot pass muster, and are either discarded as an option or assisted by the periodic insertion of cascade-proof structures to reduce the cost of cascades to an acceptable level. Otherwise, the line is invited to cascade at great cost. Thus, the load cases applied to such flat-panel 2D structures are limited to the transverse and vertical planes. Alas, it is a 3D world.

The uneven build-up of ice that is typical of a precipitation ice storm's onset and departure produces uneven tensions between spans, however modest they might be. The stiff timber arms of the standard K frame could withstand the longitudinal pull from any I string insulators that were swung by the unbalanced ice and keep the hefty vertical loads somewhat confined to the plane of the flat structure. However, the hinge-based braced-post insulators offer practically no longitudinal resistance and floated up to 1.5 m out of the plane of the flat panel structures. Their resistance was truly minimal because the framing did not even employ the recommended sloped hinge axis that is intended to develop some resistance to longitudinal movement, as we described in Chapter 4.

Thus, the large vertical load was positioned about a metre out of the renovated structures' plane, front or back. This large offset of the vertical load bent and broke the poles well above the ground. Thus, the renovation of the structures via the addition of the new hinged insulator assemblies was the destructive anomaly inserted into the line. Without that change, the line may well have survived the ice storm. After this failure was described to the insulator vendor, the arrangement disappeared from their catalogue.

The best way to understand this failure is in terms of transfer of slack from one span to the next by these insulators. If the posts had been fixed at the base and not hinged, the avoidance of a line failure would depend on their cantilever strength. Perhaps that strength would have sufficed.

Nebraska, part II

Above, we described a two-part 100 km cascade that occurred in Nebraska. It was a perfect laboratory for understanding anomalies. Only a year later, an April snowstorm visited the same utility and wreaked havoc on a whole host of lines of all voltages. But, the best part, if the word 'best' should be used in the context of a disaster, was that an unsolved mystery of the fatigue-failed shieldwire from the earlier cascade event was solved.

In this April storm, a wet snow and wind event poured into the area overnight from the north. The first indication of the storms' nature was indicated within the carnage of the distribution lines. This part of Nebraska is criss-crossed by a network of straight roads on an east–west and north–south grid. Nearly every distribution line of wood poles on the north–south roads remained intact while nearly every line along the east–west roads was flattened. So, the wind direction mattered.

There were three areas of failure worth describing. The least damage was to some 115 kV H frame lines. There, the damage was limited to a few broken shieldwires. The second area was a stretch of about 80 km of 345 kV right-of-way running east–west in which the shieldwires of both of the big latticed tower lines were on the ground, broken in every third or fourth span. Third, one 230 kV H frame line suffered a cascade failure in two unrelated locations.

The 115 kV shieldwire failures were the least useful to study as some clean-up work had removed the wires, and broken ends were not available to inspect. The plethora of shieldwire failures on the 345 kV tower lines provided a better opportunity. These two tower lines were parallel and nearly identical and about 15 years old. The shieldwires were gripped to the towers by small clamps with rubber grommet inserts. There were two shieldwires on each tower line, providing lots of wreckage to inspect within a short distance.

Each wire breakage was somewhere in-span, and upon breaking the windblown, wet-snow-laden wire tension at the nearest tower squirted the rubber grommet out of the

Figure 7.4 (a) The lightning-induced fracture break and (b) the tension break

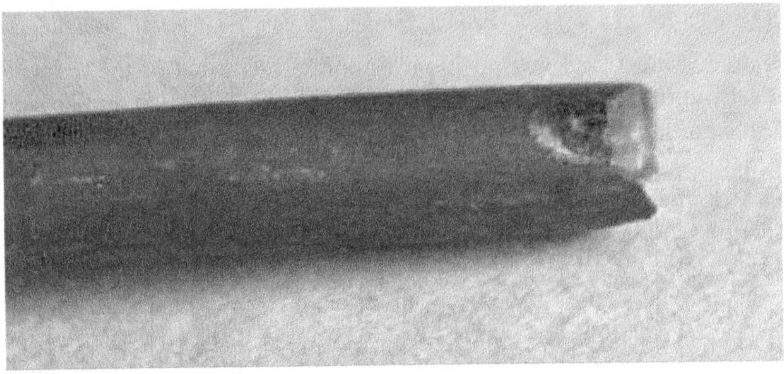

(a)

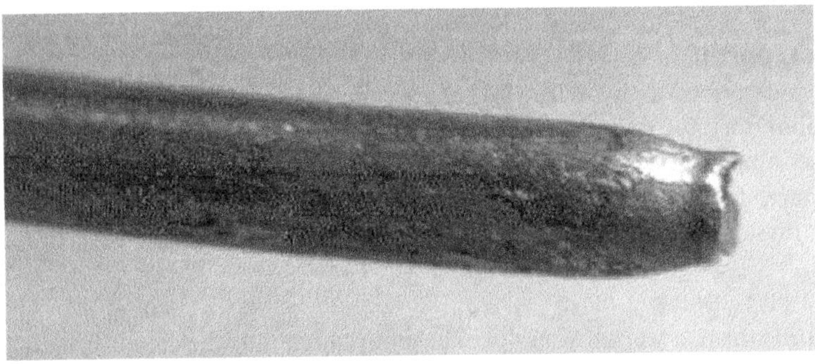

(b)

clamp's grip, and some length of wire went shooting through the clamp into the next span. The next grommet slipped sometimes. As noted, the wires were broken every three to four spans for many kilometres. On inspection, every broken wire set (pair of ends) collected showed the same pattern described above from the year before. One or more strands were broken in fatigue, and the others were necked down in tension failure. Figure 7.4 shows the tension failure mentioned above and two fatigue failures with the lightning burn mark described further above. This time, the wire ends were sent to a laboratory to understand the burn marks present on each fatigued strand.

As it turned out, these burn marks were lightning strikes that instantly made the strand brittle via heat through forming a metal structure called martensite, which is much more brittle than the steel's original state. The occasional to relentless Aeolian vibration common to shieldwires eventually cracked this brittle metal, and from that time on, the wire had the integrity of only its remaining unstruck strands. It was decided by us that in lightning-prone locations after a number of years, a shieldwire's structural integrity should be defined by less than all of its original (seven) strands. That is very disconcerting, as almost every engineer's shieldwire calculation assumes that the wire has its factory strength throughout its lifetime. We think not so.

The failures on the 230 kV line provide more lessons in anomalies. The first section of line to fail cascaded to the east from a substation location. It travelled for several kilometres across rolling hills. The trouble started in the first span outside the substation, and was triggered by the geometry that we so typically install in such a location. With H frames, the shieldwire is placed at the top of each of the two poles, and this places them above, parallel to and equidistant between the conductors. But, when we get to the substation, we move the shieldwires relative to the conductors to the station deadend structure's attachment points outside of the outer phases. To do this, the shieldwires must cross over and above both outer phases.

The weight of the wet snow this night brought one of the shieldwires down into the phase conductor, where they crossed and burned it off, and the resulting tension unbalance on that windy and snow-laden night triggered the cascade away from the station. The unique geometry of the wires in the station entry span was our anomaly and the source of the trouble.

Elsewhere on the line, a farmer had constructed a 'tank' or reservoir of several hectares in size, right on the line's path. To accommodate the tank, the once bullet-straight line was diverted around the farmer's project by the installation of four right-angle corners and sections of line that took the line direction from east to south to east to north to east. Got that? There were two to three spans on each section between the four new corner structures.

Let's orient ourselves by standing south of the reservoir looking north. To the left and well ahead, there is the line arriving to a deadend corner west of the reservoir. The line turns southwards towards us and turns east at the second deadend just to our left. It passes in front of us, turns north to our right at a third deadend until it reaches the original alignment, where a fourth deadend turns the line east along its original path. Let's number these deadend structures in the order mentioned as 1 to 4. Each deadend is a three-wood-pole affair guyed against both of the conductors' directions. The line has a shieldwire at the top of each pole, and at the top of the two outer poles of the three-pole deadends. They too are guyed against each wire direction.

Years later, on the night of our storm, a cascade started at deadend 1 and travelled west. The spans attached to deadend 2 failed towards deadends 1 and 3. The failure westward from deadend 1 was independent of the other. It was initiated by an unfortunate anomaly. Understand that the entire line renovation to allow the farmer's reservoir construction project was the creation of numerous anomalies in a section of line that was anomaly-free before the renovation.

The guys supporting the line tension to the west at deadend 1 were necessarily placed on the reservoir side of the structure – in other words, towards the water and the wet ground the water caused. Being frugal, the entire bypass was rather tightly spaced against the reservoir, so much so that the shieldwire anchor on the north pole of deadend 1 was practically going to be in the water. To avoid this, it was paired up with the north conductor guy anchor set a few metres closer to the structure. Who made this decision? The result was that the most highly loaded anchor was in the wettest soil. The screw anchor pulled out of the soft mud, with a full load of wet clay between its flutes looking like a large, muddy cork from a giant wine bottle, with the screw mechanism still attached.

Structure 2 let go of its intact state because a helical grip let go of one of its guy wires. Since these helical grips are designed to be stronger than the wires they hold, it was surmised that the grip had been installed, removed and reinstalled – a no-no! The loss of the structure caused the longitudinal collapse of its neighbouring structures falling away from corner 2.

The devilish details and other matters

The stories above centre around two types of weather events that can often harm transmission lines: wind and ice. Here, we focus more on the bits and pieces of a line and how they play a role in failures.

Insulators

The following subject is one of the few areas of transmission line work where knowledge and the statistical appraisal of data may foretell disaster and prompt the action needed to avoid it. There are two parts to the problem that follows, but we begin with a very brief reminder of a basic fact of statistical work. That is, with an assumed normal distribution of characteristics such as the strength of a component thus:

- 10% of a sample will have a strength that is less than the mean strength by about 1.23 × COV (coefficient of variance)
- 1% of a sample will be at less than the mean by about 2 × COV
- 0.1% or 1/1000 will be at less than the mean by about 3 × COV.

Our immediate concern focuses on the insulators, the cap and pin or clevis types that are used in the tens of millions on transmission lines. Throughout Europe and other places, the cap and pin/ball and socket (B&S) are not used. In their place is a long insulator that is used in a series of 1–2 or 3 as the voltage grows. When working with insulators, awareness of probability-based criteria becomes a very live issue, particularly at the higher range of voltages, where the deadend assemblies of a single 800 kV deadend tower can contain more than 900 B&S insulator units and each kilometre of 800 kV suspension towers may contain just as many.

When this quantity of insulators exists in a structure, it is most likely to be a structure in an EHV line that is most vital to the system integrity, and they are likely to be in multi-string bundles. This means that the risk of problems that we describe here is increased as the number of units increases and as the importance of the line increases. How frustrating!

Multi-string bundles of insulators exist because the strength of a single string is not sufficient to carry the conductor tensions. To ensure equal load sharing between the parallel strings, yoke plate arrangements are used. These yoke assembly configurations have differing and less-than-awesome performance characteristics when things go wrong, as described earlier. Although physically in parallel, they act effectively in series when something gives way.

'Effectively in series' means that parallel strings of insulators in the same assembly that appear to offer a measure of redundancy if there is a premature failure of one unit in one string will not reapportion evenly. When the yoke(s) rotate(s) upon the loss of tension in one string, the assembly lengthens, adding a severe impact component as the entire span of attached ice-laden conductor is set in motion away from the assembly, then stopped. The loads in the remaining units do redistribute, but unevenly with much more than the anticipated share of the tension going to one string, practically ensuring its failure as well, provided the initial failure is not due to an extremely weak unit.

Thus, a premature failure of one unit can easily lead to failure of all of the strings in the bundled assembly, causing the dropping of a phase from the tower. Although the arrangements provide for parallel tension paths, yoking arrangements can and often do result in behaviour not much different from the situation if all were in series.

In the case of a deadend tower, the failure of one unit, then its one string and then the entire strain assembly under ice load can lead to the destruction of the tower and cascading of kilometres of adjacent tangent towers. Failure of but one single insulator in a thousand could and has done just that many times. Some of the very expensive cascade failures in Quebec, Canada, in 1998, for example, were initiated by a deadend tower's insulator unit failure of just this simple sort. Thus, the importance of the statistical fact noted above. If the COV of your insulators is 10%, then it is statistically true that every single-circuit strain tower on your 800 kV system will have at least one insulator unit with a capacity of only about 70% of its rated strength.

Consider that a single-circuit deadend structure on a much less important (to the system) 115 kV line has only about 50 insulator units, and perhaps even uses only six polymer/glass rod insulators. The chances of a weak unit there are greatly reduced. Wish it were the other way around?

Some yoke plates are wide and shallow while others are narrow and deep. We discussed this in detail in Chapter 4. When the yoke plate rotates and injects this length into the span of wire almost instantly, the entire span is set in motion and then suddenly stopped. That is a lot of $F = m \times a$ stuff going on.

In Chapter 8, we will discuss the attractive principle of designing for the expectation or accommodation of failures. There, we ask you to think about the consequences of certain things going wrong. Considering the merits of your insulator choices, factors of safety and yoke plate dimensions and arrangements should be on your list of considerations.

Back-to-back angles

This is a slight design nugget that seems to wander around trying to find a home, for no one seems to want to print it. It doesn't belong in a line-loading guide or amid the formula of a tower design manual, yet it pops up every now and then to perplex all but the very few that have encountered it before.

Tall lattice towers of long river crossings can make efficient use of stitched back-to-back angles that have been found to be torsionally unstable in strong winds. The twisting back and forth usually does not do damage to the pairs of angles themselves, but attached and smaller bracing members, often bolted by coped ends or thin gusset plates, can be quickly broken.

On occasion, such failures have been found on structures before the conductors have been strung. The gusset plates can be thickened or members changed to improve resistance, but there is a very 'easy to install' baffle or wind damper that has been found to be effective.

Another feature of back-to-back angles is that they are necessarily stitched together with bolts at intervals. Being engineers, we often like symmetry, so placing one of these stitch bolts at mid-length of a back-to-back angle brace is very commonly done. Putting a hole through one of the legs of each angle at precisely the point of maximum bending stress is not the thing to do. Placing that hole elsewhere is so easy and beneficial. Make it a habit.

The ups and downs of tower 113R

Throughout this book, you have found that we return to the Kemano–Kitimat transmission line in British Columbia for a variety of reasons. In Chapter 6, we noted that Brian White said of the place, 'More interesting transmission line engineering to learn from has occurred that we can see from where we stand than anywhere in the world.' Well, there is far more to the place than just that. Here is one of the stories that belongs in this chapter on failures.

In January 1992, the twin-tower line over the Kildala Pass was again in trouble. Tower 113R was a suspension structure that was at the upper end of one of the 2500 ft (760 m) long spans coming from the 1955 catenary. It had been hit by a dry powder avalanche from the back side that hammered into a deep ravine beside the tower, and the air blast completely overwhelmed the 130 ft (39 m) tall, 35 ton (32 tonne) tower. Looking downhill with your back to the tower, the avalanche came from about 8 o'clock and from about 300 m above, where a break nearly 0.75 km long and 5 m deep was left in the snow on the slope.

Figure 7.5 shows the site and avalanche source slope above the tower 113R site. The photograph in Figure 7.6 is taken from the avalanche slope but south or to the right of the avalanche start area. The latticed tower 113R is in the foreground, before the event. The other tower – 114L, a five-legged aluminium tube structure – supports the other circuit, and sits on a knoll about 12 m above tower 113R. It was out of harm's way in 1992.

The loss of one of the two circuits puts the lines' smelter operation at a curtailed level and in complete jeopardy should the single remaining circuit be lost as well. In addition, the loss of power sales revenue is impressive.

Figure 7.5 The Kildala Pass avalanche site

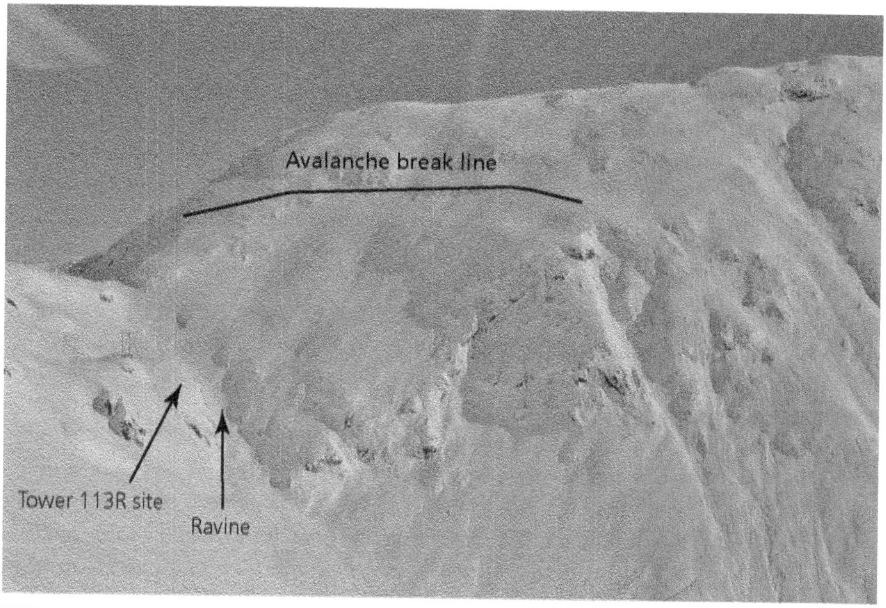

Figure 7.6 Tower 113R in the foreground and tower 114L behind it

Under that circumstance, time is of the essence to get back in service. It is important to put the event in context. Other towers on the line had been destroyed by avalanches in 1975 and 1986, with this site having been tested by a lesser event in 1988.

The site was buried under about 6 m of hard-packed snow, and January access was only by helicopter through often tricky weather. Study of the failed tower showed something a bit unusual and useful to the solution. While the air blast struck from the 8 o'clock position, essentially at the towers, hindquarter looking downhill, the tower fell uphill of its footings. This single insight into an unusual failure mechanism offered an opportunity to quickly create a replacement structure under near-impossible conditions.

The uphill failure of the tower can be clearly understood when the slack in its attached spans is recognised. The span to the tower above, on the pass, is rising about 10° above horizontal and 1500 ft (460 m) long, offering only a few metres of slack. The span to the 1955 catenary on the downhill side of the tower dropped away from the failed tower at 30° below the horizontal. The 2500 ft (760 m) span to the catenary offered dozens of metres of slack and an insulator string of more than 20 ft (6 m) in length that was backed up by a span of 4000 ft (1220 m) beyond the catenary offering much more slack. Thus, when tower 113R was pushed downhill in 1992 by the great force of the avalanche's wind blast, it was restrained by the three uphill conductors – each with a breaking strength of just over 135 000 lb (600 kN). The tower had no choice but to be pulled uphill by these conductors, as the tower on the uphill side was strong and backed up by a short 900 ft (275 m) span to a strain tower. The downhill spans offered no resistance to the uphill pull, and so the tower fell uphill in a crumpled mess (Figure 7.7).

The available heavy lift helicopter was a Sikorsky S-61, limiting the weight of all things moved to site to 6000 lb (2700 kg). Machines are always in the region for logging work, albeit not so much in December.

One of the rebuild options was to use large timber poles, of which there were many available, but they would have weighed 8000 lb (3630 kg) and more. Although they could be set in holes dug down in the packed snow, and probably develop enough resistance to remain upright, the problems as the snow melted would be almost insurmountable.

The original latticed tower had its lower leg extensions strengthened with considerable extra steel about 3 years earlier as Figure 7.7 illustrates. Digging and hacking with axes in the near-ice-hard packed snow around these very strong reinforced legs of the tower showed that the two downhill legs of the four might offer pedestals for erecting something, while examination down to the rock showed that the two uphill legs were damaged beyond repair, so the issue devolved into building something based on the helicopter's lift limit and that could be in some way set on top of these two intact downhill legs and have the height and strength to replace, at least for the remainder of the winter season, a 35 ton (32 tonne), 130 ft (39 m)-high very strong lattice tower.

The vertical conductor loads at the location were about 8000 lb (3630 kg) per phase plus about 1500 lb (680 kg) of transverse load per phase because the tower carried a 6° line

Figure 7.7 The fallen hinged tower

angle. Within 24 h, a tower concept plan took shape and a truckload of steel was ordered. Two latticed masts and a latticed cross-arm to form a modest H frame, each of nearly the 2700 kg lift capacity, were fabricated in the owner's plant maintenance shop, flown in and placed in position, and the 'little tower that could' became a reality.

> Now, I get personal. The rest of this story is about two more failures. Both occurred at this site. One is Brian's and the other is mine. We like to think – and I did offer this pearl of insight to Brian on an occasion when he was quite upset over the writing up of his failure – that, with helicopter pilots, there are those who have crashed a machine and there are those who will; with designers, there are those who have sliced open their finger with a knife and there are those who will. So it is with adventurous line engineers . . . there are those who have had a failure to their name and those who will. Let it be a measure of the company that you keep and the pushing of the boundaries that you exercise and experience.
>
> That is easy to say in light of the fact that nobody was hurt or died as a result of either of our failures. We thank the Lord for that! It would be easy to leave these somewhat embarrassing stories out of the book but for the fact that the lessons learned are so important. So, we air our laundry here in our front yard for all to ponder.

The 'little tower that could' was initially designed to be fixed by bolts at the bases of its two legs to the top of the reinforced legs of the destroyed tower. The new tower's legs were

60 ft (18 m) long, shorter than the original tower's height, but ground clearances were not an issue at the site. The maximum bending moment on a fixed-base 18 m leg, when loaded (w) along its length (l) by the very serious pressure of the next avalanche, occurs at the base connect, and is $w \times l^2/2$. However, if the legs' base connections were pinned and the top of the legs are held (reasonably) in place by the conductors headed uphill to the adjacent structure, the bending moment is only $w \times l^2/8$ – a quarter of the fixed-base options. Designing for the latter arrangement's loads was attractive due to the legs' weight restriction. Thus, the two legs were designed as pin connections to the old tower's pedestals, resulting in the tower's name – the hinged tower. The hinged tower's vertical status was dependent on the weight span and the ability of the attached spans to hold it upright in place.

That is a bit of a weird thing to comprehend. If the slack in both directions from the hinged tower is relatively small, as one finds with typical transmission line spans, and if the weight span is reasonable, the tower will wobble a bit but it is held vertical unless provoked by a significant horizontal force put onto it. There, we just gave away the problem!

The tower was erected, and the line put in service in only 3 weeks. This was a magnificent feat given that, by necessity, the work was done near the top of a mountain in January at a location where the average snowfall is about 15 m.

> In March of that year, only 6 weeks after completing the work, Brian was enthusiastically telling us of the adventure. He was also telling us of another collapse wherein a tower was brought down by in-cloud icing that loaded a span on one side and pulled the tower towards that span. I asked him, 'What is going to happen to the hinged tower if the span up to the pass gets loaded with rime ice?' Premonitions suck, for, in November, only 10 months later, the tower failed in just that manner. I was asked to investigate, and the findings were sobering because, as is written elsewhere, the devil is in the details, and the details were all against Brian on this one.

As now is well understood, the hinged tower fell uphill when it failed. Figure 7.7 illustrates the failure's pertinent points very nicely. In the photograph, the far phase position is on the ground and that conductor clamp has slipped along the conductor, stripping much of its four layers of aluminium back for several metres. The other two phases remain gripped to their conductors and are restrained by the long span back to the 1955 catenary holding the near leg and bridge above the ground.

A little history of this location is now useful. Before the 1955 catenary was installed and after only the first few months of the existing lines' service, tower 113R was a strain (dead-end) tower that spanned down to tower 112R set in the Glacier Bowl below. Tower 112R was among those lost in the January 1955 avalanche. When the catenary was built that summer, tower 113R was converted to a suspension type, and its line angle changed from about 3° right to about 6° left (looking downhill). It was built on the bisector of the original line angle, which meant that, thereafter, it sat about 4.5° off of the bisector. This is the first detail and anomaly to conventional calculations that caused trouble. If it could have been oriented normally, the hinged tower in Figure 7.7 would be rotated slightly anticlockwise, so that the two footings would appear more in line with each other from the photographer's viewpoint.

The hinged tower had two sets of paired internal guys to hold the load of the 6° line angle. They crossed within the H frame. The pair that held the angle's load was attached to the top of the near leg in the photograph and to the base of the old tower's far pedestal in the photograph. They were not attached to the leg of the tower but to the old tower's base. That was the devil's second detail. Third, the gusset plating at the pin location at the leg's bases was such that they would bind with other items there if the tower ever rotated more than about 30° uphill. This was important, since it was said that the tower could rotate up to 45°, find equilibrium and return to vertical once the ice load on the uphill wires melted away. Wild, but true!

When the lines were constructed in the 1950s, the armour rods under each suspension clamp consisted of a hand-wound set of 15 rods. When replacement rods were ordered in 1992, some 40 years later, a set of 16 preformed rods arrived. Fitting 16 rods into a space once filled by 15 rods required that the new rods' diameter and the resulting outer diameter of the assembly be smaller. The grip of the suspension clamp on the conductor via the armour rod set was important. Recall that Equation 3.14 in Chapter 3 tells us that the tension difference through these clamps is about 4000–5000 lb (1800–2270 kg). So, we have a fourth anomaly injected into the structure. Finally, remember that the location offers enormous slack to the tower from below via the long span to and long insulator strings on the 1955 catenary.

Now, let's consider the events of the failure. It was a blustery night, not a gentle accumulation of in-cloud icing. It was not the planned avalanche blast to the hindquarter. Mother Nature is very inventive and does not play fair. The tower was buffeted by wind, and there was icing on the higher span but it was not extraordinarily heavy. The inclination of the two spans attached in suspension to the tower were rising at 30° from below and rising at 10° going uphill for an overall angle turned by the conductor of about 20° and an average slope through the hinged tower's clamps of 20° off of horizontal. This non-horizontal arrangement greatly reduced any build-up of tension from the long span below as the tower rotated uphill, and yet the span above did the opposite and retained much of its tension as the tower rotated.

The fact that the tower was rotated about 4.5° off of the circuit's centreline bisector meant that the transverse load on the tower was not in the plane of the tower but angled away from it by that angular amount. This attempted to twist the tower top anticlockwise as the three phases were attached to the tower offset from the leg's centres to accommodate insulator swing in the line angle. The two guys holding that transverse load were attached to the fixed base but about 1 m apart – one at the downhill face of the hinge tower's leg where the hinge pin was located, and the other at the uphill face of the leg. When the tower rotated uphill, the uphill guy loosened quickly, and all of the guy load went into the guy attached at the pin location at the downhill face of the leg. This load transfer fed into and enhanced the moment arm that the off-bisector transverse load needed to twist the tower. As the tower rotated uphill, that highly loaded guy ripped its connection to the old tower base clean away.

There remains the question as to whether the far phase suspension clamps, loose perhaps due to the armour rod change, let go before the tower went past a point of equilibrium balance or whether the twisting of the tower and the lack of sufficient conductor tension back downhill so overloaded the clamp as to force it to slip regardless. Figure 7.7 certainly conveys the twisting action in the failure, either way. Subsequent analysis of the tower's stability, as provided by the geometry of the attached spans, suggested that its stability was not great. The span downhill was simply too long with too much slack to provide any restraint against uphill rotation.

We should say that, in the absence of all of these noted details that fed this tower's demise and are never normally present, this design concept has great value as a fast return-to-service idea. Just watch for the anomalies – the devil's details.

After this second failure at this site and the other noted avalanche failures on the line in previous years – now four high-mountain failures in 17 years, the owner was quite interested in taking any reasonable proactive action that could be imagined. It had, by the way, made the decision to replace the original latticed tower with another identical tower since, after all, it had lasted 38 years. In light of that decision, the focus for this site remained on the powder avalanche and the uphill demise of the tower as assumed future event scenarios.

While there is much more activity at the site to record, little of it falls under the heading of 'lessons learned from a failure'. Only this part of the site's story qualifies for inclusion here. Since the owner had no particular fear of out-of-the-ordinary ideas and no fear at all of things called catenaries, it was suggested that a temporary catenary could be used at the site to facilitate the return to service should the new tower – one just like the original – fail.

The premise for the plan was that the tower would again fall uphill, and the intact conductors, although perhaps scraped up and dinged, would land intact in the snow on the brink of the steep slope below the tower. If there was a cable set across the line and above that wreckage, the ruined tower could be cut away, and the conductors repaired and lifted onto that cable for temporary support and a fast return to service – letting a permanent solution occur slowly instead of in a rushed environment as occurred with the hinged tower's installation. It took a week short of 3 months to replace the hinged tower, whereas the estimate for a return to service with a temporary cable in place was 3 weeks.

This temporary cable was installed. It was a 1.75 in. (44.5 mm)-diameter, galvanised bridge strand cable 2200 ft (670 m) in length. To function, it would be anchored high on the mountain slope to the east and near the base of the aluminium tube tower sitting on the knoll beside tower 113R, as seen in Figure 7.6. This placed it high enough to be used to hang the failed line's conductors safely above the ground and any tower wreckage that would surely exist.

Figure 7.8 Pinning the temporary catenary to the temporary anchor

The line of the cable passed right by the south face of the new latticed tower, 113R. The plan required the cable to be anchored low, near the base of the latticed tower 113R, as a place for its long-term storage until such time as it might be required for service. This temporary but long-term anchorage near the base of the tower would never see great loads. If tower 113R fell again, the segmented cable would be refitted from this low anchorage to the in-service anchor on higher ground on the knoll beside the aluminium tube tower, 114L. The temporary storage anchor was a single pre-stressed high-strength anchor rod, and the cable attached to it had about 20 000 lb (9100 kg) of tension. This is described in the previous chapter as 'the KitiKat'.

Figure 7.8 shows the guys pinning the long cable to the vertical storage anchor, on which you can see a man's hand leaning.

> It was fun to design and install a cable system to carry a transmission line, albeit a temporary one of a minor nature. It was not fun when I got a phone call 6 months later (January, as I recall) to be told that my cable appeared to be missing. A helicopter pilot passing by the location, who was aware of all cables that could trap him, noticed its absence. It was not until spring that the snow reduced enough to visit the site. As soon as I looked at the single anchor at the base of the still-standing tower 113R, I knew what a bone-headed mistake I had made.

The single anchor was installed vertically into solid rock and the load on it from the cable was applied exactly horizontally, perpendicular to the anchor rod's direction. You can see the anchor rod in Figure 7.8. The man in front is leaning on it. That, in itself, is a no-no. The attaching hardware was tightened down against a small plate grouted to the rock, and the calculation made was one of shear only on the rod's strength. The anchor appeared broken off clean at the rock surface. On closer inspection, that was not quite the case. In fact, the load on the rod from the cable had moved the rod ever so slightly, and that movement caused about 50 mm of rock depth to break away. This put the rod into bending (9100 kg times a 50 mm moment arm), and it broke. The numbers did not entirely compute, in that the rod was only close to being overstressed. But, there was another issue. The location was bare, smooth rock and it sloped away at about 30°.

We have already noted that the snow at this location gets up to 5 m deep and sometimes more. About 60 ft (20 m) of the 2200 ft (1000 m)-long cable lay at the bottom of this snow along the rock surface. The power of snow creep down that slope and the grip that it had on the cable should not be underestimated. The creeping snow pulled the cable offline, and its tension increased by some incalculable amount. The load on the anchor was no doubt greater than it was without the deep, creeping snow present.

The single anchor rod's replacement was a stout six-anchor rod affair, designed in principle by my dear friend Adam Charneski (Figure 7.9). This man had 40 years' experience on this line at the time, and did nothing light-weight because he *knew*. You can see the 1955 catenary 760 m away in the background.

Figure 7.9 Remaking the failed temporary anchorage. (Inset) Adam Charneski

Missing bolts and pieces

Missing parts are a relatively common sight in some countries of this world where there is a constant battle between power utilities, who have to go so far as to weld nuts onto tower bolts to try to prevent theft of the steel angles, and those for whom acquisition of a piece of steel angle represents a successful day's reward for labour.

However, a photograph of some local engineers standing below that fully loaded angle structure with a missing brace member will bring a bit of a smile and should startle you. We are reminded of a tower designer who told a friend, 'We stand behind our towers, not under them.' In some countries, the effects of welding the nuts are destroyed galvanising, and the rust blotches should depress you.

It is more than a bit startling to note the somewhat similar complacent attitude of some engineers in less impoverished countries towards a few, or at times many, missing bolts of a lattice tower or mast construction. The small bracing members of a lattice tower, and especially of a lattice mast, have a critically important job to do in stabilising the major load-carrying member or the leg chord itself, and one missing bolt can reduce the load-carrying capacity of that member by 50–75%.

Many years ago, the investigation of the failure of a guyed television mast found that it failed under loads generated by wind to be about 50% of the design wind speed for the structure and thus about 25% of design wind load – a finding agreed on by all parties. Visits to similar guyed masts by the same manufacturer and maintained by contract with the same supplier discovered a few too many missing bolts and bracing members in each of the masts. Further investigation disclosed maintenance contracts that specified climbing inspections every 3 months to find and replace missing bolts and braces. The nuts were simply torqued to a given value, and no thread punching or other method for locking the nuts was attempted.

As the removal of a single bolt can effectively double the buckling length of a leg chord of a mast and reduce its strength to as little as a quarter, the failure of a mast at one-half the wind speed and about one-quarter the loading is to be expected.

When a failure investigator is having difficulty in establishing a failure scenario that fits all the observed facts, they will usually give thought to the possibility that defective material could possibly be the cause, but it is seldom the case, at least with metal towers (although sometimes hardware is defective).

Good maintenance, or what is now part of asset management, should not be limited to frequent inspections to find and replace missing bolts. It should include identifying critical tower sites – those exposed to strong winds – but, above all, good maintenance should seek to solve the problem of how the bolts got loose.

If bolts are becoming loose because of a lack of locking devices, or poor construction procedures, or by wind-induced member vibration, that is one thing. Routine inspection will quickly determine the problem, and it can be fixed, by retightening and retorquing the bolts, by installing locking devices or by limiting the vibration by additional bracing or other means. If, on the other hand, tower members are being stolen for local reasons, that is a more serious problem and one much more difficult to control. Sometimes, education will work, sometimes replacing bolts with highly torqued extra-high-strength bolts will work. Sometimes, replacing locking devices with better ones will work, sometimes punching the threads will work. Sometimes a combination of all of these will work, and sometimes nothing will work. In any event, those towers deemed to be more prone to severe weather should be regularly inspected and corrected, particularly prior to the severe weather season.

Regardless of the foregoing, at all times, including initial construction and maintenance, all nuts must be locked.

One of Brian's mantras during his seminars was to state that 'There is little in this business that requires the application of accuracy to greater than about $\pm 10\%$ because the input that we have and the control that we have over so many things is no better than that. The nearly sole exception is the spacing between two holes in a piece of steel.'

Figure 7.10 shows a gusset connection in which, on all three connecting braces, one of the two bolts shown on the drawings is missing. Most of that top face bracing had been bent by heavy snow sitting on top of the bridge the winter before. The tower was in its 55th year of service. Witness the thinning galvanising. The reason the bolts are missing is because the holes were too close together to have ever installed them.

Figure 7.10 Missing bolts in 'Tower Bridge'

Now, you must know that we have no interest in ever pointing a 'gotcha' finger at anyone because such actions always yield repercussions. But, we find it funny at this point to note that Brian designed this tower! He never did answer the question: 'Is your mantra above regarding the need for accurately spaced bolt holes the result of learning the hard way?' Whether it is or not, we all know that we learn best by our mistakes and infrequently by our successes. And despite the comments above regarding the perils of missing bolts, we will lean on another statement: 'Give these creations some room for being misunderstood.'

Blowout

This is the short story of two parallel lines that were strung with different-size conductors. The line with the lighter conductor – as measured by the conductor's density (weight/diameter ratio) – was placed close to, parallel to and upwind of the line with a heavier (denser) conductor where they crossed a large valley. The lighter line occasionally blew into the heavier line, and flashed over because in the steady, synoptic breeze that can impact a span high above the ground across a wide valley, its blowout angle and lateral displacement was greater. The lesson is to keep an eye on and understand your surroundings.

One man's garbage is . . . or garbage in – garbage out

In Austin, Texas, an investigation was undertaken to determine why the relatively new concrete pier foundations under steel poles appeared to be deeply cracked as if by structural overload, even though there was little reason to think that such a loading had ever occurred. The knowledge of a concrete expert was sought.

Our expert had seen this before, and gave our party a tour of a multi-storey concrete parking garage that showed the same problem. He also relayed the story of an investigation of concrete railway ties that were 'falling apart' based on the day of the week that they were made. As it turns out, the folks who make cement from limestone by burning the rock in a kiln find the kiln to be a handy place to enhance their fuel source by disposing of other peoples' garbage – burning rubber items such as disposed surgical gloves. However, this increases the sulphate content of the cement powder that comes out of these kilns. Sometimes that matters.

The chemical formula to describe the setting of cement and water into concrete is described as being too complex to write down. One action that takes place is that sulphate in the mix is absorbed into the mix, and all is good. But, if there is excess sulphate, that is not absorbed in the normal way but is left to do other things. If moisture exists, it joins with the excess sulphite to form crystalline ettringite. The process is called delayed ettringite formation, and amounts to expansive crystal formation in the concrete. That expansive crystal formation in concrete is tearing apart concrete foundations within 5 years in substations and under line poles in places where the cement is so formulated. The foundations' structural service life is cut very short.

This was an investigation made in the mid-1990s, and we have not heard of the phenomenon since. Has it left us due to changes in cement production? Or have we not looked?

A mechanism waiting to fail

Alaska is a special place in so many ways. One of them is the challenge the people there face dealing with permafrost. The trouble with permafrost occurs when we do something that ruins its permanence. When building something on permafrost, the name of the game is to keep that ground frozen. The transmission line engineers in Alaska developed a type of structure that has as its purpose, first the retention of frozen ground but also the ability to deal with ground that goes through the expansion and contraction actions associated with freezing and thawing. The structure is called the X tower.

The X tower in Figure 7.11 is guyed ahead and back to single anchors set on the centre-line of the line. The right-hand photograph shows that the two legs of the tower are set in pins attached to brackets clamped to piles driven into the ground or bolted to bedrock. The idea is that the structure can hinge back and forth on the pins, restrained only by the guys, and, if frost should raise a footing, the yoke as part of the guy system (as seen in the photograph) will fold, inserting enough slack into the guy to permit the structure to rise without damaging anything else. Thereafter, the clamp can be slid down the pile, and the structure's original condition can be restored without incurring damage.

The success of this mechanism depends on a number of things, including the attached hardware pieces and the detailing of the yoke. The version installed on some lines will

Figure 7.11 The tubular X tower (138 kV), Alaska

never permit success. In detail, the vang on the end of the yoke is placed in a vertical plane aimed at the tower. As the yoke folds in a scissor action, the plane of each vang will rotate away from aiming at the tower, and the light-weight thimble eye pinned to it and holding the guys towards the tower will be pulled out of its intended and capable alignment and tear apart unless the vang plate bends easily and maintains its directed angle towards the tower. It won't. Details matter!

Fortunately, structures set on rock and others set in areas with no frost action will not have their yokes tested, and of those there are many, because these towers are installed across Alaska regardless of the soil and frost conditions as a matter of state pride.

Flexible and tubular towers

Speaking of Alaska, renovation of the line just discussed, 15 years after its original construction, addressed some of the original line design's shortcomings. It suffered a number of unplanned outages when heavy, wet snow stuck to the conductors and carried them so close to the ground as to short out. The nature of the line's design was the problem, and the fix addressed them.

The line used fairly long spans. It operated at 69 kV but was designed for 138 kV. The structures, as described above, are very flexible, being pinned to the foundations and guyed as they were. When large snow loads – up to 125 mm of radial snow – on the wires accumulated, and more so as it fell off unevenly, the tension unbalanced, and easily swung the insulators along the line and even rotated the tower cross-arm, readily feeding enough slack into the heavier span to send it towards the ground.

The solution was to insert new structures into many of the low-clearance, long spans. This raised the conductors and reduced the available slack. The new towers were tubular guyed-V designs. These were flown to the line by helicopter, and lowered into the line (Figure 7.12). The legs of the towers were open to slide between the phases, and were closed onto a common pin. The legs of these towers were quite slender and up to 80 ft (24 m) long. To mitigate vibration of the legs, three chains about 6 m in length were installed internally at the top of the leg. These three chains were fastened to one face inside the tube and to the opposite face about 6 m below. As the tube would try to vibrate, the chains would dampen the motion.

On inspection of the shop drawings, we noticed $\frac{3}{4}$ in. (19 mm) nuts welded all over the outer surface of the tube. The fabricator explained, 'These are so we (you) can attach sections of angle iron to the outside of the tubes to break up the wind – in case the chains don't work.' So, that answer is both humorous and enlightening. The understanding and management of the Aeolian vibration of long, thin tubes and the mitigation of the same is, within our industry at least, not well understood and is as much art as science.

Fast forward about 15 years to near today. A colleague had just returned from Alaska and reported that the larger, 345 kV versions of these guyed-X structures on another line were having weld failures at the ends their tubes due to the tubes' vibration. They too were equipped with nuts all over the surface, but the associated angle iron pieces had never

Figure 7.12 Flying the guyed-V tower into the existing line

been attached. The question for today is: what good is a back-up plan if some process for deciding its implementation is never in place?

It was told to me some time ago that the non-tapered aluminium tube structures that are used for the higher-elevation Kildala Pass crossing of the mountain on the Kemano–Kitimat line vibrated badly upon initial installation in the relentless wind high on a mountain. The immediate and successful solution was to helically wrap the long legs with a length of fire hose to break the symmetry of the cylindrical surface – much as a wire is spiral wrapped around modern car antennae for the same purpose.

The ultimate solution that is still in place on the aluminium structures is the placement of a set of dampers halfway up and inside each leg. These dampers consist of a set of three automotive-type shock absorbers that hold several hundred kilograms of weights in place, suspended inside the tubular leg. These tower legs are about 1 m in diameter, and access to the top of the tower is by a ladder inside each leg. Access to the interior of the leg is by a small trap door at the leg's bottom and top. Every so often, inspection reveals a pile of dampers and weights heaped up in a mess at the bottom of the leg. Maintenance matters!

Distribution or transmission?

In 1999, a nasty little wind event in the slightly rolling farmland of Minnesota destroyed a 34.5 kV double-circuit wood pole line that took the juice from a widespread wind farm to the grid. It was a typical mid-summer weather front that spawned a local but intense wind event.

The damage occurred to the pole line along a pair of gravel farm roads, one running north–south and intersecting with the other running east–west. The wood pole line ran along the west side of the north–south road, and turned to the east at the road intersection, running along the south side of that road. From the road's intersection, the entire pole line was down for 14 spans to the north and for 14 spans to the east. The spans were short, so the overall damaged distances were about 800–900 m in each direction.

At the north end of the damaged section along the north–south road, the last structure damaged was a guyed deadend structure that was in place to turn the line to the other (east) side of the road before heading further north. At about ten spans north of the line's turn at the road intersection, there were three lonely evergreen trees within 150 m of the line, just to the west. One of them was uprooted and lying on the ground. The other two were undamaged and now lonelier than before.

On the north-east corner of this country road intersection there was a farm – a house, a large barn and a few out-buildings all surrounded by a dense grove of large trees. It is typical to surround a farm's housing and outbuilding location in the US Midwest with a thick grove of big trees to protect it from the cold winter winds and summer sun. All of this was essentially across the road and within 25 m of the fallen pole line's first several spans at the road intersection. Yet, there was not a single damaged branch of any tree or damaged building.

About 0.5 km or less to the north-east of the farm house and barn, several very tall wind turbines stood in the farmer's field. It was reported that the turbine near the east end of the fallen line's damage limit recorded a wind gust of 110 mph (175 km/h) by the wind gauge on its top. Another wind farm anemometer located a few hundred metres from the turbine recorded only an 87 mph (140 km/h) wind speed.

So, we have one tree of three uprooted and the others undamaged, measured wind gusts of 87 and 110 mph, no damage to the farmhouse tree grove or buildings and 10 of 13 poles broken along the north–south road and lying to the east on the road and 14 poles broken and lying to the south along the east–west road. Finally, in the field south of the damage, the new bean crop was growing. The young plants were only about 0.2 m tall, but all lay flat on the ground, bent to the south-east. The deadend at the north end of the damage was destroyed. The deadend at the road intersection was also destroyed, and the damage ended to the east at suspension structures that were damaged but not fallen.

The line was an unusual construction. It was a double-circuit 34.5 kV line with a neutral conductor below the circuits that were each mounted on a wood cross-arm – one circuit above the other. On each cross-arm, one phase was set on one side of the pole and the other two were set on the other side of the pole. Each phase was made from 336.4 kcmil (170 mm^2) Linnet ACSR/T2 conductor. The neutral was also a T2 conductor. The design wind speed was reported to be 120 mph (190 km/h).

T2 conductor is composed of two standard ACSR conductors wound around each other on a lay length of about 2.5 m. So, the pole line was supporting 12 Linnet conductors in six T2 pairs plus the T2 neutral. The poles were very stout and buried deep in response to the unusually high design wind speed and number of supported conductors. They were class H1 poles set 15 ft (4.6 m) in the ground. This is about 5–6 ft (1.5–1.8 m) deeper than usual. The poles that were broken were very strong new poles. They broke at the ground line or several metres above the ground. The upper sections of all the poles along the east–west road were thrown 10–25 m south into the bean field. There they lay with all the wires intact between them.

The wind came from the north-west, striking the north-south run of the line first. An analysis of the high wind on these structures says that the pole should break at the ground line, not higher up on the pole. To break the pole higher up from the ground line requires a vertical component of load on the pole. So, interested in which pole broke first, we looked for a pole broken at the ground line surrounded by poles broken higher up. On this north–south run, we found that pole near the midpoint of the damaged set of ten poles. The three unbroken poles on the north–south run of the line were the last three poles near the road intersection where the line turned to the east and which were across the road from the grove of trees surrounding the farm buildings.

That these three poles were standing suggests that the brunt of the wind force passed to the north-east of the farmhouse grove of trees and missed impacting those trees and the three poles near the corner.

The deadend structure at the north end of the damage was broken above the lower cross-arm guy attachment point. The deadend pole across the road, on the east side, was not damaged, perhaps because it was placed very close to the dense grove of trees that prevented a very high wind from striking the line. The deadend structure at the road intersection where the line turned east was also destroyed. Both deadend structures were of the same design.

Each circuit deadended to a steel cross-arm – one phase on each end of the arm and the middle phase to the arm at the pole. Each arm was backed up by two guy wires set on a 1 : 1 slope to separate anchors in the ground. All the guys were attached to the poles by steel wrap-around pole bands. At the upper cross-arm, the pole bands were attached 0.3 m or more below the cross-arm. At the lower cross-arm, the pole bands were attached to the pole 15–23 cm below the cross-arm. With this arrangement, the back-up support for the conductor tension is not aligned with the conductor tension at the pole, and the pole is put in bending by the mismatch.

With poles breaking and falling to the south of the north deadend and to the east of the road intersection deadend, the conductor tension was enough to break these poles at the lower cross-arm. In detail, the higher guy assembly on the upper cross-arm overloaded first. At the road intersection pole, that top guy broke in the wedge connector to the anchor rod. At the north deadend, the top guy's steel pole band broke. After these initial triggers, all sorts of other bits failed as well.

The deadend pole at the road intersection broke because the wind struck the length of line to its east and caused a serious tension increase at the pole. The loss of the deadend and the concurrent wind front along this length of line allowed the 14 suspension poles to the east to fail and be thrown to the south into the bean field. The damage ended along this length of line because the wind front was felt no further to the east. Without that direct load on the wires, the big poles had enough strength to remain standing, albeit with damaged cross-arms.

Review of the line's design criteria showed us that the pole stresses were at a maximum under the 120 mph design wind, and that the overload factor on that wind load was 1.00. Recall that the published strength of wood poles to use for analysis is an average strength, not a high value meant to capture the varied strength of the vast majority of poles. If the usual overload factor of 1.33 had been used, the actual wind speed capacity of the poles on this segment of line was closer to 95 mph (145 km/h), not 120 mph (190 km/h). So, we find a design calculation flaw in that it was not compliant with the national/state standard. If the wind speed and load applied to the poles is matched with the average strength of a set of poles and the poles have a reasonably large range of strength, as wood is known to have, a failure of some poles is practically certain.

Let's look at the design of the deadend structures a bit more. At each cross-arm, there is a set of three T2 conductors, each with a rated strength of 28 200 lb (12 780 kg). The guy strength backing them up is 25 000 lb (11 340 kg) per guy at a 1 : 1 slope. If we de-rate each guy by 10% for fitting abuse, the capacity of each in the direction of the conductor load is 15 900 lb (7200 kg). But each guy supports 1.5 conductors, so the ratio of the conductor tension capability to the back-up guying strength is about 1.77 : 1. This means that if something happens to the system to generate huge conductor tensions – say a bunch of poles flying into a bean field – the guying at the end of the line section could easily be broken.

Recall the guiding rule that says that, unless economics disallow, it is a very good idea to have the strength of a deadend and all of its bits and pieces of insulation and guying hardware exceed the breaking strength of the conductors attached to it.

On this line we find a design error, a guying attachment geometry that reduces its intended strength and a failure to adhere to that guiding rule. What else happened? The line was designed by the utility's distribution engineering department, not its transmission engineering department. We offer you a generalisation based on personal and observed experience. Distribution line engineering is generally all about the development and application of standards whereas transmission line engineering can have standards to employ but the work is typically based on the first principles of physics and engineering. We suggest that the three errors noted above that contributed to the failure were set in place due to the use of standards-trained engineers and designers in lieu of engineers who are used to working from first principles.

This is not meant to say that the use of engineers knowledgeable in first principles would have prevented the event, but we are suggesting that the chances would have been reduced. Designers and engineers who have been exercising their craft in the standards-based environment of distribution line engineering often have no understanding of how lines behave structurally – that is to say, they have no experience with the insights that this book is trying to convey. The use of T2 conductors and the high design wind speed put this line's design needs outside of the boundaries defined by a utility's distribution design standards.

> While wandering around the site on the first day I investigated the event, the facts leading to it occurred to me, and I felt a wave of panic, because the error of not realising that the nature of the line was outside of the designer's usual experience was the primary error, and I was well aware that we are all capable of not recognising that fact. That is to say, this could happen to any of us. From time to time it either has or will.

The perfect storm

This last event illustrates very well the problems that short spans create, and we tell the tale in terms of slack. Having virtually no slack can mean big trouble. The story is also an example of the common phrase, 'the perfect storm', in which a number of poor choices combine to create a disaster when any one of them in isolation would not have been much of a problem at all.

The line was a double-circuit wood pole design for 12.5 kV – a typical run-of-the-mill distribution design, you would think. But, like the last story, this distribution design was put outside the normal box by a single decision – the conductor choice. The line ran for about 80 spans in a very straight line with 250 ft (76 m) spans from a powerhouse wall to a river crossing, where the line turned and had a very long span before wiggling to its destination. In other words, the line was a short-span design for a long distance without any corners and then became quite different in its final few kilometres of length, turning a few angles and crossing a river with a 1270 ft (387 m) span. The short-span section used single wood poles with three insulator posts mounted on steel stand-off brackets back to back up each side of the pole. It was real standard stuff. All poles and spans were nearly identical in length.

The exception to this boring arrangement on this part of the line was the two spans at the powerhouse. These two spans were both deadended at each end, and the first span was half the length of other spans in the line at 34 m. Barring that exception, this was a section of line without anomalies until it was decided after the line was constructed that it should have two transpositions put into it at the third points for power flow balancing. So, two simple tangent pole framings at poles 30 and 60 were converted to deadends that allowed jumpers at these two locations to swap phases between various conductors. This was done by cutting the conductors and replacing the insulator posts with steel arms with deadend assemblies attached at each end.

All of the conductor deadends were terminated with bolted wave seat clamps, not compression units. The conductor was a large 1192 kcmil (604 mm^2) ACSR with 54/19 stranding, meaning that it had a 19-strand steel core 11 mm in diameter. That conductor

Figure 7.13 The wave seat clamp, intact (left) and broken (right)

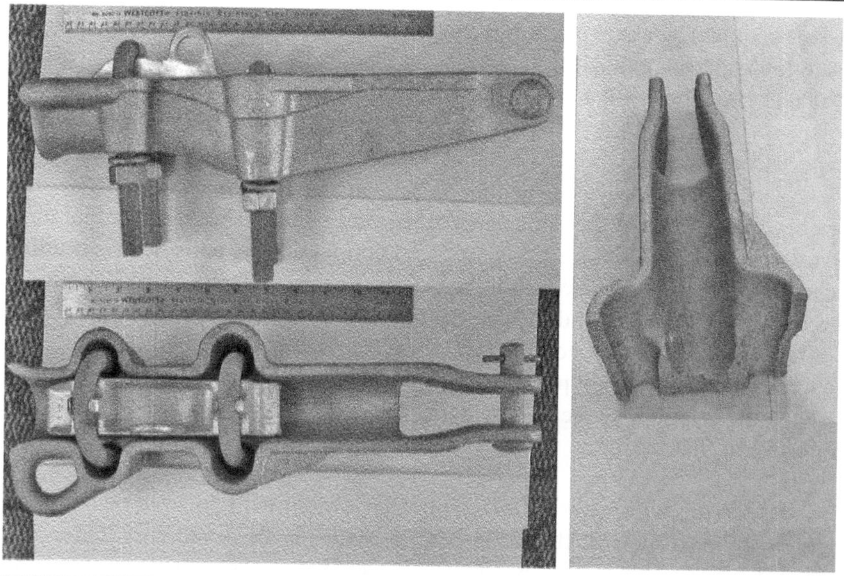

with its big steel core had no business being used on this short-span line. But, it was a standard conductor to the local transmission company, and handy to use. The line was located in a place that gets quite cold, so the sag–tension calculations were done down to −30°F (−34°C).

One winter night, when the line was in its first year of service, the temperature dropped to −42°C – a full 8°C lower than the temperature that the line was designed for – and the line fell apart. More accurately, one or more of the wave seat clamps pulled apart in tension at the powerhouse wall, at the second structure from the wall and at both transition poles that had been inserted many spans away (Figure 7.13).

- Pole 1 at the powerhouse span: 3 of 12 clamps broke.
- Pole 2, a small angle pole, guyed one way: 3 of 12 clamps broke.
- Pole 30, the first transition pole: 6 of 12 clamps broke.
- Pole 60, the second transition pole: several of 12 clamps broke.
- In the long span area of the line: nothing broke.

At each of the poles when one clamp broke, the load sharing, most notably to the unit on the other end of the steel arm, was overloaded and broke. So, at each pole we had a triggering break followed by numerous victim breaks. At the transition pole failures, several adjacent poles were also broken after the line's intact tensions were released, creating longitudinal loads on the poles.

The obvious question was: why did these hardware pieces break? In Figure 7.13, you can see that the clamp is a U-shaped trough casting. The conductor is laid into the trough from the top and clamped down into the trough by two U bolts. One of the U bolts presses the conductor down into the low point of the wave-shaped trough bottom. This bending and pinching of the conductor gives the hardware its grip on the wire. We call it a wave seat clamp. The clevis pin at the end of the unit is attached to the pole hardware. The catalogue data provided by the vendor declared that the unit has a rated strength of 12 000 lb (5440 kg or 53.4 kN).

But, we calculated that the tension that night rose to only about 7000 lb (3175 kg or 31 kN), or less than 60% of the unit's declared strength. We had a unit tested in a laboratory with a piece of the same conductor held in it and, sure enough, that unit also failed at 7000 lb. It was our assertion that the clamp was overly stressed by bending the stiff conductor into the wave seat with the U bolt. As the conductor resists the bending, it causes the cast unit to respond by also bending. In doing so, the top side of the clamp goes into tension. That top side of the casting is an open flange with large waves in it as it curves around the U-bolt positions. The point of maximum stress will be in the inside face of the flanges right at the U bolt doing the conductor compression. Indeed, every clamp broke in that location.

The reason for the high stress in the clamp that caused it to fail well below its stated strength was the stiffness of the big conductor with the 19-strand core. On asking an experienced lineman if he would ever use such a style of clamp with a big, strong ACSR conductor, he said, 'No, only with all-aluminium conductors.' To which conductor size and type should this clamp design by limited so that its rated strength can be trusted? We do not know, and we also believe that this particular vendor does not know either, nor did they seem to care very much during the investigation.

The unplanned temperature drop of 8°C below the design temperature on these short spans and especially on the very short span at the powerhouse wall caused a significant tension spike. It did not exceed the rated strength of the deadend clamps, but would the designer have used the choices of clamps, conductor and design tension if they had used a lower tension limit for the design? The designer should have, because the temperature that night broke no records.

Finally, this event is a great example for describing why short-span lines must be strung loose. If the sag had been increased by 0.5 m or so, there would have been enough slack in each span to absorb the thermal shrinkage of the conductors on a night like the one described without the problematic increase in tension. Yet the line's ground clearances could be maintained at a very modest cost.

Finally, spans with practically no slack in them cannot be accurately set correctly by a contractor. If a contractor can set sag by sighting a line to within 50 mm and the sag itself is only 0.60 m, then they cannot set the sag or tension within better than 10% or so of the target. If the slack on a short span is only 15 mm and the contractor is cutting in dead-ends, as needed at the transition poles, the accuracy of their work must be beyond possible.

Give the contractor a fighting chance by understanding slack.

Summary

This has been an important chapter if you want to understand the structural behaviour of transmission lines and then to do things to reduce the likelihood that you will be responsible for a failure yourself. It is not possible to put down all of the stories and words to reduce the odds of not understanding to zero. As a follow up, read about failures and investigate line failures to your best ability whenever you get the chance. You will always learn something useful by doing so if the investigation is done well. Many are not.

If all you ever do is design only to the local codes and local standards, your value to the industry will be very limited. To go well beyond that is personally rewarding and of great value to the industry. You decide.

Understand anomalies and the effect they have on the line. Understand slack and its importance to managing tension when the line is intact and after something fails and loads want to transfer along the line.

As an example, a common scenario threaded through several of these stories and through other stories not told (they are not ours to tell), is the placement of structures on a line running up a long slope. When a line runs up a long slope, it is common and attractive to have long spans in the mix. When ice accretes on the conductors above the long span, the tension in the iced span increases and the tension in the lower span does not. If the lower span is long with a great deal of slack in it that can be transferred to the upper span by the along-line swinging of insulators on the common structure between, that structure needs to be able to handle the longitudinal load.

In a story of the failure of such an intermediary tower that we cannot tell, the recommended solution was to install another tower in the long span to reduce the available slack. Elegant yes, but not always practical or possible given the terrain that can be there. The solution that was used to replace the failed suspension tower was to install a deadend tower. That was unnecessary. Sufficient but unnecessary, and not elegant. A more elegant solution is to put the deadend tower up in place of the failed suspension tower but fit it with suspension insulator assemblies. After all, the swinging was not the problem. The problem was the strength of the tower.

If you are designing a new line up a slope and want long spans set into it, do not use deadend towers with deadend insulator assemblies attached. That will cost a great deal of money. Install the stronger towers but use the suspension insulator on them to keep the installation costs reasonable.

A saving grace of the often-mentioned Kemano–Kitimat line is that most of its towers in the mountain area are very strong where avalanches keep destroying towers – even the suspension towers. Of the many towers that have failed, the adjacent tower has never been destroyed. When a line is placed in a location and terrain that makes damage recovery very, very expensive, this inherent strength of towers is a very good thing.

One more time: understand anomalies and slack.

REFERENCES

Peyrot A (2009) Wind loading: uncertainties and suggested simplification. In *Electrical Transmission and Substation Structures 2009. Technology for the Next Generation* (Vogt MW (ed.)). ASCE, Reston, VA, USA.

Wong CJ and Miller MD (2010) *Manual of Practice 74. Guidelines for Electrical Transmission Line Structural Loading*, 3rd edn. ASCE, Reston, VA, USA.

Structural Engineering of Transmission Lines
ISBN 978-0-7277-5988-7

http://dx.doi.org/10.1680/setl.59887.263

ice
Institution of Civil Engineers
publishing

Chapter 8
Projects

Once a transmission line is understood as an electrical and structural entity, it is time to work with other people to do something with it. You can study it, you can revise it or you can create it from scratch. Doing such things with a transmission line is called a 'project'. Executing a successful project is an art and science all unto itself, and is therefore worthy of a chapter in this book. Here, we mention some of the infrequently discussed subjects related to project engineering, and specifically subjects related to transmission line projects.

A preface to getting busy

Only part of the work on a project is engineering, so we discuss here the context for the engineering and try to paint a picture that will show you a much broader list or range of tasks and responsibilities than you may realise from the context for your work. We hope it helps.

Due diligence and best practices

A project is an event where the engineering facts and principles that you understand, such as those in the preceding chapters, are put into practice. It is where the rubber meets the pavement, and it is important to understand your legal obligations as an engineer.

As a professional engineer or a staff member of a consulting or engineering service group, you are obligated to exercise 'due diligence' or 'a standard of care'. To my untrained ear, these two phrases have the same meaning, but to a lawyer they are different. Take a moment and go talk to your favourite lawyer about the difference between them. However, both mean that you are obligated to have an understanding of the work that you are executing that is at least as good as most others who execute similar work. You are not expected to be a superstar, but you are expected to be reasonably competent.

Some contracts are just a few pages long, while others make up a mountain of paper. Taking the 'few pages' approach avoids spelling out all the intricate instructions that would direct the execution of the work. In place of all the intricate directives embedded in a lengthy contract, the sparsely worded contract asks that the engineer performs the work in accordance with the 'best practices' of the industry. Perhaps this is done to attract the best that the contractor and engineer have to offer, and perhaps it is to avoid writing about things of which the author does not know. But, there is a problem.

On asking a consulting company's legal counsel about the meaning of 'best practices' and asking how the term is to be understood and addressed, he flipped through several pages of one of his grand old books and then declared that: 'Best practices amounts to only avoiding negligence.' This should not give comfort to anyone using the phrase with the intention of getting the 'best' applied to your project. It would appear that, if you seek excellence, you need to do the hard work to attract it. It cannot be bought as easily as simply asking for 'best practices'.

You will find too that the 'standard of care' rises over time. A normal instinct for many engineers in a competitive environment is to distinguish themselves by attempting to be the best and the first. Many engineers want to stand in front of an audience and say 'Look what I did', 'Look at the ingenious manner by which I was able to solve this problem', or 'I believe that this is how this should be done'. Each time such a thing is said to the industry and that which was said or done is not refuted or discredited, the bar that defines the 'standard of' gets nudged upwards.

> I once said to a group of engineers who were embarking on a large set of projects involving a fairly new type of calculations that were undefined by the client and to be defined by themselves that they should think carefully about doing the work in a way that sets the bar no higher than necessary, because they would have to step over it with such work ever after.

Communications

> There once was a consulting company that paid a lot of money to a former client upon the failure of a significant component on a line that the consultant had engineered a few years earlier. The money was not paid because the consultant had made any errors but because it did not have the records to prove that it had not.

Communications, both written and spoken, are used to record actions of the past and to convey intentions for the future. When a relationship is founded on love or even friendship, a lack of good communication can usually be forgiven and overcome without much pain – think of your friends and lover(s) here. However, when the relationship is founded on the exchange of money for work, poor communications will cause you grief. It is not news that the requirement for good communications between parties and persons on a project team will never be deemed excessive.

This is the age of communications, and there are tools aplenty to use for conveying information and recording events. You can never record too much information, and you can never express your needs and intentions too often. Communicate or pay! Despite all our efforts in this electronic age, it remains much easier to lose records than to create them and file them away. Records tend to last in a retrievable form for a surprisingly short time. Pay attention to this, and be careful. Paper burns or mushes in floods. Computer storage units fail, and the authors of software upgrades have limited interest in these being functional retroactively very far into the past. All these factors are important, and good luck with that.

> Some of my most important and precious information from past projects resides in boxes under my desk and has been with me for 25 years. I have pulled files from those boxes more often than I ask for a recent project file to be retrieved from the official electronic storage medium.

Remember too that we each have our own brain and mind. Each of us lives in a world unto itself – some more than others – wherein the centre of the universe is not shared with others. What you think requires little explanation, because it is central to your understanding of things, but you must realise that your view will usually be foreign to the minds of others. If it is necessary that others understand what you are thinking or doing, it can require considerable care on your part to make your message clear. That level of care and effort is necessary to success. Practise clear and thorough communications, whether written or spoken. Your success depends on it. When the other person is being unclear to you, do not guess, ask for clarification.

> There have been times when I was told that I said something that I was sure I had not. For example, I was recently told: 'Do you realise that you just called that girl by another girl's name?' But, there are times when I will remain convinced that the error was in the listening, not in my speaking. When I recently asked a contractor why he took a certain action, his answer was 'Because you said to do it.' I know that I had said no such thing because, technically, it was something that I would never have done or recommended. In this case, I am convinced the listener got it wrong. A colleague suggested that the situation may have required a diagram – they say a diagram is worth a thousand words.

> Isn't it odd that when a bird in a flock chirps, either all the birds flee or they all do not. But when a human in a crowd yells, the crowd reacts with a wide variety of responses. Personally, I am quite convinced that we are such a diverse collection of hearts and minds that clear communications between us is a real chore.

Project engineering guiding principles

Sometime around 1980, the world went to hell in a hand basket. Well . . . it got more complicated. Before then, transmission lines tended to be owned and operated by utilities with years of practice in and understanding of their service territory. They had engineering staff or a consultant who engineered a line after the alignment had been established and before a contractor was put to work and before materials are purchased. The engineering process involved very few computer aids, so the individuals tended to understand the work and their responsibilities by virtue of the lack of a 'black box' computer.

Since then, much has changed. Transmission lines, large and small, are being proposed and owned by developers – organisations with much less or no experience with transmission line development, design, installation or operation. These owners and many utilities are guided by lawyers or accountants, and funded by financial institutions or wealthy investors. They have reduced the number of engineering staff or have none at all, and they necessarily rely on consultants, who often come from afar and have little knowledge of the service territory.

Project execution is often via design–build or engineer–procure–construct (i.e. turnkey) style contracts, where the engineer is working for the contractor, not the owner, and is required to produce the design as a parallel activity to identifying the alignment, and easement and permitting rules development, in parallel to the procurement and construction itself. When the engineer is acting in service not to the facility owner but to a contractor who has limited knowledge of, or interest in, the essence of the engineering work or role in the project, things get difficult.

In addition, line engineering itself has largely disappeared into the 'black box'. The use of black box computer programs has allowed inexperienced design engineers – or even salesmen – to provide services to inexperienced owners with errors or inefficiencies going unnoticed. Sometimes, unnecessary problems are generated and efficiencies are rarely found. The present situation often couples inexperienced engineers with an inefficient project design process. Overall ... what a mess!

> As many people have, I was long ago pointed to Ayn Rand's books and read *The Fountainhead* and *Atlas Shrugged.* As an engineer, I found it quite agreeable that persons, such as engineers, who make contributions via their creative minds to society for its benefit and gain were considered by her to be an essential part of a society. The bad guys in her writings were the non-producing bureaucrats. The changes in our working environment since the 1980s seem to comprise a situation where the bureaucrats have won the recent battles. Sadly, from my point of view. If it is not actually sad, it is at least frustrating.

Every once in a while, we have been requested, presumably by some of these inexperienced persons or organisations, to provide a flowchart or, similarly, a description of the execution process for a transmission line project. We try to oblige but, in this new environment of diversely different system owners, financiers and operators, one or several project process examples cannot provide useful, detailed guidance. There is simply too much diversity among the players, schedules, constraints and so on.

It is not valuable to the transmission line engineering community to offer guidance that supports the idea that this work is easy, or that inexperienced engineers should get the idea that they can do this work well with their black boxes, for owners who know no different or have limited interest in the process, as expressed by their willingness simply to follow such 'flow-chart guidance'. Project processes must be learned on the job, as no two projects follow a common path to completion, and engineering needs can cover an astonishing array of issues. Design engineers must do the work of getting experience by working with people who have the experience, and then they must 'think on their feet'.

That said, we can take the request on at a higher level. The following are suggested guiding principles for successful project execution:

- Identify all components of the engineering effort required:
  - routing support
  - permitting support
  - contracting format guidance

  - legal support
  - engineering
  - procurement support
  - construction support.
- Identify resources for all required efforts:
  - manpower,
    - in-house
    - external
  - materials,
    - understand alternatives – pros and cons.
- ID the stakeholders and understand their objectives:
  - financiers
  - internal management and other departments
  - landowners and neighbours
  - vendors
  - contractors
  - government bodies and regulators.
- Write a management plan:
  - roles and responsibilities
  - communications plan
  - quality assurance/quality control plan
  - document management plan.
- Execute all the required and resourced tasks as soon as possible. Do not procrastinate.
- Maintain and understand the schedule and anticipate your client's needs. Try to put yourself in the client's shoes.
- Maintain control over stating your delivery capabilities.
- Keep an open mind with regard to all options, at all times.
- Challenge decisions:
  - measure alternatives for their relative merit.
- Understand the iterative nature of design development and work. Be prepared for the tide to change at any time.
- At all costs avoid hearing 'You are holding up the contractor'.
- Protect your own interests and help your client.
- Document everything.
- Communicate relentlessly.
- Be a team player.
- Have a win–win objective with everyone else.
- Rely on others ('trust but verify').

Try practising these principles and you should stay out of trouble, actually be useful and enjoy your results. A further thought: success comes from working with people who share your beliefs and goals. This is true in all relationships, and is necessarily true on projects. The more you can understand and be supportive of the goals of all the stakeholders listed above, the better your project will go. If you cannot get well aligned with some of these folks, you will have trouble.

My current employer uses a process to help decide whether to pursue a project for which it has been invited to bid. It is called a 'go–no go' process/decision. All the questions that the process requires the persons contemplating a bid to answer are effectively asking the team to consider three things: the value to the company of pursuing and doing the work; the ability to perform the work suitably; and our the knowledge of and empathy with the organisation and persons seeking the services.

These three components of an engagement between two parties – value, capability and empathy – are the key to success. In fact, for a successful adventure together both parties to the proposed contract must answer positively to all three components at both the personal and corporate level. Please understand this, or suffer through many miserable project events.

Cost control and people

Sometimes it is stated, or can be observed, that 'cost is no object'. The project is stated to be, or appears to be, driven by other factors. It may seem freeing when this happens but it never lasts. Eventually, the cost of the project and of many of its elements will matter a great deal. This is always the case, so do not be surprised when that day arrives. In fact, prepare for it.

We have already described the extent to which an engineer can affect the cost of a project, and shown that it is not a great factor in the scheme of things. But your client/customer or boss tends not to know or believe this. So, the engineer can try, and under certain circumstances must try, to influence cost. Point out where savings lie, while knowing that there is a frustratingly powerful yin and yang force at work causing increases here due to savings there. Point this out and then let the decisions of others be what they are, because their decisions are based on the centre of their universe, which lies elsewhere from yours. This approach makes neither of you right or wrong, and when the decision is not yours to make there comes a time when arguing about it or being upset with it is effort or angst not worth exercising.

When a large financial institution, be it the World Bank or a billionaire, finances a project, the overriding objective is for there to be a low to no risk of the investment being overrun. The primary interest of the financier is that the price given is the final price. There is less interest in what the price is. Couple this with the fact that many transmission line projects since the 1980s have been paid for by a guaranteed return-on-investment formula.

Back when a fully integrated, full service utility generated, transmitted and distributed the electricity to your home and business, the transmission component of the process was not seen as a profitable component of the process. It was paid for by the generation side of the business. After these fully integrated companies were dismantled into separate companies, the new companies whose business was only the transmission of power had to find a way to get paid. Some of those components of payment are a return on the money invested in new transmission line facilities, and even the right to charge the customer base for projects pursued but never constructed. If this is what it takes to attract the money necessary to grow and evolve the world's transmission line networks, so be it. But, inherent in this system is the lack of an incentive to be cost-effective.

Consider too that the operating and maintenance (O&M) costs of a facility are treated as an expense to a facility owner, whereas the capital cost incurred to install the facility is treated as an investment that is used to justify the rate the owner can charge for the electricity delivered. If the design engineer offers a design that can be installed for a lower cost, the owner may be inclined to notice that the engineer is reducing the size of the pot of money that he uses to justify the rates charged for his product. If the design offered also incurs a new O&M cost to train or equip the O&M staff to deal with something new, the argument for adoption is diminished further. Under these circumstances, advances in design are not appealing to the owner. Something there should change.

It seems that the World Bank and the others do not have a clue as to what a cost-effective, efficiently designed transmission line looks like. The wealthy investor is more likely to prefer to invest $2 billion at a healthy return than the $1.5 billion that is required to be efficient with the dollars spent. What does all of this mean for the design engineer?

I expect that most engineers are very nearly 'genetically programmed' to provide value – to provide a lower cost solution that does not degrade performance. What I am telling you is that you will not always have a partner in this goal. This may frustrate you when it occurs, but *c'est la vie*.

When it comes to your own work, remember the principles above, and remember the target of the day. Stay focused on what matters. There are two things that commonly occur. Every once in a while a project comes along that is just what you have been waiting for. Its needs are right in your wheelhouse – you think. It is a human characteristic to hang out and dwell on what we love to do and to ignore what we do not enjoy. When this happens to you or your staff, the budget can drain quickly while you are having so much fun. Oops!

Second, have the right person do the right task. Who is more expensive to the project? You, or the other person who costs half what you cost or twice what you cost? Generally, a person's charge rate trends towards being proportional to their productivity and the value of their ideas. Often, the expensive person gets the work done at a better cost to the project than the less expensive person. But, you can't have all the old, experienced hands doing all the work. They have to take the opportunities to teach the young and inexperienced, or else they'll be stuck with the work until they die, and they will be on a team with no younger people because those folks are not going to hang out with them waiting forever for their own opportunities. Part of the expense of executing work is the training of the inexperienced. Pick your people and your moments, but do it always.

Consider too the technology changes that take place with our design tools. As a person ages, he or she tends to lose touch with and be less efficient in using new tools, but gains experience outside of the use of tools and with the non-technical aspects of the work. Thus the roles played by individuals must change as they age. Let the younger people run the tools and have the older persons guide their work. Have the older people tell the younger people what to do, but not micro-manage how they do it.

With regard to 'what matters' ... understand that there are people who do very well in executing tasks but who are also quite incapable of discerning which tasks matter. These people will drill and drill down into a subject to understand it completely, only to eventually be told that all of that drilling was for naught and all of the cost of that drilling was wasted because the subject is not what matters. Learn to identify these people and task them accordingly.

Finally, at a technical level, cost control requires a tracking method and something trackable. Define the work in trackable units and keep records. No two projects are the same, and no set of people interested in the project is like the set of people interested in a previous and seemingly similar project. Think!

Operations and maintenance

Parents conceive a child and, in the early years, they are the primary developers of the child's character and knowledge base. As the years go by, the child encounters other children, then other adults, and at some point the parents find that they are no longer in control of that child's character or knowledge development. At some point, the child is full grown and a contributor to the society in his or her own right. An engineered and constructed entity is not much different. During the design stage, a transmission line is, in many ways, very much the child of the design engineer. Eventually, the line's future development and contribution to society is necessarily defined by the work of others. It is useful to know this, and even to prepare the line for that.

It is not useful if you design a line that does not perform according to the expectations of the operator or cannot be maintained according to the practices of the maintenance entity. Understand the operational needs. This requires learning, up to a point, that business, or interacting with the operator, is possible, whenever you can. We call this 'designing for maintenance'. Maintenance practices vary widely between organisations and between lines in a network. Is maintenance to be live-line or not? Will the line be inspected intensely, or ever? Will access be by climbing, bucket trucks or helicopters or some other means? Find out and design accordingly.

Technically, learn what goes wrong over time. Learn about erosion, corrosion, oxidation, wear, fatigue, ultraviolet degradation and so on. Learn this from technical literature and by attending meetings with people who study these subjects. Try to incorporate what you learn into your designs, remembering that the devil is in the detail. Avoid the nasty, risky and unknown whenever you can.

The design process

The design process is a subset or only a part of the already described project engineering process. Being a smaller subject, it can be better described than the ever-changing project process. The design process involves a collection of issues, and the attention to be paid to each varies from project to project. Basically, you set up the criteria, invent or select the components, lay them out along the route, check them for compliance with the criteria and make adjustments until it all works, optimise your effort (perhaps set up contracts to get the work done), and then deal with the people who will build and later own, maintain and operate the structure.

Designs: standardise or specialise?

Some power line owners (not all are electric utilities) have no interest in using standard structure designs on multiple projects, while others cannot imagine trying something new – something that wasn't adopted maybe decades ago. Both good and bad results can arise from either *modus operandum*.

The use of standards achieves the following:

- cost savings through expediting choice
- cost savings through mass purchasing
- cost savings through simplified maintenance and construction
- safety through simplified/predictable maintenance and construction.

The use of standards prevents the following:

- cost savings through thoughtful choices that address the present and unique challenges
- cost savings through optimised purchasing
- risk avoidance as a design challenge steps outside of the standards' presumed scope of use.

There are organisations that exercise their line engineering with standards that were written up to 50 years ago. There are organisations that are very much anchored to their designs even when the project at hand has rather new constraints. One has to wonder what could ever compel them to revisit the nature and content of the standards. One has to hope that their standards were near perfect when created, because all cost-increasing or safety-reducing features are embedded in the work done in accordance with the standards for the whole time they are used. Technologies advance with time, and while not all changes are improvements, many certainly are. Organisations stuck with very aged standard materials and ways of doing things do not remain efficient or cost-effective.

Why do organisations stay with their long-standing ways? Some organisations offer no incentive to be creative and offer new solutions, and they may even offer disincentives. Some organisations are forced to comply with methods and materials dictated by the local jurisdiction laws. Their engineers may know very well of better solutions but the lawmakers do not, or they have other agendas. If you are an engineer because you like clarity and valued solutions, I am sorry. You live among human beings!

Standards allow designers and engineers without suitable knowledge to perform work – presumably successfully. When a line is designed successfully by such people, this can be called a good thing. However, it is a risk, because, over time, excessive dependence on standards leads a person or a group to complacency, and that is clearly a bad thing. Left without any guidance but the standards, people lose understanding of the intent and the intended limits of application of the standards.

Standards tell a designer *what* to do. Very often, they do not tell the designer *why* it is to be done. There are often many ways to skin a cat, and a standard has selected one choice from the many. In many organisations, the documentation explaining why the choices expressed in the standard were made from among the many choices available does not exist. When the time comes to revisit the standards, it is very difficult to revise them in the absence of a description of why they are what they are. A standards document should have an accompanying rationale document to facilitate its eventual revision.

Whether a project justifies breaking ranks with standards is always worth discussion. Some projects will and some won't succeed in breaking away. The factors are generally not easily valued, so anyone who needs a monetary-based argument may need therapy after the discussions end and the 'wrong' choice is made. Certainly, engineering standards should be under near constant review, and that is nearly never the case. Good standards are rare, and they are pieces of artful work.

Design criteria

One of the first things a designer must do when embarking on a design effort is to establish the performance requirements of the final 'product'. View this as describing an envelope outside of which the design must not protrude. The envelope has many more than three dimensions – imagine that if you can. It is a complexly shaped envelope, and to feel most confident that its boundary will not be violated somewhere define its shape using as many points as possible.

This is a simple way of saying that you should be in the habit of making a list of criteria that is as extensive and elaborate as possible, even if you are basically convinced that many of the criteria will not control any aspect of the design. To do otherwise is to presume that you understand all aspects of the problem at hand well before you have given it much thought. It also does not give you much room to properly address shifting requirements and new ideas as they show up during the design process.

> Some years ago, we designed a tower for a project, and listed 15 load cases. Some were essential, in that they addressed the fundamental code requirements of wind and ice and broken wire, and so on. Others were our choice, including the already discussed tornado 'wind-on-tower-only' load case. The tower was full-scale tested and failed at 3 min into the 5-min hold at full design load. It did so high up in the tower at a complex junction of five members in the plane of a tower side under the maintenance-load case. Did that matter? The answer is 'yes', but only because we said it matters. It did reveal a poor alignment of member work lines, and their adjustment perhaps improved the integrity of the tower design against a host of real-life loads.

It is commonplace to see tower design criteria that consisted of a list of several hundred load cases. Many are the attachment of loads to less than all attachment points in every combination possible, multiplied by the loads being applied over a range of all possible directions. I have listened as very experienced designers ridicule this incredible list as 90% irrelevant because they knew what the controlling loads will be and the rest was a waste of time.

That attitude is commonplace and dangerous. In this age of computers, the analysis of hundreds of criteria takes a few seconds more than the analysis of a preferred few. Unless you are very sure that you are right about your expertise, load up the list. Design criteria should cover more than the classic technical issues of electrical and structural concerns, and include the sustainability criteria described in Chapter 9 if you want to call yourself 'green'. If you are a structural engineer, do not forget the electrical constraints, the environmental constraints, or the construction and maintenance criteria.

Electrical considerations

The first constraint to be understood is that of the electrical performance requirements. After all, this is an electrical transmission line. System operators will judge the success of a line in terms of its planned and unplanned serviceability rate (hours available for service per year compared with the hours in a year, expressed as a percentage). As lines become increasingly important to the performance of a network as a whole or to a key customer, that serviceability rate target is increased.

The target is also increased as the system as a whole becomes challenged, such as when new components were not added when they should have been. If you have only one line but would benefit from a second line that does not exist, the one line you do have takes on an incredible importance. The pushback against the installation of new lines and the upgrading and uprating of existing lines forces the existing lines to perform better and better, even when faced with their own ageing.

The difficulty is that the correlation between the serviceability rate and the actions the design engineer can take to affect it certainly is unclear and very difficult to quantify and manage. In addition, the serviceability rate is also affected by many things outside the control of the design engineer, such as a changing role of the line in the system, revised conditions along the route as of the line (as set by humans' decisions) and the financial conditions that affect the maintenance of the line.

Now that we have this pointed out, and although once again you are not in control of the situation, you are required to carry on and do what you can. Obtain and understand the line's key performance requirements: normal and transient ampacities, definition of transient, voltage drop limit, radio and TV interference, audible noise limits, electric and electromagnetic field limits, and, for long lines, the surge impedance loading. These criteria will feed into the selection of conductor size, phase bundle dimensions and phase spacing. Do not forget that trading-off is the name of the game. If you are a structural engineer, learn a bit about electricity, and, if the other experts involved also learn a bit about structural engineering, discussions with the experts in each arena will be more interesting and fruitful.

Electrical clearances

In Chapter 3 we noted some factors associated with the electrical clearances, and we have more to say here.

> When working on developing the line design criteria for some proposed lines in Kenya, we came to the humorous, but basically true, statement that 'phase spacing is determined by the reach of a baboon, and ground clearance is determined by the height of giraffes'. In Canada, ground clearance must take into account the fact that a person on a snow machine (a) loves to travel along a well cleared right-of-way, (b) likes deep snow and (c) may have a big antenna on the machine to talk to his buddies. In dry farmland, such as the middle and western states of the USA, some irrigation systems are made of aluminium pipe sections that are moved around by farm labourers. The easiest way to carry a long metal pipe is over your shoulder, with one end probably higher than the other.
>
> In Panama, I once saw a sloth hanging from a 12 kV phase wire about 10 m from the nearest pole. How did he get there, and how will he get back without, you know … bBzt!? Perhaps the long claws of sloths are good insulators. I wished him well and moved on.

Build a tower and birds will hunt from it, or nest in it, and poop from it. Or, snakes will climb it to get the birds. The issues are regional and varied.

It is also a well recognised, but poorly quantifiable, fact that conductors move around due to thermal changes, wind and ice, and even with time (creep). They may move around at the structure, depending on the insulator assembly design, and they certainly move around in the span, especially near mid-span. Quantifying these movements or displacements from a defined position of rest always involves acceptance of varied amounts of approximation and some sense of what constitutes acceptable risk. To begin, the position of rest is only approximately known. The thermal expansion and creep characteristics of the conductor are known only reasonably well. The amount of ice that might ever accumulate on the conductor is always a rough estimate.

Most notably, the wind force on a span of wire and the reaction of it to that force are poorly modelled in our business, in part because the wind environment and nature at each span is unique and there are so many spans to deal with. Approximations and assumptions that render the work cost-effective while mitigating risk to our satisfaction are necessary, and the name of that game.

But, a good thing happens as spans get longer. While you might think that bigger sags would allow a span to move further in accordance with a describable relationship, other facts come into play to mitigate the movement. Short spans move easily, in part because the load sources are local and fit the dimensions of the span better than they do on longer spans. There is also an $F = m \times a$ issue in play. If you hit a light mass with a force, it will move easily. Hit a larger mass, it will not.

Generally speaking, the nature of wind is that a background, synoptic wind speed of a moving mass of air carries with it the feature of small events of higher wind speed. These events are gusts, which are almost always present, and downburst, micro-bursts and so on, which are less prevalent to rare. In general, the faster the small event, the shorter its duration and the smaller its dimension. So, as span length increases, the smaller is the effect of these faster and faster events. As you move through the spans on your project, or that are under consideration for your project, you need to adjust the effect that the wind has on the span, and you can adjust the clearance required accordingly. The design guides floating around the industry use various formulas to deal with this. There is more than one formula out there because the subject is poorly understood, and we are unwilling or unable to properly represent this piece of rather obviously complex physics with a truly correct formulation.

When spans are short, parallel wires are not likely to act in unison in response to a wind or ice load, but when spans are very long they do act in unison. When you get the chance, look along a very long span of line when a strong crosswind is blowing. The wires hardly move, certainly not quickly, and they seem to be parked in the sky as a unit – parallel but curved in transverse displacement. So, phase separations and clearances to other things become more critical as the span length increases. The clearance requirements for short spans tend to be specified rather directly by codes and standards because these had their genesis in long-ago systems that are now our low-voltage systems. With very long spans, you may need to find your physics-based guidance elsewhere.

The best guidance comes from finding an existing installation that has comparable dimensions and is subjected to comparable environmental loads. Look for this basic rule: a long span horizontal separation equal to 1% of the span length plus the insulator length of the line (to account for the voltage differences). Expect examples of vertical separation to be more difficult to find. CIGRÉ has published a document that contains the parameters of many of the world's long spans. Remember that a span that has never caused trouble may be overdesigned, but you don't know that. Shrinking your project's dimensions in the name of bravado or to seek recognition requires better evidence of what is possible than a review of successful installations. Find one that failed and was fixed. We point you to Davis *et al.* (1963).

Clearances express risk. For example, the arm or body face of a support tower is always close to a conductor that it supports but a very tall truck passing under a conductor is in proximity only momentarily, probably not at the tower nor when the conductor is likely to be at maximum sag. The line's nominal voltage is always present, as a minimum value and with the probability expressed by the above-noted serviceability rate. A momentary voltage occurs in a fraction of a second due to a short circuit or switching surge. So, using our example, the passing truck is at a very low risk of encountering the voltage spike, and the tower face is not. Accordingly, different separation values can apply. The separation to the tower body should be greater than to the passing vehicle, except that the vehicle is likely to pass under the conductor where the conductor's position is not confined or even well known. So, the vehicle gets a big clearance value as well but for a different reason.

A conductor hanging in suspension on a tower will be at rest most of the time and could be blown for several seconds or minutes towards the tower face by extreme wind. Accordingly, different separations can be applied for the two weather conditions of no wind and extreme wind. The insulator length basically represents the spike or surge voltage separation requirement of a line, as determined by the physics of the ability of a spark to jump to ground at that spike voltage.

One final comment on clearances is useful. Increasing the clearances will almost invariably mean increasing the cost of the project. If you like working on the cheap, at least consider this. A dimension increase applied to many spans will cause a measurable cost increase. Increasing the clearances in a few spans or that one big one may not. It will be very difficult to renovate an entire line if a mistake is made but it will not be so expensive to renovate a single and unusual span.

Those are two opposing concepts to work with, so here is a hint to help you decide. If your client is a contractor, don't ever try to tell him that coming back to renovate a location, however discrete, is your suggestion, because you are asking him to put much of his profit from the project at risk. As he probably has no feel at all for the risk you are suggesting, and because he will assume it will be his problem to solve, his response will be to reject the idea. Remember, all the project's stakeholders have their unique centre of universe coordinates and goals.

Environmental considerations

Once you have understood the electricity and the air space its conductors require, you can then look at the strip of land that either you need or is given to you. The situation is often the latter, and your challenge is to work with a right-of-way that has features you would rather did not exist, such as a point of intersection (PI) in a hole or a wicked side slope, and/or has a width that you invariably wish was greater. The more onerous and persistent problem is insufficient width. Ah, but again, if this were easy, why would you be earning such a handsome pay cheque? Let's talk about locations, noises, blowout clearances, working space and guying.

> Two stories ... I was once told of a design and construction interaction that I hope is true because it is just too funny. After the construction crew went out to survey tower locations, they complained to the engineers that they had done a lousy job because they had put all the PIs in low spots. But the complaint was a little misdirected. In fact, the designer had laid the line out on the surveyed profile drawn on a long roll of paper – this event preceded the present-day computerised environment – but had laid the paper upside down on his table.
>
> That said; I did once review the profile drawings for a project that I eventually was pleased to avoid, and in this case also the PIs were placed in all the holes of very rugged terrain. One can only guess as to why, and then run away. I have a guess, but ... never mind.
>
> Related to the first story, because it was the same organisation that told it, a construction crew arrived at a particular tower site, which on paper was on a very

well defined mound of ground in the middle of a very flat agricultural area. How fortunate! But, the mound was not to be found. A close examination of the aerial photography from which the profile had been made showed the mound to be a giant pile of hay that the farmer had long since fed to his cattle.

Recently, I was involved in a project that was in the middle of nowhere, and there was only one landowner – the government – and there were no obstacles except cliffs. But the alignment had a few silly kinks in it because it was laid out by a surveyor using a surveyor's definition of the 'centre of the universe'. He had his PI coordinates and knew his next, end point coordinates. So, he set off on the 3-mile run as if it were a fired a bullet. This line of sight took him generally along and parallel to the top edge of a long, deep ravine.

Oops! He came to a spot unworkably close to the edge of the ravine, so he planted a PI before the trouble spot, beside and offset from the spot, and then again back onto his original line after this spot. Further along, he found that a curve in the project's access road was encroached by the alignment of the line, so he added another three PIs to dodge that spot. All this was the result of the surveyor believing that what mattered above all else was his original long bearing from the first PI to the end. All this all occurred before the line engineer was brought on site. We campaigned and prevailed, reducing the surveyor's six PIs to two, one at the side hill offset and the other at the road offset, even though the clearing was well underway.

As this project was located in the middle of nowhere, we were planning to use a flat phase arrangement, with the two circuits beside each other on separate structure sets, with long spans to reduce the number of structure locations where the contractor would have to set up and do work. After all, we had all the room in the world. Not so, as it turned out. Somebody decided, before the engineers were involved, to provide a right-of-way width of only 30 m to accommodate two 230 kV circuits. The eventual design was forced into vertically stacked phases sharing a common two-pole structure with short spans. Too bad!

The problem with a PI is that it forces a structure to be placed exactly there unless the angle is miniscule, in which case its structure can be moved forwards or back a bit. If the spot is not a great location for working, or for guying if needed, or requires a tall or deadend-type structure because it is in a hole, then that PI will cost you money. When given the opportunity, select the alignment PIs while bearing in mind with tower cost, access, work area and guying. There is a considerable amount of money at stake. Remember that optimisation via a computer program is done after the PIs have been set, so their suitability, and therefore the efficiency of the project, is up to you. A forgotten feature of significant angles at a PI location is that it is an attractive, and perhaps necessary, place to set up stringing equipment. Setting a PI at a location where stringing equipment will need to be placed requires that a few hundred metres of space be available in line with the conductors in each direction beyond the actual, permanent easement. Remember ... design for construction!

Not all line owners pay sufficient attention to the need to keep the noise of a line away from populated areas. Some have quite stringent requirements, while others have none. Noise can be addressed in a number of ways, including conductor choice, bundle dimensions, and phase spacing and arrangements, but also by the obvious – keeping the line further away from the populated area. By this I mean offering to pay for a wider right-of-way so that the edge of the line's domain abutting the populated domain is further from the noisy wires. Ignoring any arguments about the need for such action, we look at the calculations for deciding an acceptable arrangement.

To understand audible noise, you need to understand decibels, ambient or background noise and its variability over a 24-h period, and the effects of weather on wire-generated noise. A more diversely managed issue is the strength of electric and magnetic fields, which is viewed as a health, not annoyance, issue. As the presence of a person within a right-of-way is a transient in both time and location, there is a growing understanding that any calculations should be based on the average height of the conductors, not their lowest height, and should be an average within the span, not a maximum, such as is found closer to the structures. The closest persistent exposure is certainly no closer than the edge of the right-of-way, and is still based on the average conductor height. The average height of a conductor in a span is one-third up from its low point. As magnetic force is based on amperes, not voltage, the sag of concern is the average sag, not the maximum, which relates to maximum amperes CIGRÉ (2005, 2007, 2009, 2011).

The conductor blowout action almost always plays a role in defining the width of the right-of-way. However, how this is calculated varies considerably. It is a clear indication that, in this business, we get to make stuff up. In the USA, the widely used default calculation is to blow 6 lb/ft^2 (20 m/s or 46 mph) of wind transversely onto the wires on an average sunny day and ensure that a separation, as specified in a safety code, from that displaced wire to an object at the edge of the right-of-way exists as a function of voltage and of what the object may be. This amounts to a significant but modest weather event coupled with a suitable clearance between items. If the span is sheltered such that a wind of this speed cannot occur, the wind force is reducible to 4 lb/ft^2. By the way, tree cover does not count, because trees are considered more transient than the transmission line. A comforting thought! But what if it is a National Park rainforest?

There are organisations in the USA that couple a 16 lb/ft^2 wind (35 m/s or 80 mph: a hurricane, even though no such things reach the organisation's domains) with a zero separation requirement to objects outside the right-of-way. This is a coupling of a very unlikely event with a higher risk clearance. So, a coupling of guesses is expected to provide suitable results. Only a history of failures with the use of a rule will ever reveal its suitability, but nobody is checking.

Very long spans over deep ravines present an interesting situation. The blowout could be quite large, except in the case of long spans, which are relatively immune to motion, barring high synoptic winds as described above. But, if the wire is higher above the ground than any structure to be built or occupied nearby, and high above any trees, ground-based occupant or activity, what would be the reason for the wider land strip, except an excuse to give money to the landowner? Barring that reason, which is not altogether ridiculous, we have suggested that a typical rule, such as the one described above, tops out its application at a rational height above around, say, 15 m, and that no rules apply above that height, barring very tall trees.

> I recently worked on a long, high-voltage project in the middle of nowhere in the western USA. One could stand anywhere on about 80% of the route and see no trees and no human activity or facilities. Yet the owner insisted on separating each line from other lines by at least 450 m, so that a fire could not destroy both.

> It is not difficult in many congested places to see lines having multiple circuits packed very close together on each tower line. I saw this years ago in the Toronto area, in Japan and Korea and in Switzerland. You have to love Switzerland. The towers are all painted green, and some lines are set on tall towers above the forest with no clearing whatsoever having been done. In Tokyo, numerous circuits are stacked so high on single towers that the North American ideas of circuit security and blowout issues that could affect adjacent (much shorter) buildings means nothing. We do what we can, not what is universally agreed or established.

A contractor needs room to get things built. Certainly, it only requires more money to fit the men, equipment and materials onto a thinner strip of ground or to step outside of it occasionally, but large structures or big sets of conductors require space for installation. Corners and deadend tower locations are particularly useful as work sites. If you have the opportunity to affect the areas available, be kind to the contractor. If you do not, your recourse may be to employ structure types and construction methods that require less room.

Guyed structures are pet preferences of the authors. Structurally, they can have the greatest integrity, and they certainly offer a lower cost solution for self-supported towers. However, the problem they bring is fitting the guys and anchors onto the right-of-way. Steep guys add expense, and properly sloped guys require a large footprint for the structure, especially at corners. At corners, seek extra guying room, if necessary. Steep downhill side slopes aggravate the issue. If you are using guyed structures in rugged terrain – a type of situation for which they are well suited because rough terrain usually means tough subsurface conditions, and guy anchors are the most cost-effective solution – and if you have a say in the alignment, choose it wisely, especially at the corners. If you do not have a say in the alignment and the width of the right-of-way, consider carefully the viability of large guy footprints in these locations.

Aesthetics

> Twice in my career I have been accused – no, recognised (about 20 years apart by two people who have never met me) – that (1) I am 'not an engineer but a frustrated architect' (said by an architect who was either biased or insightful), and (2), said by the other guy, 'you are not a transmission engineer, you are a transmission architect'. Granted, I try endlessly to make my profession's structures as attractive as possible. It's a battle.

> Recently, a friend who knows this about me asked if I had time to look at a recently constructed high-voltage line in the city I was visiting. The reason he wanted to show it to me was because the people around him in the business all agreed that this was a terrible line. He said that people not in the business – members of the general public who know he was in the business – were stopping him to ask why that line was built the way it was. He said that 'When the general public notices something about a transmission line, something about the line is out of the ordinary.' So we went on a tour.
>
> The line was alongside a multi-lane divided highway and ran for many kilometres around the city. Before the tour was done, I had to agree and the word that would not leave my head was 'hideous'. Every few structures along the way, the structure design changed from one odd arrangement to another arrangement that was even stranger. Many of the features were unlike anything I had ever seen before and were not particularly understandable. The loading on the poles was lopsided and the poles all bent into the load, and the very long tubular arms drooped under the weight of the wires. It was aesthetically hideous!

Barring such extreme examples, beauty is in the eye of the beholder, and, collectively, we will never agree on what looks good. While the designer may have an opinion on what looks good, it is the opinion of the people, who will have to look at his work day after day, who count. The good news is that those people are as likely to be without an opinion as not, and can be swayed. Even some of those who do have an opinion can be swayed, unless aesthetics is just their excuse to hate you. So, here are some guidelines.

People think poles look better than latticed towers but they do not realise that, while poles do present a cleaner look when they are close by, at a distance latticed towers are very hard even to see while poles can stand out like long, bright neon sticks. In other words, the visually pleasing nature of a structure depends on where it is relative to the viewer – close or distant. The aesthetic value of a structure depends on its simplicity. A large pole or a large latticed tower can be clean or cluttered with appendages and details.

People want a structure to look like it is doing its job successfully. Long parts or members, such as cross-arms or the entire pole itself, should not appear to be losing the battle with gravity. Pre-cambering poles and even long arms angled away from the applied load are appealing. If these are bent towards the load, as if pulled there by the load, the result is not visually appealing.

The top diameter and taper of a pole should be limited. Limit taper (the reduction in diameter with increasing height) to less than 1 : 30. An economical pole is a thin-walled pole with a large diameter. If you seek a thin pole, you will pay extra but it can look much nicer. Poles shaped like pyramids and that block the sun or are large enough to allow an eagle to nest on top are not attractive. They are a sure sign of inexperience. Given the chance, hide the poles behind natural features such as tree groves or hills. For up-close towers, use landscaping to hide them at ground level.

There is a lot of silly work going on to make towers be sculptures. The result seems very structurally inefficient, and therefore costly. Or it may be just costly – some art is not particularly durable in the fashion sense, or is not likely to please many, unless a marketing campaign can be structured to reward you for trying.

Do consider that you are placing your entire project on someone else's land and playground. If you know that you are not particularly aesthetically minded, you should try to be 'kind to their eyes' (an architect's phrase). I have one pet goal with aesthetics, and that is to apply the Fibonacci series (the divine proportion ratio) to my tower dimensioning. A structure that employs this divine proportion in its shape is naturally pleasing to many people, and they won't even know why. Try it!

Design load cases

Consider environmental loads, construction events, maintenance events and post-failure events. Here are some suggestions.

Weather events:

- code required case(s)
- extreme ice (or snow)
- extreme wind (synoptic)
- combination of ice and wind
- microburst, gust or tornado event on the tower only
- unbalanced ice (ice shedding or rime ice on inclined spans)
  - include a temperature for each of the above
- moderate or minimal ice and/or wind for extreme situations
- hot summer ambient temperature
- warm winter ambient temperature (when snow can still be deep)
- everyday condition, no wind
- vibration temperature, no wind
- annual average conductor temperature for creep
- galloping temperature (with ice?)
- stringing temperature range
- construction weather range.

Construction and maintenance events:

- loads on structures and wires during erection and stringing
- wind loads on towers with wires not yet strung
- less than all phases attached
- structure lifting loads when lifting by crane or helicopter
- personnel and equipment on structure (or wire) loads
  - rigging points
- wear points for components that can move back and forth cyclically.

Post-failure loads:

- This issue was addressed in Chapter 7. Ask yourself the question: 'What will happen if (name a part) fails, and what do I want to do about limiting that damage?' At the very least, think about containment.
- If you can ask a 'What if...?' question about enough appropriate line components, and answer the damage-control question, then you will develop a few load cases dedicated to controlling failure events. This is the source of broken wire and most other longitudinal load cases.

The point of the above list is simply to have you understand that there is much more to designing a line than doing what the code tells you to. The load cases that are suggested by most items in the list are to account for the realities of what goes on out there when you build the line, maintain it, watch it age and have a century of weather beat on it. Remember the point made earlier – we design for certain load cases and events but lines fall down or misbehave for other reasons. The reason for the more exhaustive list of things to be concerned with is to try to capture these other reasons for trouble occurring. Use the weather and construction and maintenance load cases to capture these events, and then use the post-failure loads as a stop-gap process to mitigate cost-effectively the consequences of having failed to capture them fully.

The actual version of the list and the detailed description of the line items will vary depending on where you are in the world. In other words, a list from another country, or climate or era may not be the best guidance. Think!

Strengths versus deflection

Some tubular poles, whether of wood, metal or a composite material, that are long and thin and lightly loaded may meet all the strength requirements but be very flexible and 'whippy'. Such structures draw the (negative) attention of neighbours and linemen. Typically, steel poles designed via a competitive bid process are thin walled and have a large diameter. These designs win in the bidding by weight, as steel is priced by the pound, more than on any other basis, when options within the narrowed parameters of a project are used. If for any reason you want to limit the diameter of pole bases, you need to ask for it and you may pay a premium for it.

If you want a pole that is not easily deflected at the top, you need to ask for this specifically. Competing designers will not offer such in their bids if this is not specified. When you design for deflection control, remember the following:

- Use 'everyday' weather. Nobody looks up when it's terrible outside. Specify a modest deflection for modest weather, and express it as a percentage of the pole length. A deflection of 2–4% of pole length is reasonable.
- Ask for camber to be built into corner structures. People get concerned when a pole is bending into its load. They like to see it winning the load battle by leaning away from the load (a little!).
- Never use factors other than 1.00 on loads used in a deflection load case. This, of course, means that the deflections that a computer calculates when the loads are factored are wrong and overstated. Ignore them.
- Don't sweat the details. This whole deflection-management gig is 'approximate'.

Many codes require that the deflection of structures be considered in calculations of clearance to adjacent items such as buildings or the edge of the right-of-way. It should be obvious that the deflection of structures is not calculated accurately, especially for wood poles. What should you do? The best answer is be conservative, but also simply to do something, because that will satisfy the powers that be better than doing nothing.

Safety, performance and survival

You should be aware that the load cases you select and the scenarios of loadings and failures you formulate are associated with certain issues that are not all the same. Some are personnel and public *safety* issues. Some are system operating *performance* issues and others are structural *survival* issues. Each of these three labels can vary between organisations, and you should settle on classifications that work for you. The International Electrotechnical Commission (IEC) through its standard 60826 (IEC, 2003) offers widely held definitions of safety, performance and survival.

So-called 'code' loadings are usually safety issues that are defined in the code's statement of intent or purpose. Other construction and maintenance load cases are clearly *safety* related, in that personnel are at risk.

The remaining load cases tend to be *performance* related. The distinction between performance issues and survival issues is whether the line is operating normally or not. There are ice, wind and temperature combinations, and other events, through which you intend that the line remains in operation.

You may want to adopt the philosophy that, while there are such weather events that should not interrupt normal electric operations, there are also more severe and less frequent weather or catastrophic events during which you are willing to sacrifice electrical operation but do not want to suffer a costly structural failure.

The point of making the distinction between these three issues is to assist in developing a more comprehensive and cost-effective set of design criteria, by assigning loads and strength combinations and buffers that are appropriate to each issue: safety, electrical performance and structural survival.

Designing for failure management

We hope, as you can see from the failure stories in Chapter 7, that the focus of typical design exercises is not on matters that actually impact the performance of a transmission line. We have a saying: 'We design for certain loads and events, and lines fall down for other reasons.' Synoptic events and basic component characteristics (tower members, conductors, foundations) are not what cause most line failures. To be sure, synoptic events and the nature of basic components need to be considered.

To truly succeed, engineers must be aware of and deal with matters such as unique and incidental matters, that is, anomalies and freak events, matters *not yet known*, construction decisions and their effects, maintenance programmes/practices, material properties in detail, limits of design methods, and so on. As you can imagine, this level of knowledge and effort is essentially an impossible expectation to place on you. Such knowledge improves over time and with experience but, even so, the oldest and best of us are still and always learning!

There are people who actually believe that, at some designed strength, all failures will be prevented, that is, no failure can occur if the line is strong enough. The reality is that, even if every structure were an Egyptian pyramid and every conductor was supported every 50 m or was insulated heavily and buried in a trench, or if every customer had a power plant at his location, *something* would eventually cause the lights to go out. You might as well ask that vehicles and the highway system that they travel on be such that there is *never* a crash – ever! Some things are unachievable. Everything that humans engineer and construct is done with the understanding that it has some risk of failing.

We have also discussed that the engineer has only a very modest ability to affect the cost of a reasonably well thought out project but has the ability to impact the cost of the eventual failures significantly by addressing the kinds of things we have mentioned throughout this book. Failures can be costly in several ways: loss of life, loss of a valuable power delivery and extensive structural loss, and even loss of reputation. Each of these costly failure types can be mitigated by the design – not entirely but the risks can be reduced. The question remains: Is the designer given the freedom to attempt the mitigation, or is the designer stopped by other interests? The answer is: freedom is rare but to be fought for.

Mitigating loss-of-life events

In general, reducing the number of failures or the extent of a failure reduces the chances of a person or persons becoming entangled in the failure. The greatest chance of a loss-of-life event occurring is during construction, when trained and untrained people are on, beside, over and under partially constructed facilities. The engineer's role in reducing loss-of-life events lies in choices of material strength and quality, understanding all manner of construction loads on the materials, and informing the construction people about the loads and stresses that are imposed on line components by their actions. Only extensive time spent with construction crews will bring useful knowledge to the engineer. Get out there.

> That said, I can tell you the lesson learned from my roommate in my final year at university. We graduated in May, and he went to work for the railway company where his father had worked for years and where he had worked every summer up to this time. One day in September, 4 months after graduating and 3 months after getting married to his long-term girlfriend, Norm was standing alongside a railcar that was being unloaded. He was just a young engineer who was on site to learn and play his role. The hook on the load let go and the free end of the lifting cable flew through the air and smacked Norm in the midsection – hard. He bled to death from internal injuries. A charming and talented life ended.
>
> Since then, I have told all young engineers who go to a construction site to be very careful and learn the ways that construction workers stay safe. You are not paid to be in harm's way.

Learn what 'in the bite' means and why it matters never to stand there. Learn about where things are going to go when they fall or start to roll, and don't be anywhere near there. Learn that you are a danger to others when you don't know the environment and others don't know who you are, where you are or what you are doing. The most terrible, and even life-changing, feeling that you can experience is being responsible for harm to another person.

Mitigating loss of valuable power delivery

Some transmission lines are an essential connection between a generator and a customer. If the connection is not in a serviceable form the customer's economy will be seriously curtailed or terminated. If a line delivers 300 MW of irreplaceable energy at $80/MW h, every day that the line is out of service amounts to $576 000 in lost revenue. If that lack of energy curtails the customer's business production, puts the company at risk of shut-down, or does shut it down for the outage duration, and if that business is the heartbeat of a town's economy, the cost of the outage is staggering and practically incalculable.

There are two things to do with lines delivering a very valued service. First, make them stout against the environmental loads, and, second, avoid cascades and other very expensive events at all cost. Structures in very remote locations should be the most secure. Series of structures in remote, expensive to access locations should have foundations that will not be damaged should the structures fail. At the most basic, the need is to avoid cascades.

> The Kemano–Kitimat double-circuit transmission line in British Columbia, Canada, is one such valuable line. More than 50 km of its 80 km length is supported on double-circuit latticed steel towers that traverse two narrow valleys in close proximity to avalanche paths and fast-moving rivers. Several of these towers have failed for one reason or another but the line has never cascaded. Why?
>
> The line was designed for quite heavy ice, which means that the conductors are strung quite loosely. The line is also in a location with low lightning-strike levels and it has overhead shieldwires only at its few end spans. The saving grace of the line is these two features, despite the designer's frustration needing to add more towers along the way due to the low tension of the conductors.
>
> The remaining near 30 km of these circuits traverses very high and very rugged mountain terrain. The design ice load was doubled, the conductor size was doubled, span lengths were often doubled, and each circuit was placed on separate structures. This part of the line has suffered six separate structural failures over its more than 50 years of service life. None of these failures caused anything but minimal damage to any other structure. That is an extraordinary record. Why?
>
> Again, the very large conductors are strung quite loosely and have long spans, and there is considerable slack in the spans to allow large displacements to occur

without a damaging increase in tension. But, in addition, the towers are simply very strong against vertical and longitudinal loads, and thankfully so.

Understand which design features create security and which trim the fat from the meat and leave little room for error, and thus little room for accommodating the unforeseen. This is particularly important for facilities on which the economic security and safety of many people depend.

A reminder on anomalies

An anomaly is a deviation from the common rule, type, arrangement or form.

We have said, and we hope we have shown you, that anomalies in a transmission line are necessary both to trigger a failure and to stop that failure. So, you need them and you don't need them? No ... you need to *understand* the behaviour of a transmission line so that you can recognise the problematic anomalies and can put the beneficial ones in place as part of your design. Let's review some common anomalies and talk about them in these terms.

Corners are anomalies; that is, they are a potential source of 'slack' if the corner fails. When a corner structure fails, it dumps an enormous amount of slack into its adjacent spans. The demand arising from the increased tension put on the adjacent structures is severe. The optimum line angle, where the increase in tension is offset completely by the correct amount of slack being introduced, is somewhere between 5° and 10°. Larger angles can reduce line tension immediately and totally.

You should now understand the hidden danger built into corners. The next question is: Why would a corner break? A deadended corner can only break and produce the foregoing scenario if the structure or the guy system (if present) fails. As most deadend structures (angle or tangent) are designed to withstand the highest tension loads expected, this should not happen unless there is faulty material, excessive corrosion or wear, a design/construction error or a natural disaster. But that is a healthy list of sources of trouble. The rule for corners is: give them structural integrity because they matter.

Deadend or strain structures are anomalies unless every structure on the line is one of these. Strain structures simply have more components on them, especially hardware that can develop problems. The same rule applies as applies for corner structures, but there is more to it.

Our first rule for deadends: never use dummy deadends! If it looks like a deadend it had better be one, at least for load conditions applicable to the presence of personnel. Recall that the extensive carnage associated with failures is usually caused by the fact that the conductors don't break and that they haul things down. If problems can start at deadends simply because there are more components on the deadend that can be a source of a failure, then up the ante at the deadends. When deadend structures and their complex hardware assemblies are a low percentage of all structures on the line, it is not expensive to spend a modest amount there to ensure structural integrity.

Our second rule of deadends, and for that matter running corners that could act as deadends if asked to do so, is: make the guying, hardware and insulator assemblies that support conductor and shieldwire tensions as strong as or stronger than the breaking strength of the wires they support. If a failure triggered elsewhere on a line wants to challenge the strength of a deadend or selected running corner structure – which, by the way, being an anomaly is therefore an opportunity to stop the progression of the failure – then make the failure break the conductors. For the sake of a few dollars spent on the next category of hardware and insulators, don't let them be a structural fuse that gives in before the conductors are fully challenged.

Long spans within a series of shorter spans are anomalies. Chapter 7 eluded to the problems that long span anomalies can invite. It is because they contain an unusual amount of slack, and the real problem occurs when that slack is transferred to adjacent and shorter spans. In this computer age, it is fairly easy to explore the scenarios of long span and short span interactions under extreme cold or hot temperature events or uneven ice load events.

Related to the long span anomaly is the supplemental aggravation of placing long spans on *inclined spans, which themselves are anomalies* on many lines. Except in tropical locations, consider the loading of high spans with rime ice, especially when coupled with the long spans as described in Chapter 7.

The engineered features of a long span in Tacoma, Washington, USA, were all aimed at addressing the anomaly that the span itself was, and mitigating possible failure events. Figure 8.1 illustrates most of these features. The single conductors for this 230 kV,

Figure 8.1 Failure-mitigating features of a large water crossing

double-circuit, 1900 m span between its very tall towers were very strong, having a rated strength of just over 68 000 kg. They were continuous conductors with no splices permitted between the deadend structures. The total conductor lengths were 2800 m per phase.

In this line, every conductor was terminated on its own guyed mast, and the masts were positioned such that the loss of one mast or the breakage of the insulator assembly on it would not impact and jeopardise any other. The suspension arms of the tall towers were staggered outward, with the lowest arms being the shortest. This allowed, in the case of a failure of a suspension assembly or a mistake during construction or maintenance, any 10 ton phase to drop from the tower without striking the support arm of the phase below. As a result, the tall towers were designed for only one broken phase. Before installation, every insulator assembly was tension tested to its maximum design load to root out any faulty components. Each assembly used a three-bundle of 25 ton insulator strings, even though the maximum design load on the span was expected to be only 40% of the conductors' 75 ton strength. All these unusual features were rational choices and considered cost-effective because they applied to only one span and the cost of its failure would be enormous.

Controlling failure sequence

If you accept the fact that a line can fail and you want to manage or control that failure somehow, then you will want to give some thought to what might fail first, what the consequences will be versus what you would like them to be, etc. Here are some guiding thoughts.

1. Wire systems are generally very strong and will not likely break first or ever. Just as easily as they will hold a system together, they will tear it down.
2. Recognise the anomalies, and design them not to be a source of failure *and* to be a stopper of a failure, that is, make your corners strong!
3. Don't depend on things breaking in any particular order if their strength coefficients of variation are large or not understood/known.
4. Let arm systems fail first to save whole-structure and foundation replacement costs.
5. Have foundations fail last, especially when the line is located in a remote location and the replacement of foundations is a time-consuming and costly task.

It is said that those who design for failures in some way – it's not important exactly how – don't experience them, while those who ignore the issue do.

Failure containment

As it is often very uneconomical to design each structure to handle anything close to the magnitude of the potential longitudinal load, even at everyday tensions, a cascade potential of some distance is inherently acceptable within the line design. This is true for all free-standing high-voltage designs – single poles, steel lattice lines and, particularly, H frames. H frames are most susceptible because the capacity of the span against wind is easily increased with no meaningful accompanying increase in longitudinal strength.

If such a line design is economically attractive, then the best protection against a costly cascade failure is the periodic insertion of a structure that can withstand the necessary loads. This is 'failure containment', and has long been a popular concept. Structures located between containment structures, and perhaps the containment structures themselves, depending on their nature, are there to be sacrificed when there is a failure in their 'zone'.

The first thought is to say 'stick in a deadend structure every so often'. Depending on the line, a full-blown deadend, with all its associated construction costs and inherent anomalies, may not be necessary. A stout suspension design can provide reasonable security. A light or medium angle structure that can carry the loads of a light or medium angle may also have the geometry and strength to withstand reasonable longitudinal loads. This is not an idea, it is a well-used tactic. Use corners as opportunities to provide containment strength. Make use of swinging insulators to reduce load in a controlled manner, if the structure design permits.

Guyed structures become very economically attractive for any location where real-estate prices are still reasonable and landowners allow. Guyed structures come in all shapes and sizes. Some are wonderful and some are not. Some of the good ones are, or can be, perfect containment structures – each and every one! These lines are inherently awesome in the face of a failure. If you get into high-voltage and extra-high-voltage line design, get to know your guyed structures.

Analysis methods

Twenty years ago, there seemed to be very little guidance on how a wood pole structure should be analysed or designed. Well, we have been saved.

Power Line Systems (PLS) has written a variety of computer programs for power-line structures, including PLS-POLE and PLS-TOWER PLS-POLE, which can analyse wood pole structures, concrete pole structures and steel pole structures. The program relies on libraries of material properties that can be edited. Therefore, it can be made to analyse other metals such as aluminium and fibre-reinforced plastic. PLS-TOWER is written to analyse latticed towers. Both programs are tailored to the electric utility industry, so the analysis methods and calculation formulas within the programs are specific to the industry, relying on industry standards such as those of the American Society of Civil Engineers (ASCE) and IEC for direction. In other words, the programs will not analyse a building or bridge properly – only power-line structures, and only in accordance with the software manual's description of the calculations.

It is necessary to study and understand the different nature of steel pole analysis compared with that for wood or concrete. Each material or product line is viewed, defined and assessed in its own way. This is a classic example of having to understand what the black box is doing. Otherwise, you will not be its partner in a successful outcome. In effect, the computer always assumes that you know what you are doing because it knows that it does not. It is just cranking out numbers for you.

These programs analyse, but do not fully design, single-pole and multiple-pole structures, both guyed and unguyed. They are widely accepted within the industry, and are reasonably priced and well supported. The programs contain enough analysis method options that users need to be aware that intelligent choices within the programs are necessary to produce the desired results.

The PLS programs offer the option to use linear or non-linear methods. But, be aware, the only reason why the linear method is provided is so you can compare the computer's results with your hand calculations that you filed away years ago. The linear method is not there to be used as a tool for present-day work. The difference between the results of linear and non-linear methods for flexible, indeterminate structures such as tall poles guyed in several directions by a multitude of guys can be very significant, and the non-linear method will not be the conservative result.

Latticed towers are trusses with minimal or no frame or beam components. They are much stiffer than pole structures, which are primarily frames with some truss components. Their deflection is relatively small, and linear analysis results tend not to differ much from non-linear results. This is true for self-supported latticed towers but not when guys are involved. Guyed towers and towers with internal cable elements demand the use of a non-linear analysis method to produce useful results. Remember the point that it takes strain to produce stress, and linear analysis methods ignore strain.

Just as the PLS programs are written specifically for the electric utility industry's methods for designing and analysing power-line structures, other programs are not. It is possible to analyse transmission line structures using software designed for general structures but you had better understand the differences in the calculations within the generic software to know whether you are remaining compliant with the industry's way of doing business.

> We once ran an exercise to compare the results obtained using PLS-POLE and a respected vendor's own design software. That vendor had designed the poles used in the exercise, and we wanted to know if PLS-POLE understood his product as he did himself. It did not. The two programs differed by about 10%. The reasons for the difference were subtle, and neither was in error.

In this business there is always a list of good reasons why you should never rely on the results of calculations to better than about 10%. This means that if you declare your answer to be correct and a competing answer that differs by 10% or less to be wrong, you are wrong.

Design tools

In case you have not noticed by now, this book does not bear down on detailed descriptions of how present-day software and other tools for doing our work should be used. We offer few exact calculations to describe the way something must be done. To do so would shorten the useful life of the book to the useful life of those ever-morphing products and processes. What we offer is a description of what is important. In fact, we ask you to do the research to find out how calculations are best made at any given time. If you are interested in the work, you will do that and be rewarded for it.

Design tools are transient but two things are not – the challenges and you. The very best and essential tools that you have for dealing with the relentless myriad of challenges that the business brings are your brain, your curiosity, your enthusiasm and your colleagues. Without 'you', the technical tools and processes of the day are useless. And, unless you are Einstein, you will flourish only when you collaborate with others of your calibre.

Thayer (1924) is a great example of the fact that the subject matter that we write about in this industry is, to some degree, unchanging. We write and we write. So many papers presented today at technical conferences break so little new ground. Have we got it all figured out, or are we stymied? There is a lot of room in this industry for bright, energetic minds to make a difference and in so doing achieve great satisfaction.

The title of this little section tricked you, didn't it? Go get 'em, tiger! You too, tigress!

Line layout (structure spotting)

Once the line designer has a right-of-way plan and profile, structures and a set of criteria to follow, he or she has the pleasure of travelling (mentally) down the right-of-way, placing structures and stringing wires. This little chore is one of the sources of gratification for the designer. It involves a stream of *important* decisions, and it feels kind of like you just built the whole thing by yourself. It is a fun chore!

I highly recommend that the designer become as familiar as is economically possible with the real right-of-way before building the line at his or her desk. You ought to consider at least one trip down the alignment with experienced eyes before doing the work. The knowledge will pay off.

A golden rule: don't ever design a line for a right-of-way that you have not actually seen. You must go to see the alignment to capture in your mind and memory the essence of the place. If you do not do this, you will have an incorrect rapport with the place and you will get things wrong.

Goals and constraints

The goal in selecting and spotting structures is to meet or exceed the minimum requirements of the design criteria at a minimum constructed cost. It would be nice to include some measurement of minimum lifetime operating cost in the process but the opportunity for this tends to be defined by the content of the criteria.

We also point out that it is far easier to simply measure the material costs of a line than it is to include the labour or construction costs. However, to omit labour in a cost assessment of a design yields a false measure of the cost of the line cost because the labour cost often outweighs the material costs, and the two are not in a constant ratio along the line.

If you are using a computer program that includes a cost function to help design the line and that cost function speaks only of materials, then you might consider throwing some labour costs into the fray under the guise of 'material assemblies'. Call it a 'loaded material rate'.

The constraints imposed on structure spotting are topics outside the design-criteria type topics. Simply put, they are a list of places where structures cannot or should not be placed. You *cannot* place structures on roads or railway tracks, etc. You *should not* place structures in places that are difficult and expensive to access, that require horrendous foundations, at the base of large hills, near river banks, and so on.

Conversely, you must, by definition, place structures at PIs (corners) and, given a choice, you should locate your PIs with the criteria mentioned earlier in mind. Implicit to the requirement that a structure be placed at a PI is the powerful suggestion that the next structure be a rational distance away. When you set a PI location, keep in mind where the next structure is likely to be as a consequence. This is a clear example of the point: no structure can be efficiently located in isolation because the choice affects the quality of the next structure's location.

Two approaches

There are two very basic approaches to line design and layout. One is to place structures of undefined strength wherever you wish or must, then record the resulting wind and weight spans and line angles created, and then design the structures to fit that layout. The other method is to design a rational family of structure types and place them on the alignment where they work, that is, where good use is made of their span and line angle capabilities. The latter method is the one necessary to achieve an optimised design when using the conventional computerised approach.

The first approach, in which structure wind and weight spans are not input values but are discovered during the process, only works well if you are going to use structures that do not have inherent span limits. The approach works well when using fibre-reinforced plastic or wood poles that are readily available in a full range of lengths and strengths. The approach works poorly when the poles are custom-designed steel poles or latticed towers, as both these structure types are only efficiently produced in large quantities and on demand. Steel poles and latticed towers lend themselves well only to the second approach, in which the wind and weight span limits are known beforehand and are input to the process rather than discovered through it.

Wind span uniformity comes from equalising spans, which becomes increasingly difficult as the ground profile becomes more rugged. Weight span uniformity is achieved by levelising the wire attachment points on adjacent structures, which also becomes increasingly difficult as the ground profile becomes erratic.

The message is to keep the wire profile level on vertically variable terrain by varying the height of each tower. Think of it as setting the structures' wire attachment points on a smoother profile than the ground profile at a practical distance above the ground, and then calling out the structure heights so that the base of each structure reaches to the ground, wherever it may be. This is best done with wood or fibre-reinforced plastic poles, and latticed towers with an array of available leg extensions. It is difficult and costly to take this approach if using steel poles.

Figure 8.2 Tower spotting before the computer age

Until the early 1980s, the only way to spot a line was by *hand* on paper using templates (Figure 8.2). Then, the computer program TL-CADD by John Bates came into its own and became very popular. Around 1992, a program by Optimal, England, became functional. However, it has not developed a large user base in North America. In 1994, Alain Peyrot's program PLS-CADD matured, and its use has been flourishing around the world, relegating most of the other software programs to the history books. Designers who are still working on paper with templates are nearly as rare as dinosaurs.

All the computer programs provide the option of optimising the line by selecting a combination of structures and locations for them that achieves the lowest cost of all options. Unfortunately, the results are only as good as the information input, and that information, we will argue, is suspect at best. They also do not consider wire tensions other than the single value you provide. They do not consider PI locations or alignment adjustments. Iterations on wire tensions and alignment adjustments are useful to control the cost of a design but must be managed manually.

Another gentle reminder is that the straightness, or lack of straightness, of an alignment should, in principle, define the optimum wire tensions for the line. If the percentage of structures – such as corners and deadends, which are defined by wire tension, and structures, which are defined by broken wire load cases – is low, the line tensions should be tight to keep spans long and/or all the suspension structures lower in height. The savings at all the tangent and suspension towers will be greater than the higher cost of high tensions at the few towers defined by tension. If the percentage of these types of structures is high, the line tensions should be lower to reduce the cost of all those expensive structures.

Survey data

Every time I turn my back the surveying business seems to undergo a technical revolution. Today, there are helicopter-borne lasers, GPS data, Google Earth, satellite-based DTM files and the like, and young engineers might assume that it is necessary to have such data available as the 3D platform on which to design a line. This is not so!

> Back in the 1990s, we engineered the renovation of a transmission line that ran across very rugged ground in southeast Alaska. There was timber debris up to 3 m deep on the right-of-way because the very large trees that had been cut 10 years earlier for the original construction of the line were never removed from where they fell. The ground was so poorly definable that you could hear small streams gurgling away under your feet, although you could not see them.
>
> The vertical information for producing our profile came from aerial photographs and stereoscopic tools used by an equipment operator who I am sure went cross-eyed. I trusted the profile so little that our drawings include no elevation, and we used a vertical buffer of 1.5 m in our clearance calculations.

Survey data of any quality can be used as the basis for a project, and the essence of the process is to pair the quality of that data with a suitable buffer to account for its nature.

It is necessary that the surveyor you employ understands what kind of data you want and why you want it (i.e. what you are going to do with it). If you send someone into the field, it is useful that the person knows why the information is being gathered. It improves the chances that the information gathered will be appropriate. When you send a LiDAR provider into the field, the same applies. When you get data for free or on the cheap in DTM form from the internet, it is useful to understand its accuracy and its ability to help define the quality of the engineering that is done on that platform.

Presentation of results

Part and parcel of the computerisation of the line-design process is the spotting of the line ‘on screen’ rather than ‘on paper’. On screen we can zoom in and out, look in 3D and plan views, and call up a plethora of data. When working on paper we had to draw the blank plan and profile to a scale that allowed us to discern things and make decisions with suitable accuracy. If you owned a set of sag templates, there was incentive to draw all projects at their scale.

Now that we don’t engineer the line on paper, it can be drawn (printed) at any reasonable scale we like, or not at all. It is possible to produce a construction package for a line that has no plan or profile drawings at all but that relies on structure lists, material lists and the written technical page. Some people do this. The point is that you might think about ‘letting go’ of certain drafting standards that apply to things such as plan and profile drawings. On the other hand, ask what the owner and contractor expect.

Optimisation

> I recently requested a colleague not to sweat the third significant figure in his calculations even though such finely tuned values had been provided to him as calculation input. He is a wise fellow and responded with a great comment: 'Not to worry, I will not polish the cannonball.'

I am not a fan of line optimisation, as it is known and shaped by the computer software vendors and in technical dissertations in the industry. The obvious objective is to lower the capital cost of the project, because that is how the value of the solution is measured. When you consider that conductor choice, conductor tension and alignment adjustments are not part of the automated process, that labour costs are both poorly knowable at the design stage and poorly represented in a calculation, that nearly 50% of a project's capital costs are controlled by matters beyond the control of the design engineer, and that operating costs are not part of the equation, you will recognise that to optimise in this fashion is to 'polish the cannonball'.

That said, in 2014 CIGRÉ will to publish a technical brochure by Dr Rob Stephen of ESKOM in South Africa on the subject of line optimisation. The document will describe a much more useful and higher level, bigger picture approach to optimisation than has been in place for the last 40 years. Once instilled in an organisation's work methods, better transmission lines will result. Look out for this publication, and watch it take hold. Become a champion of it.

Contracts

Remember what we said about the impact that the engineer can have on the cost of a project by addressing the technical content and issues. It was not often very much. A poorly written contract or a contract of an inappropriate type will have a much greater impact on the cost of a project. Why is it that, after all these years, we have not settled on good language and a good form of party relationships, and understood responsibilities for the work that we do? Why is it that we are still fighting to plug the loopholes and contain problems with new language requirements? Why is that we have a seemingly endless stream of train-wreck stories about project cost overruns, failed schedules, personnel replacements and the financial ruin of companies as a result of the approach taken on a large project? One of the reasons is the nature of the contract format selected.

Goals

> I once spent the day travelling around a potential job site with a contractor. As we drove along, he told the story of a bidding process he'd just been involved in called an 'online reverse auction'. Bid values were posted for a project to a website, and all bidders could see the last, and presumably lowest, bid. If you wanted the work for even less money, you needed to post a lower bid. Two things happened. First, this man's contracting firm figured out that it took about 20 s for a bid to post for others to see it. He wanted the work, so he waited until about 30 s before the closing deadline and posted his price just below the last one. No one else had time to respond. He won the work.

> Later, one engineer who helped run the bidding process asked how he liked the process. To the engineer's surprise the contractor's response was 'No good at all'. The contractor explained that contractors are in the business of taking risks, and are willing to lose on occasion in an environment where they are also allowed to win big, hopefully more often than not.
>
> Therein lay the message. If the owner and his engineer – folks who do not typically have risk-taking – offer a process to bidding contractors that denies them the possibility of winning big, while other owners with other projects leave the winning big opportunity open, the contractors will not play with you. They will go to gamble with the other guys because they tend to be gamblers at heart.

The owner wants value and no financial or performance risk. The engineer wants value. Both often confuse low capital cost with value. The contractor wants profit. The contractor needs profit to stay in business, because next week he may lose a bet. There are honest and capable contractors and there are contractors who are less so. Honest and capable are not traits that are necessarily joined at the hip although being capable allows one to be honest and being incapable can breed dishonesty. When you find an honest contractor, cherish him. When he is honest and capable, use him. Let the others learn elsewhere. The cost of cleaning up the work of an incapable or dishonest contractor is far too high to consider their presence on your project. Unless, you too like to gamble.

Our office once found itself without a decent specification for the design, fabrication and erection of latticed steel towers. We called a variety of sources and received their offerings. All appeared different but, amazingly, all contained a common clause – to the letter. We began to suspect that, once upon a time, some genius wrote a very nice specification and we have been abusing it ever since. Wouldn't it be nice if we could all come closer to speaking a common language with bidders and suppliers. We might gain something. In our business, there is no such unifying force.

Contracts needs to be written in a certain language. You need to cover all topics, you need to provide an 'out clause' for all parties for all obvious and not so obvious scenarios that might occur. The more events on which you can place a known price, the less argument you will have from the contractor. If the contractor starts to lose money, he is not your friend until you pay.

A useful point to understand is that it is at the points of interface between parties' responsibilities where contracts run into trouble. So define the responsibilities at party interfaces very clearly. Obviously, the unknowns are sources of trouble. Think subsurface conditions and undefined work. Every interface and every unknown is a place where the contractor can insert his pry bar and start digging at the owner's bank account. Manage these relentlessly.

Writing good contract language to achieve a successful outcome is both an art and the result of a dedicated education. Try thinking like the contractor!

Types

It is a widely held view that the engineer–procure–construct (EPC) contracting format puts all the risk related to interfaces, the unknowns and the unknowable onto the contractor, thus shutting down the pry bar activity. Good luck with that.

> I asked a contractor if he liked EPC contracts. This individual said 'no'. He claimed they don't care to engineer the project and they are not particularly interested in the procurement duties. They like to build things.

At the top management level contractors do not mind EPC contracting as much because risk taking simply means adding more money to the cost of the work. If the owner does not want risk but does not mind the cost of the work as long as the pry bar does not come out of the contractor's toolkit, all will be good. But, understand this, transmission line projects are lousy candidates for EPC contracting. This is because large projects have inherently large unknowns compared with large projects in other fields of endeavour.

Consider this comparison. A long transmission line project consumes about the same amount of land with its right-of-way as a major airport consumes for its use. Suppose the transmission line were constructed on the airport property – back and forth, back and forth filling the property with towers and wires. The risk to the contractor would then be minimal compared with placing that transmission line in its actual location. On the airport property, there would be practically no geotechnical risks. Access to each structure would be simple, and dealing with the single land owner would be a one-time challenge. In reality, the geotechnical unknown of a long transmission line cannot be understood cost-effectively, access is very complex and landowner issues can be staggering. That is why a transmission line project is a lousy candidate for EPC contracting.

Many owners don't take much risk, and the larger the project and the greater its cost, the more this is the case. The reason for this is that very big projects cannot be funded from the coffers of the average utility. They need a financial partner, and it is *those* guys who don't take risks with their investments and yet don't care what the cost is as long as they know the number going in. Why don't they care – sometimes?

The simplest way to get a lower price for a given package of work is not to place the responsibility on the contractor/supplier. Responsibility costs money! This is the distinction between a 'fixed fee' and a 'time plus expense' contract. The middle ground is the 'unit price' contract. That said, I expect that on every major project a discussion is warranted to help decide what type of contract is best and where the delineation of responsibility should lie.

Packaging

A contract *document* consists of the boilerplate, quasi-boilerplate, technical specifications and drawings. The actual names of these document parts vary between organisations. The so-called boilerplate and quasi-boilerplate come from management and legal counsel, who are rightfully reluctant to edit any of it, the technical specifications come from the engineer and is more project-specific, and the drawings often come from other parts or persons in or near engineering and are often a mix of standards and project-specific efforts. Therefore, the battle is to build a coordinated and complete package. Ask yourself these important questions:

1 Is *everything* addressed?
2 Are the points of interface all identified, described and managed?
3 Are the responsibilities for the unknowns assigned?
4 Is *anything* addressed more than once, and are the instructions consistent?
5 Are the drawings required in the field and are they suited to field use?
6 Is the written document required in the field and is it suited to field use?

These questions may seem easy to address but the words *everything* and *anything* cover a very lot of ground.

Boilerplate

'Boilerplate' refers to the general and supplemental conditions describing the responsibilities of the parties to each other. It does not describe the work, nor is it about how to execute the work. It tends to be useful project after project, and is typically not in the purview of the engineer to shape. Nonetheless, the boilerplate needs attention to attempt to make it bulletproof.

> We provided the engineering for a line project and put our estimate for the contracted work that was to be bid at US $4.8 million. The owner was a US state agency that was required by law to award on low price. We received prices ranging from $4.6 million to $5.4 million, and one of $2.8 million. In a rational world, we would suspect that the offering of $2.8 million was from someone who did not understand the scope of the work. He got the job.
>
> After a suitably short period of time on site, the contractor submitted a request for extra payment to the amount of $1.6 million. The claim was based on the nature of some particular language in the contract boilerplate, not the technical specifications. The owner's project manager was no slouch. He knew what was going on. He asked that we, the engineer, battle the contractor on his behalf to reduce the $1.6 million claim. Have you ever seen engineers try to battle a contractor? It is laughable!
>
> It was also painful. The contractor got his $1.6 million. The owner got the project for a cost slightly lower than the engineer's estimate. The engineer got a headache.
>
> The following year, phase II of the same project was put out to bid and another contractor was awarded the work. We had closed the language loophole in the boilerplate but this contractor found another one. It gets tiring.

Technical specifications

The technical specifications are the engineer's responsibility. Here you have to describe everything about the contractor's interactions with the right-of-way, the tower sites and the materials, from purchasing, handling storing and installing. The contractor may not know how to exercise these interactions, or may not want to exercise them as you want, even though he knows he should. Regardless of the reason, you must provide a complete set of directives on every subject imaginable. If you leave a matter unspecified, a contractor will someday squirt through the gap either intentionally or inadvertently, and when he does eyes will turn to you.

The good news is that the vast majority of the content of the technical specifications necessary to cover a project's concerns is applicable to project after project without editing. This means that careful attention to the development of a technical specifications package will, over time, produce a pretty good product. That's the good news. The bad news is that many organisations require their staff to multitask so much on a daily basis that attention to this specifications package development tends to lag and, well . . . never get quite finished. Give it attention, you will be rewarded.

Contractors, consultants and manufacturers

Two anecdotes: I have worked with a fellow who seemed to operate by the credo – if you want something done right, do it yourself! He was quite sharp but eventually 'crashed and burned' due to work overload, and the employees around him found it difficult to learn anything from him. In the first few years of my career, I worked with a few consultants, and quickly decided they weren't all that I expected them to be – now I am one. Karma?

Every contractor, consultant and manufacturer is a human being or a collection of human beings, and that is both their failing and their redemption. As people come in all manner of greed, grace and grump, and have a unique centre of universe, as described earlier, I suggest you search for individuals and organisations with whom you can have a satisfying relationship. The array of personality types available to you is staggering. If you find someone to whom you relate well, you can accomplish almost anything together.

Remember the personal and business entity requirement for a successful outcome – all key individuals and both companies must:

- See value in the relationship.
  - Is this venture something we want to spend money on?
  - Is this venture the way we want to earn money?
- Have the capability to execute their role.
  - What services do we need and want? Can we provide what the other party cannot?
  - Do we have the time and expertise to deliver?
- regarding
  - Can I put myself in the shoes of the other party, and am I willing to deliver what he/she wants and needs?

If everyone cannot answer 'yes' to all – and we mean *all* – these questions, you will have a less than stellar experience.

Their agendas

Companies, be they contractors, consultants or manufacturers, are required to be profit driven. Historically, not all utilities have been so driven by profit. Even if an organisation has been a profit-oriented utility, it may have operated, with relative safety, effectively as a monopoly in a service area. Accordingly, any contractor, consultant or manufacturer that has been around for more than a decade or so is probably quite efficient with its time compared with a utility. Sometimes the difference is astounding.

> I recently read an article authored by a senior manager of a major electrical utility that is actually a branch of a federal government. He claimed that his internal staff was undoubtedly more efficient than any consultant that they could hire. If that is true – and it may be – it is because that organisation has exercised a business plan that has intentionally or inadvertently made it so.
>
> In his case, the organisation has retained a significant technical staff and kept them well educated in their duties. They have a very bureaucratic approach to the business, thus reducing the incentive and ability of many service providers to be attracted to working with them. Finally, the culture in the organisation is such that outsiders are not embraced and informed as to how the work is to be executed, thus inviting failure when tried.
>
> There is no way that the internal staff is naturally more useful to the organisation. It is that way by choice.

You will find that consultants (and, sometimes, contractors) get hired for one of two reasons: they are hired for their expertise, or they are hired only to provide needed manpower. The latter is less enjoyable work, it is generally easier work and not worth as much to the client. It is easier to find consultants to provide man-hours than it is to provide genuine expertise. Not all consultants are deeply or broadly experienced. Select carefully, unless all you need is man-hours.

While not all organisations have valuable expertise to sell, it is not likely that they have two rate schedules – one for man-hours and another for brain hours. Shop carefully, and get to know the person who walks in your door.

The in-house staff of a utility undoubtedly know a great deal about their system and their policies and practices. They may well know very little about how business is conducted anywhere else, and their technical knowledge is quite limited. Consultants tend to be the opposite if they are any good. They may know nothing about your particular system, policies and practices but they can bring a lot of new ideas to the table. Both parties need to recognise the nature of the other, and be patient with knowledge transfer between them if a fruitful outcome is to be achieved.

Over time, you will encounter many failed relationships and the ugly projects that result. Hunt for and cherish the great ones.

Getting what you need

Every good contractor, consultant and manufacturer is a busy person/place. Be firm with them when you need results, and be lenient when you don't. Clients who act like squeaky wheels get the attention only because often there is a lack of hours in a day and tasks need to be prioritised.

Be fair, and communicate explicitly and clearly. Unless you have (unfortunately) hired someone who simply cannot do what is required (in over his or her head!), you will get what you need, and it will be provided happily and readily if the relationship is one of mutual respect.

It is our *modus operandum* that, even when hired simply to provide man-hours, we offer any knowledge that we might have on a subject, just in case the client might find what he didn't know that he didn't know is useful. We then consider it the client's right to take action on the information or not.

That said, we do think there are as many bright, intelligent and energetic individuals working at utilities as there are in consultants' offices. Conversely, there tend not to be many mediocre folks in consultants' office. We can't afford them!

Line 'life' after engineering

In recent years, considerable work has gone into writing computer programs that manage transmission line maintenance programmes. This is a very smart move – not because they are computer programs but because it is a plan to pay attention to the line after the design engineer walks away.

A line design engineer uses a set of design criteria to guide his work. Essentially, all matters described or defined in the design criteria are a snapshot in time. A snapshot doesn't look far into the future, if it looks ahead at all. Typically, a line operates for at least 40 or 80 years after the snapshot is taken. The designer is long gone and is powerless to look after his 'baby' in its morphing surroundings and environment change. Who will take care of the engineer's 'baby'?

Operations

Transmission systems grow, and as they do so 'member' lines play different roles within them. Operating practices change. Operators change. Owners change. Voltage changes, ampacity requirements change, the surrounding vegetation comes and goes, contamination comes and goes, maybe the annual weather changes. Conversely, the wires, insulators and structures age but don't change, and operating performance will actually or apparently change.

Maintenance

If you really want a transmission line system that runs like a Swiss watch and makes its owners happy for decades, then we suggest that the design engineers can do little more than give the line(s) a very good head start. To repeat an analogy – the design engineers are the hospital team that brings the line into the world and sends it home in a healthy condition. The real parents of the line are the maintenance team and maintenance programme that chaperones it through the next five to ten decades of its service life.

Accordingly, I suggest that the maintenance programme that you have (or don't have) is far more relevant to your line's success than is the initial engineering.

> I once helped my father-in-law assemble a steel panel door for a storage shed. We got it a bit wrong, by putting the translucent panel at the wrong height. But, it would have worked fine. I said to him, 'It's just a damn barn.' He responded with, 'Yeah, but it's my damn barn.' We reassembled the door.
>
> Some days, I think 'It's just a damn transmission line.' But more often I think, 'Yeah, but it's my damn transmission line.'

In this chapter we have tried to illustrate that you work among other people to do your engineering, and that you work with them, not in isolation. You cannot accomplish your work alone, both because you are unable to and because others will not let you.

It's fun and gratifying to do a good job. Embrace the chaos!

REFERENCES

CIGRÉ (2005) *Use of Corona Rings to Control the Electrical Field Along Transmission Line Composite Insulators. CIGRÉ Technical Brochure 284*. CIGRÉ, Paris, France.

CIGRÉ (2007) *Characterization of ELF Magnetic Fields. CIGRÉ Technical Brochure 320*. CIGRÉ, Paris, France.

CIGRÉ (2009) *Technical Guide for Measurement of Low Frequency Electric and Magnetic Fields Near Overhead Power Lines. CIGRÉ Technical Brochure 375*. CIGRÉ, Paris, France.

CIGRÉ (2011) *Electric Field and Ion Current Environment of HVDC Overhead Transmission Lines. CIGRÉ Technical Brochure 473*. CIGRÉ, Paris, France.

Davis DA, Richards DJW and Scriven RA (1963) Investigation of conductor oscillation on the 275 kV crossing over the Rivers Severn and Wye. *Proceedings of the Institution of Electrical Engineers* **110(1)**: 205–219.

IEC (International Electrotechnical Commission) (2003) IEC 60826: Design criteria of overhead transmission lines. IEC, Geneva, Switzerland.

Thayer ES (1924) Computing tensions in transmission lines. *Electrical World* **84(2)**: 72–73.

Structural Engineering of Transmission Lines
ISBN 978-0-7277-5988-7

http://dx.doi.org/10.1680/setl.59887.303

Chapter 9
Sustainable development

This book so far has been aimed at educating you on how things work and how to go about your work. It has dealt with things as they have been and how they are. Here, we offer a subject that will suggest that you do something novel. With this subject, if you should feel it is important, you have a very serious challenge and the opportunity to try really to get something new done.

> At a recent conference, I described the world in a certain way to have the audience understand its fragility and finite nature. I said that if the world was the size of a basketball, the air and water around its surface would be like a coat of paint on the ball about 1 mm thick. The Earth's crust that floats on the molten mass of rock that lies beneath would be as thick as the vinyl and rubber hide of the ball. All of the air inside the ball is molten rock. We live on the skin of the ball within a layer as thin as a hair.

Over the time that the Earth has existed, it has at different periods been covered completely in water, frozen solid all around, and hotter and drier than a desert, and we have had many ice ages blanket large portions time and time again. Our planet keeps changing, and is fully capable of changing again. While these changes take so long that you could dismiss the matter as not a problem, it is also clear that those changes that we tend to foresee and predict have a habit of taking place earlier than ever expected – to our naive surprise!

The global view

Living sustainably had been humanity's way from the beginning of time until about 150 years ago, when certain societies – most notably those originating in western Europe – took a different path. These societies headed down an unsustainable path when they stepped into the industrial age, figured out how to produce much more food per hectare of ground via seed genetics and fertilisers, and discovered antibiotics. These three developments permitted a population explosion on the planet unlike anything else in mankind's history. If you think that the way things are today is how they were through the ages until the late 19th century, think again. For many centuries, humans did not live unsustainably, and there is no reason to believe that we can continue with our current practices from this point forwards, because our present consumption rates of the Earth's resources has no long history to suggest our present path can be sustained.

Today, we find ourselves in the unpleasant position of admitting that our societies cannot continue to consume resources in the way that we are accustomed to without jeopardising our neighbours and future generations, and the Earth's plants and animals. We are recognising that the piper will be paid, and if we do not take serious actions quickly, the payments will be painfully high.

CIGRÉ, in its *Technical Brochure 340* (CIGRÉ, 2008), has suggested that the definition of sustainable development (SD) be:

> Development that will guarantee to the next generations the same levels of opportunities, welfare, prosperity and availability of resources that we have today.

There are some problems with this statement, namely:

Guarantee?	This is too strong a word. An acceptable path to a guarantee will never be envisioned. But, we can try.
Next?	There are gross inequities between nations and people within nations today. Fixing that is part of SD.
Same?	As technology and other factors change the world, there has never been a generation that had, or even wanted to have, the same opportunities, welfare and prosperity of a prior generation. The goal is to have it be better. The outcome is that it is simply different.
We?	The definition was written by persons who are reasonably well off and living in rich countries. It is a skewed perspective from which to write.

In simple terms, the definition assumes global continuity and seeks the impossible. We suggest a modified definition of SD. The highlighted words above drive the changes:

> Development that improves the possibilities for current and future generations to have a more globally uniform and ever-improving level of opportunities, welfare, prosperity and availability of resources than people around the world have today.

The United Nations (UN) and other organisations that operate at the global level have put their minds to SD. The UN published a report to standardise how SD efforts are reported by organisations. It is called the *Global Reporting Initiative* (GRI). In work by CIGRÉ (in its *Technical Brochure 340*) and the Canadian Standards Association – example international and national (Canadian) organisations that have also put their minds to the subject of SD within the realm of power delivery – we find the UN reporting system being used.

In this reporting system, SD requires addressing and making improvements to a long list of issues that fall under three main headings: societal, environmental and economic. 'Societal' includes looking at things such as labour practices, human rights and product responsibility, to name a few. 'Environmental' examines an organisation's impacts on living and non-living natural systems as measured by inputs (material, energy, water) and outputs (emissions, effluents and waste). 'Economic' examines an organisation's impacts on economic conditions of stakeholders at local, regional and global levels. When you look at attempting to change our behaviour so that our presence on this planet has a better chance of being a pleasant experience for more years than not, you come to realise that all of these issues do require addressing.

SD is not simply about consuming less energy and material. It is about seeing that the way we do things does no harm to anyone, present or future, and would have us repair the grosser inequities that have already developed. The charts shown in Figure 9.1 are from the *Living Planet Report 2012* (WWF, 2012). Each bubble is a country, and is sized by population. The further to the right that a country is placed, the more highly developed it is. The further up on the chart a country is, the more resources it consumes to be as developed as it is. The two bubbles that are high and to the right on the charts are USA and Canada. The two largest bubbles are India and China. Over time, most low and left bubbles are moving to the right and therefore upwards as well. This expresses pretty well the juggernaut of change that is taking place and the strain on resources that is coming. The array in the top chart places the spotlight on the enormous inequities between countries. Supplemental work in the World Wildlife Fund (WWF) report speaks to the inequities within countries, as per the second chart. Notice the shifts to the left of various bubbles. The magnitude of the shift from the position in the top chart to the same country's position in the lower chart is a measure of the disparities in human condition within that country.

SD actions are only meaningful if they address all of these disparities and the global excess resource and energy consumption, because serious inequities and excess consumption are both unsustainable. There is a lot to do if you care to address SD. You could say that the industrialised countries need to change their ways, and the developing countries need not to do what the industrialised countries did. In other words, the industrialised countries need to say to the developing world: *'Don't do it that way we do, develop a better way. Otherwise, you too will have to undo your ways.'*

SD within the power delivery industry

In the electrical energy business, most attempts to achieve sustainability are directed at the generation side of the business and at the demand side – consumers' use of that generated energy. Conference topics, government regulations and incentive programmes point us away from fossil fuels to less energetic and more technically challenging sources such as geothermal, wind and sunlight. We are trading smoke for mirrors. Couldn't stop myself from saying that! Consumers are prompted to trade in their gas-guzzling vehicles, turn off the lights in vacant rooms, turn down the heat, buy higher-efficiency light bulbs and turn off the TV in the spare room. One component of the energy enterprise that is not mentioned often is the bulk power delivery system between the generators and the consumers. What can be done there?

The classic answer seems to be 'not much', because losses (energy produced but put to no good use) in the bulk delivery system pale compared with the losses at the generator or within the consumer's domain. We offer the suggestion that this is not so, because there is more to this than energy losses along the wires. We believe there are significant gains to be made with the bulk energy delivery system. Those gains will require real paradigm shifts within the systems' owner, designer and operator camps, and that makes the subject worthy of attention.

Figure 9.1 The human development index from the *Living Planet Report 2012*. (Courtesy of WWF, 2012)

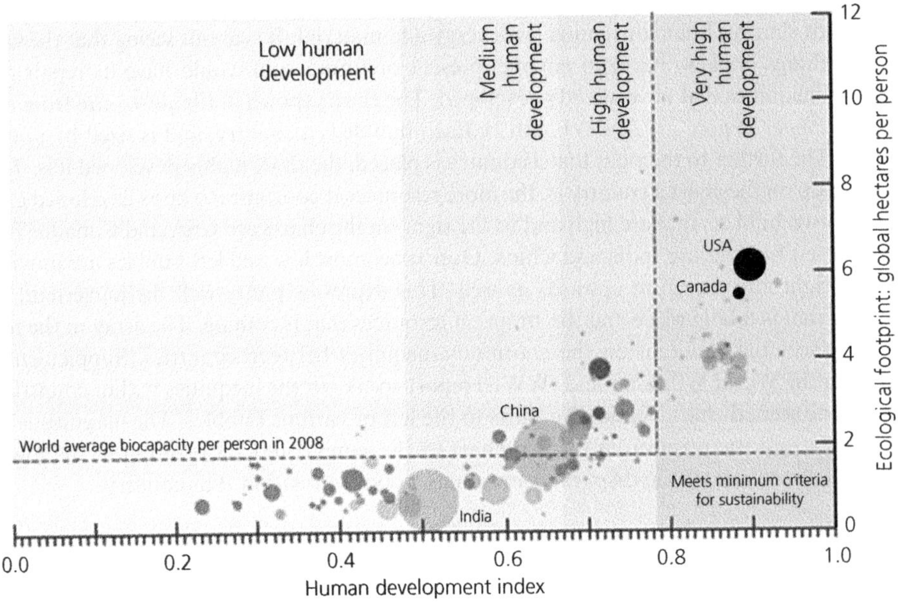

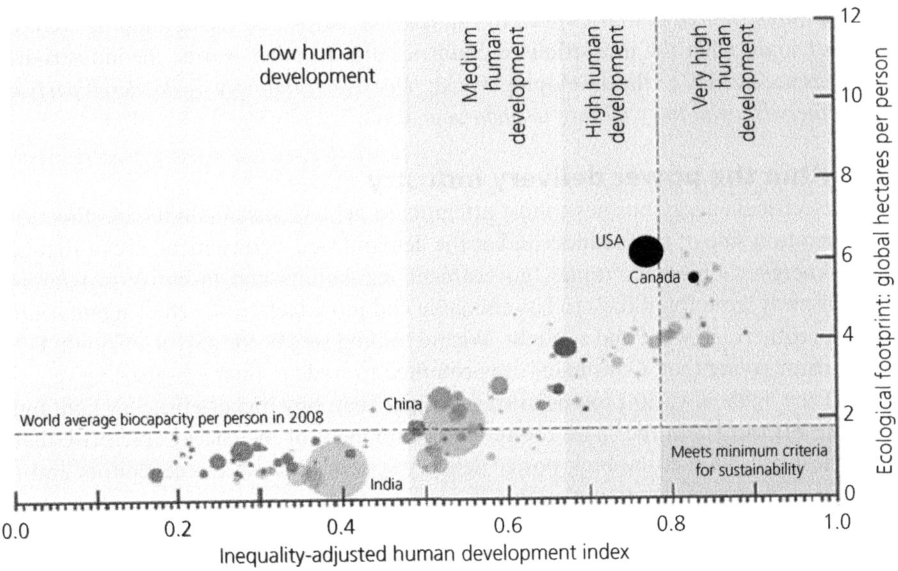

Waste not, want not

There are two subjects of interest for improving a bulk delivery system's contribution to sustainability – what we install as components of the system and how we operate the system. When we delve into these two subjects, we find that current criteria for the locating, design, construction and operating of transmission lines within the bulk delivery system are sometimes in conflict with criteria that target sustainability improvements. Embracing these conflicting criteria, shifting decisions towards honouring those that address sustainability improvements and away from honouring the classic criteria, is the difficult paradigm shift that must occur if the owners, designers, constructors or operators of the system choose to embrace and contribute to our societies' improved sustainability.

The engineer's role

As engineers, we can see what we could do within the context of power delivery. We should be able to suggest to owners and developers of transmission line facilities that they could include design criteria for their project that address SD in addition to addressing cost and operational integrity as we normally do. If a facility owner is serious about contributing to SD, they can then let us, as experienced engineers in the industry, bring SD-valued ideas to the project. If that sounds like an outrageous idea, consider the following. In January 2014, the Association of Professional Engineers of British Columbia, Canada, rolled out its new and improved sustainability guidelines for its members. The guidelines include an obligation that its members discuss the opportunity to incorporate sustainability into the project/work with their client. If the client declines ... fine, the engineer's obligation is completed. So, we seem to have other people thinking as we suggest. It is nice to not be alone!

As engineers of transmission line facilities, there is a limit to our ability to affect many of the GRI SD reporting items. We find its very long list being shortened considerably when you consider where transmission line design engineers can and cannot have an impact. *CIGRÉ Technical Brochure 340* points us to certain areas. In CIGRÉ's analysis of actions taken by surveyed facility owners, the three main 'pillars' of SD – the environment, the economy and society – are acknowledged, making *Technical Brochure 340* compatible with the global conversation. This technical brochure summarises the approach to SD that is being exercised by the reporting utilities. The list below outlines that work, and the italic items are those that we believe design engineers can easily affect.

Environmental aspects:

- *reduction in the environmental impact of operations*
- greenhouse gas policy
- renewable energy sources
- *resources consumption and energy efficiency.*

Actions dealing with the society:

- support of key social programmes
- stakeholder consultation
- support of ethical business practices
- promotion of health, safety and employee welfare.

Actions on the economy side:

- *support of R&D*

- support of business development
- policies to improve the supply chain.

Technical Brochure 340 noted that the key area of improvement in utilities, practices toward SD can be summarised as follows:

- Greater awareness and reporting of climate change:
  - *improving resources use and reducing the impact of current operations*
  - *improving environmental performance*
  - *working in partnership (with stakeholders and/or other businesses) to find solutions to environmental problems*
  - *investing in R&D in new technologies*
  - *supporting increased access to affordable electricity.*
- Strengthening the relationship with the community where the company operates:
  - donating time and resources to social causes that resonate with the organisation.
- Promoting ethical business:
  - promoting the well-being of employees.

Technical Brochure 340 also noted that there is a need for improving and harmonising the reporting about environmental, social and economic performance. The main topics are:

- Greenhouse gas emission data: the information should be quantitative and allow a comparison with past periods (for the same company) and other companies.
- *Renewable energy source generation capacity and production data should be given. More emphasis should be placed on items such as the use of resources (water use and consumption, use of materials, etc.).*
- The effects on the biodiversity of operations should be dealt with.
- Social and economic issues should be given higher attention; the main issues should be clearly defined (e.g. ethical business).
- Social and economic appropriate performance indicators should be developed and used; to this end, existing guidelines could be used.

The italic items in the lists above can be summarised as follows, and are the subjects that the design engineer can embrace to address SD:

- reduce environmental impact
- reduce resource and energy consumption and energy
- put effort into R&D – new technologies, and wider education of existing better ideas.

Grid design and operations

This discussion should look into how we manage power flow if in fact we do; how we decide what lines should be added to or deleted from the grid; and whether these decisions can be based on new, additional criteria that would improve sustainability (i.e. lower delivery losses, not install materials that do very little work or show preference for development of more sustainable power sources rather than simply cheaper power sources). Some thoughts:

First Is electricity itself a catalyst for sustainability, or is electricity's presence in our homes and businesses a hindrance to sustainability? One could argue well that the developed world's portion of the 6–7 billion souls on this planet need to get back to the lifestyle of the past to get humanity to a truly sustainable condition. Well, that is not going to happen voluntarily. It may be worth planning for, but it is not something to pursue. Let's assume that electricity is here to stay.

Second Do we assume we will cut back – way back! – on electricity consumption as a necessary step towards sustainability? This is not a concern. If electricity consumption is reshaped by society's concern for and efforts towards a sustainable lifestyle, it will take time. It is likely to take a sufficiently long time to allow the power delivery system to morph along the way at a suitable pace, given that components of the grid do age and are retired on a 50–100-year lifecycle. The grid will adapt naturally. Our job is not to obstruct the adaptation with personal agendas.

Third Should the grid be shaped to support the sustainability principles outlined in the discussion below on grid component location, design and construction? The principles below suggest basically that 'whatever you have decided to build, build it with less effort and materials'. At the grid level, this means, for example, that, if you have decided to deliver 4000 MW of power between distant source and load points/areas, then doing that with less effort and materials must be a primary decision driver in designing the facility. This can mean, for example, one 800 kV extra-high-voltage line being built instead of a collection of 500 kV lines, consuming less steel and less land. It can mean removing and recycling the materials from existing lower-voltage lines on an existing right-of-way and replacing them with fewer higher-voltage lines. It can be done, but it will take the understanding of the issues by the affected public, and their acceptance, political will and corporate commitment to sustainability as a cause equal in stature to cost-driven profit. That is a lot to ask.

Fourth Should the grid design favour preferred energy sources – sources that support sustainability more than others? The argument against such a simple statement is that 'the market must decide'. Certainly, in many countries, the market is God, so it becomes necessary to recognise how to redirect the market forces.

If you care to become a champion of sustainable development, there is a ton of opportunity and a ton of challenges.

Grid component location, design and construction

Location

Typically, new and large transmission line projects take many more years to site than to construct. It seems that every issue imaginable is included in the discussion/debate/

decision. The beetles, birds, beasts and bushes all have representation at the table, as do the landowners, neighbours, corporations and political jurisdictions. That's all good, since many of the siting conversations are designed to ensure the sustainability of the interests of each of these groups. But, there are some overriding criteria in place that limit options and prevent meaningful advances towards sustainability.

Regulating bodies specify rules regarding the placing of new lines relative to the location of existing lines. Some jurisdictions want large separations between circuits even though one can point to many lines with four major circuits or more on common structures. The reasons for circuit separation relate to the causes of outages and the ability to separate in light of other land use demands. Certainly, these reasons and abilities vary within a country and around the world. We will see later that certain sustainability gains can be made by piling the circuits onto common structures, while others are only available with single-circuit structures. Since the arrays of choices one can make are limited by the regulator's constraints, it is incumbent on the regulator to provide constraints that can also embrace SD.

The same can be said for government agencies that develop criteria for line configurations and constraints. As an example, one county jurisdiction in the western USA declares that guyed structures cannot be used in its domain because they cause fatal collision problems for a precious bird species in the area. If we drill deeper into that subject, we can point out that the ever-present low-voltage lines use guy wires of about 9 mm in diameter and they may not have brightly coloured guy guards on them. Very-high-voltage structures will use guys wires two to three times that diameter, and can have all the guy markings one desires placed on them. Does the jurisdiction's decision account for the relative visibility of the guy wires? What time of day do the birds fly, and how high above the ground? Do these birds slam into other things such as fences regardless, living a doomed existence despite our intentions? If this level of thought has not been put into the jurisdiction's decision, then the opportunities for reason-based design changes and line location that address sustainability are seriously hindered.

It is important to have the regulators and the controlling government agencies as partners in sustainability goals by informing them of how sustainability gains are made with the locating and design features of transmission lines, and asking them to drill deeper into their understanding of their stakeholders' problems with a line so that they can make smart decisions and be useful partners in achieving improved sustainability.

Design and construction

When we first contemplated the subject of SD, we looked at the carbon footprint as a measure of a new project's sustainability. It quickly became apparent that this method of measure was not useful for several reasons. For example, in different parts of the world the sustainability of using a wood pole is not consistent, depending on availability and on a pole's ability to survive the environment for a period of time. These vary dramatically with location. A wood pole is a good solution for a certain voltage or capacity of line but not so for another. The same can be said for steel, concrete and, indeed, all materials. Wood is a good solution where wood exists and where it will survive for a long time. The sustainability of wood varies for many reasons, and the same can be said for most materials and resources, including labour. In different parts of the world, the balance between labour and material costs varies dramatically, affecting a sustainability measure. Finally, lines are constructed for very different purposes, disrupting the meaningfulness of a comparison.

The best way to measure the sustainability merits of a design is by comparing it to a competing design for the same project.

SD design principles

We propose three guiding principles aimed at improving a transmission line project's contribution to sustainable development. These principles relate to the italic items above from the CIGRÉ work. They are:

1. Whatever materials are to be used, use less.
2. Some materials are fatally flawed (poisonous) to something important and must not be used.
3. The exception to 'using less' is for the conductors' aluminium content and conductivity, because low losses equate to fossil fuel not burned.

Each of these three principles is discussed below.

Principle 1: SD requires using less of whatever it is that you are going to use

Material not consumed is material made available for something else. Making, transporting and installing material requires energy. Material not used is energy not spent. Energy not spent is energy retained for something else.

There is a real opportunity here, because our industry is too often not efficient with materials or effort. That may seem harsh but, when you consider the magnitude of the SD challenge, everyone who cares to contribute to SD must consider improvement, even when you might think you are presently doing a pretty good job. The use of long-standing practices, preconceived or incorrect notions regarding design features and materials, and the absence of knowledge of alternative solutions all diminish the possibilities in this area.

We are confident in claiming that there are many opportunities throughout the network of lines and stations to make adjustments to our practices that will reduce the amount of materials and the effort that we put into the installation and maintenance of these facilities. R&D and education play a major role here.

As an example, a recent, large transmission project in the western USA used a structure type that weighed considerably more than an alternative that should have been considered as being very compatible with the location. The selected design was also a very heavy version of its own style. A calculation showed that the alternative choice of structure for the project would have released enough steel and concrete to supply all the structural needs for another project of the same size. In other words, the material and energy waste was equal to the materials needed.

We would hope that if the consideration of project sustainability were entrenched in our design and construction choices process, this unnecessary waste of raw materials would not happen. Not every project can find such large opportunities, but even savings on a more modest scale are useful towards supporting SD.

We suggest that SD design and construction criteria for projects require the measure of the amount of materials on a 'per unit length/per MVASIL' basis (MVASIL is defined below). The materials to measure are the important large quantity items, such as:

- tonnes of steel/km/MVASIL in structures and wires (conductor core, if steel)
- tonnes of other metals or wood/km/MVASIL
- tonnes of aluminium in conductors/km/MVASIL
- tonnes of foundation materials/km/MVASIL (concrete and moved earth)
- tonnes of earth moved/km/MVASIL for access roads and site preparations
- hectares of access road and work sites/km/MVASIL
- hectares of land removed from alternative service/km/MVASIL.

All of these things are easily measured with the necessary accuracy at the design and bid stages of a project. Again, while one project can be compared with another of similar nature, use and environment, the very best use of the measure is to compare design alternatives for a project during its development stage, and that is a simple exercise that a design team can execute.

The MVASIL value used in the denominator was selected because current work at CIGRÉ on a proposed method for optimising transmission line designs uses the same value for similar reasons, so compatibility is good. MVASIL is the capacity of a line to deliver power in units of MVA (megavolt-amperes) at that line's SIL (surge impedance loading). We would prefer to use the power actually delivered rather than this declaration of optimum capacity, but MVASIL is an easy value to calculate at the early design stage whereas the power actually delivered will necessarily be a guess.

A few guiding hints to get your SD score improved are:

- latticed steel towers use less steel and concrete than tubular structures
- guyed latticed towers use less steel and concrete than self-supported latticed steel towers
- access road building creates far more moved earth and removes much more land from alternative use than do the activities and footprint of the line's support structures
- lines constructed for system security (N-1 service) purposes will score poorly, so they deserve their own criteria to avoid unnecessary consumption.

RELATED CONSIDERATIONS

It is *possible* that the SD winner in a design choice competition is not also the winner on price or other measures, and it will be incumbent on the owner to decide the degree to which sustainability will be honoured. But, it is *probable* that the winners of these sustainability measures are also the winners on cost, because less energy expended and less material consumed should generally translate into lower capital cost. It is worth remembering that large projects financed by bankers and the like are not incentivised to develop low-cost solutions. Bankers prefer a no-risk price over a possible low price, and that makes the cost higher. When the banker is an investor who will receive an attractive return on the investment, the incentive for a low cost is gone, and attraction to a high cost may even exist.

Recall the nature of utility financing where capital expenses support the rate base and the operation and maintenance (O&M) costs do not. Therefore, capital costs are preferred to be high in order to pay for the O&M expenses. These days, many large projects have these roadblocks to low cost, to low material consumption and to low energy expenditure solution.

Language embedded in contract terms and conditions and performance metrics are often designed to move risk away from an owner who is willing to pay for the shift, as noted above. This is a driving force behind the use of engineering, procurement and construction (EPC) contracting. Yet, watching the application of EPC contracting by a variety of owners shows that the method is understood differently by different owners and clearly misunderstood and misapplied by many. The money spent is large and mostly unnecessarily spent. Money is a resource.

There are significant SD obstacles and opportunities embedded in contract language and environmental constraint choices.

Design criteria can create great costs. The use of inappropriate design criteria is common. It should be incumbent on designers to understand the impact on cost of the criteria of choice and to weigh the choice against alternatives. In other words, design criteria must be made more appropriate to the project's requirements so that the costs incurred by them create only appropriate costs.

Principle 2: SD cannot apply to certain materials, products and technologies

The first principle requires the caveat expressed here that there are materials, products and perhaps technologies that are fatally flawed, meaning that in some way they are 'poison' to something in the environment, and the quantities of such things should not be reduced but eliminated completely. In other words, find an alternative to these things. The complexity may be that whether a choice is poison or not can be location dependent.

'Poison' can also refer to a material or product that is such only on a comparative basis, but so clearly that its elimination or seriously reduced use is easy to accept. For example, some wood pole preservatives are more detrimental in certain vegetation-rich and wet locations than in arid, rocky and more vacant locations. But, global action may be preferred over selective action.

Principle 3: SD must include reduced operating losses

The principles above cover capital cost matters. The other opportunity for developing a transmission network that contributes better to SD lies in the operating savings that arise from a planner's and designer's choice of voltage and conductors. A network with lowered power delivery losses translates directly into fossil fuel not burned, and that is very much the point of the discussion.

The classic calculation used to select conductors for a new transmission line plots rising capital costs against reducing cost of losses as the amount of aluminium is increased in the conductors. This was described in Chapter 4. When the two plots are summed for a range of conductor sizes, the optimum conductor is identified at the low point in the classically U-shaped plot. In our experience, the plot is not a deep U shape but a very shallow U shape. Review of several such plots shows that the sum of the capital and operating costs changes only about 2–3% across a range of aluminium content of 20–25%. The net present value (NPV) cost of a project is usually quite insensitive to the aluminium content of the conductors. This is increasingly true as the voltage increases, because other base costs are large and independent of the conductor choice.

If the NPV of design options is nearly immune to the conductor choice, then the choice has little to do with lowering the NPV and everything to do with shifting the cost from the present to the future – or not.

Given (1) this insensitivity and (2) an all-too-often lack of concern for high capital costs as described above and (3) a desire to lower the electrical losses for the sake of fossil fuel not burned, we suggest that the best conductor is one with an aluminium content well above the classic NPV calculation optimum. Given that the capital costs are more predictable than a 40-year projection of the cost of losses, this choice also hedges the bet on being wrong with the very fuzzy cost of losses projection.

The advent of new low-sag conductor types (at present, only some with a competitive purchase cost) or the use of trapezoidal strands of aluminium to pack about 20% more aluminium into a diameter than can be achieved with round-strand conductors may offer bangs for your buck in selected situations if explored. Notice that we did not say low-sag, high-temperature conductors. High temperature use means higher losses. Whenever possible, run conductors cool.

The strength of a conductor is defined by the loading criteria, most notably ice. The tension to which a conductor should be pulled – relatively loose or relatively tight – should be defined by the alignment. Since the strength and weight of corner and strain towers are dependent on line tension, and the height of tangent towers is defined by tension (sag), you will find savings by abandoning a globally applied tension rule and making relatively straight lines tighter and relatively crooked lines looser. As tension rises and falls, the strength requirement – the core content requirement – will rise and fall accordingly.

At the end of the day, intelligent attention to design criteria and the construction method choices can have dramatic impacts on the project capital, maintenance and operating costs, and therefore on the SD contribution of the facility to the world. But making changes that benefit SD is likely to be seen as a dramatic shift in present habits. It will be hard! So be it, because not making the change will eventually be harder and not fun.

And that, boys and girls, is what we have to say on the subject of structural engineering of transmission lines. We hope it was a pleasure to read and is helpful to you for years to come as you exercise a career in this field.

Peter and Buck

REFERENCES

CIGRÉ (2008) *Utilities Practices Toward Sustainable Development. CIGRÉ Technical Brochure 340*. CIGRÉ, Paris, France.

WWF (World Wildlife Fund) (2012) *Living Planet Report 2012. Biodiversity, Biocapacity and Better Choices*. WWF, Gland, Switzerland.

Structural Engineering of Transmission Lines
ISBN 978-0-7277-5988-7

http://dx.doi.org/10.1680/setl.59887.317

Chapter 10
Further information

CIGRÉ technical brochures

The following is a list of most of the CIGRÉ technical brochures for overhead lines. Most of them are very good. CIGRÉ offers an intelligent stream of technical information, and you would be well advised to access these documents – you will be well educated if you do. Go one better and become a member of CIGRÉ; the documents are then free of charge and available online.

If you can get into the habit of attending CIGRÉ events and being involved in their work, you will meet many of the best minds in the business, and some of the characters.

072	*Guidelines for the Evaluation of the Dielectric Strength of External Insulation*
201	*Maintenance Outsourcing Guidelines*
206	*The Design of Transmission Line Support Foundation*
207	*Thermal Behavior of Overhead Conductors*
230	*Assessment of Existing Overhead Line Supports*
244	*Conductors for the Uprating of Overhead Lines*
255	*Material Properties for Non-Ceramic Outdoor Insulation*
256	*Life Cycle Assessment (LCA) of Overhead Lines*
265	*Life Cycle Assessment (LCA) of Overhead Lines*
273	*Overhead Conductor Safe Design Tension with Respect to Aeolian Vibration*
274	*Consultation Models for Overhand Line Projects*
277	*State of the Art Survey on Spacers and Spacer Dampers*
278	*Influence of Line Configuration on Environment Impacts of Electrical Origin*
281	*Design and Installation of Micropiles and Ground Anchors for OHL Support Foundations*
284	*Use of Corona Rings to Control the Electrical Field Along Transmission Line Composite Insulators*
289	*Reliability Based Design Methods for Overhead Lines: Advantages, Applications and Comparisons*
291	*Guidelines for Meteorological Icing Models, Statistical Methods and Topographical Effects*
294	*How OH Lines are Re-Designed for Uprating/Upgrading*
299	*Guide for Selection of Weather Parameters for Bare Overhead Conductor Ratings*
306	*Guide for the Assessment of Old Cap and Pin and Long-Rod Transmission Line Insulators Made of Porcelain or Glass*

308	*Foundation Installation and Overview*
309	*Asset Management of Transmission Systems and Associated CIGRÉ Activities*
316	*Defense Plan Against Extreme Contingencies*
320	*Characterization of ELF Magnetic Fields*
322	*State of the Art of Conductor Galloping*
324	*Sag-Tension Calculation Methods for Overhead Lines*
331	*Consideration Relating to the Use of High Temperature Conductors*
332	*Fatigue Endurance Capability of Conductor/Clamp Systems*
340	*Utilities Practices Toward Sustainable Development*
344	*Big Storm Events*
348	*Tower Top Geometry and Mid Span Clearances*
350	*How Overhead Lines Respond to Localized High Intensity Winds*
353	*Guidelines for Increased Utilization of Existing Overhead Transmission Lines*
361	*Outdoor Insulation in Polluted Conditions: Guidelines for Selection and Dimensioning*
363	*Reliability Based Calibration of Foundation Strength Factor Using Full Scale Test Data*
369	*New Developments in the Use of Geographic Information as Applied to Overhead Power Lines*
375	*Technical Guide for Measurement of Low Frequency Electric and Magnetic Fields Near Overhead Power Lines*
376	*Cloud To Ground Lightning Parameters Derived from Lightning Location Systems*
384	*Comparison of General Industry Practices for Lattice Tower Design and Detailing*
385	*Management of Risks Due to Load Flow Increases in Transmission OHL*
387	*Influence of the Hyperstatic Modeling on the Behavior of Transmission Line Lattice Structures*
388	*Impacts of HVDC Lines on the Economics of HVDC Projects*
389	*Innovations and Standards*
395	*Investigation on the Structural Interaction Between Transmission Line Towers and Foundations*
396	*Large Overhead Line Crossings*
399	*Tower Testing Methodology*
402	*High Impedance Faults*
410	*Local Wind Speed Up on Overhead Lines for Specific Terrain Features*
416	*Innovative Solution for Overhead Line Supports*
416 Annex	*Innovative Solution for Overhead Line Supports*
418	*Status of Development and Field Test Experience with High-Temperature Superconducting Power Equipment*
419	*Treatment of Information Security for Electric Power Utilities*
420	*Generic Guidelines for HV Assets*
421	*The Impact of Renewable Energy Sources and Distributed Generation on Substation Protection and Automation*
422	*Transmission Asset Risk Management*

423	*Technical and Commercial Standardization of DER MicroGrid Components*
424	*New Trends for Automated Fault and Disturbance Analysis*
425	*Increasing Capacity of Overhead Transmission Lines – Needs and Solutions*
426	*Guide for Qualifying High Temperature Conductors for Use on Overhead Transmission Lines*
427	*The Impact of Implementing Cyber Security Requirements using IEC 61850*
428	*The Effect of Fabrication and Erection Tolerances on the Strength of Lattice Steel Transmission Towers*
429	*Engineering Guidelines Relating to Fatigue Endurance Capability of Conductor/Clamp Systems*
440	*Use of Surge Arresters for Lightning Protection of Transmission Lines*
448	*Refurbishment Strategies based on Life Cycle Cost and Technical Constraints*
471	*Working Safely while Supported on Aged Overhead Conductors*
473	*Electric Field and Ion Current Environment of HVDC Overhead Transmission Lines*
477	*Technical Brochure on Evaluation of Aged Fittings*
482	*State of the Art for Testing Self-Damping Characteristics of Conductors for Overhead Lines*
485	*Overhead Line Design Guidelines for Mitigation of Severe Wind Storm Damage*
496	*Recommendations for Testing DC Extruded Cable Systems for Power Transmission at a Rated Voltage up to 500 kV*
498	*Guide for Application of Direct Real-Time Monitoring Systems*
508	*HVDC Environmental Planning Guidelines*
515	*Mechanical Security of Overhead Lines Containing Cascading Failures and Mitigating Their Effects*
516	*Geotechnical Aspects of Overhead Transmission Line Routing – An Overview*

Papers by H. Brian White

Finally, we offer you a list of many of the papers written over the years by Brian White. They illustrate the mind of the very clever, practical and passionate man. They are worth seeking if you want to read refreshing material on the subject of transmission line engineering. These papers were written for a wide variety of organisations, and some were quite informally presented. Therefore, the information listed is sometimes incomplete.

1 Cross Suspension System – Kemano–Kitimat Transmission Line.
Presentation: Annual Meeting, Engineering Institute of Canada, May 1956.
Printed in: *The Engineering Journal* **39(7)**: 901–911, 926 (1956).

2 Design Problem: How to Maintain Tower Clearances at Low Temperature.
Electrical World, 9 December 1957.

3 Chute-des-Passes Transmission System.
Presentation: Canadian Electrical Association, Quebec City, 27 January 1959.

4 Chute-des-Passes 345 kV Transmission Line.
Presentation: EIC Winter Meeting, Winnipeg, 1960.
Printed in: *The Engineering Journal*, October 1960.

5 Bibliography on Bundled Conductors.
Presentation: American Institute of Electrical Engineers, June 1961, Paper No. CP-61-871. [Chairman of Task Force]

6 An Investigation of the Design Factors of Safety against Uplift of the Footings of a 345 kV TL.
CIGRÉ Study Committee Report, 1961. CIGRÉ, Paris.

7 Guyed Mast High Voltage Transmission Structures.
Presentation: American Institute of Electrical Engineers, Transmission and Distribution, October 1961, Paper No. CP-61-1100.

8 Guyed Mast High Voltage Transmission Towers.
Presentation: Canadian Electrical Association Regina, Saskatchewan, 20 March 1962.

9 The Use of Helicopters in Line Construction.
Presentation: Canadian Electricity Association, Eastern Zone Meeting, Quebec City, 22 January 1963.

10 A New Concept of Tower Loadings.
Presentation: Canadian Electrical Association, Montreal, Quebec, January 1964.

11 A Guide to Transmission Structure Design Loadings.
Presentation: Institute of Electrical and Electronic Engineers, New York, February 1964, Paper No. 64-62 (in six parts). [Co-author]

12 Limitations of Stringing and Sagging Conductors.
Working Group of Towers, Poles and Conductors of IEEE. Author of section on 'Sagging Offsets'
Presentation: IEEE WP Meeting, New York, February 1964, Paper No. CP-64-146.

13 An Investigation of the Design Factors of Safety Against Uplift on the Footings of a 345 kV Line.
Presentation for discussion: Working Group on Factors of Safety, CIGRÉ Study Committees 6 and 7, Madrid, May 1965, Paper No. SF23.

14 Bibliography on Transmission Tower Foundations.
Presentation: IEEE, Detroit, MI, June 1965, Paper No. 31-CP-65-710. [Co-author]

15 Conductor Selection for Co-ordinated Transmission Line Design.
Presentation: Rendiconti Della LXVII Riunione Annuale, Sardinia, 1966, AIE Paper No. I-37/1966 (in Italian).

16 Loading and Random (Casual) Failures of Transmission Lines.
Presentation: CIGRÉ Study Committee 22, Working Group on SFs, Richmond, VA, May 1969, Paper No. 22-69 WG06-4.

17 Special Report to Study Committee 22, 1970. CIGRÉ, Paris.

18 Report to Canadian Electricity Association on Line Loadings, Factors of Safety and Security.

19 Special Report to Study Committee 22, 1972. CIGRÉ, Paris.

20 Some Aspects of Transmission Line Security.
Presentation: Reliability Conference, 1972.

21 Structural System for the James Bay Transmission Lines.
Presentation: Hydro Quebec Symposium, Varennes, Quebec, October 1973. Transmission of Electrical Energy at EHV and UHV AC (First Disclosure of Cross Rope Suspension, i.e. Chainette Tower).

22 Use of Helicopters in Transmission Line Construction.
Presentation: 214B Electrical Construction Panel, September 1974.
23 Transmission Line Reliability.
Presentation: American Society for Quality Control, Portland, OR, April 1975.
24 735 kV Chainette Towers.
Presentation: CIGRÉ Study Committee 22, Working Group 08, Mannheim, 1975, Paper No. 22-75 (WG 08) 01.
25 735 kV Chainette Towers – Part II – Construction & Testing.
CIGRÉ Study Committee 22-76, Working Group 06, August 1976. CIGRÉ, Paris. [Co-author]
26 A Modular Design System for Guyed V Towers.
Presentation: IEEE, New York, January 1978, Paper No. F78-151-3.
27 Special Report to CIGRÉ Study Committee 22, 1974. CIGRÉ, Paris.
28 Special Report to CIGRÉ Study Committee 22, 1976. CIGRÉ, Paris.
29 Line Reliability – Progress and Problems.
Presentation: Reliability Engineering Conference for the Electric Power Industry, Portland, OR, April 1978.
30 Failure Containment for Overhead Line Design.
CIGRÉ Study Committee 22, September 1978, Paper No. 22-13. CIGRÉ, Paris.
31 Special Report to CIGRÉ Study Committee 22, 1978. CIGRÉ, Paris.
32 Some Observations Resulting from the Study of Seventy-Four Line Failures.
Presentation: CIGRÉ Study Committee 22, Working Group 06, September, 1978, Paper No. 22-78 (WG 06) IWD. CIGRÉ, Paris.
33 Some Destructive Mechanisms Activated by Galloping Conductors.
IEEE PES WM, New York, February 1979, No. A79 106-6.
34 Position Papers, CIGRÉ Study Committee 22, Working Group 06 – Line Security – 1979:
(a) Notes Regarding Controlled Sequence of Failures Missing
(b) Failure Containment for Multi-Circuit Towers
(c) Failure Containment – Energy Dissipation with Time.
35 Towers Without Crossarms.
Presentation: CIGRÉ Study Committee 22, Sienna, September 1979.
36 Building for Tomorrow. Ohio Brass, Mansfield, OH, 1979.
37 Draft of document on special loadings (failure containment and C&M).
International Electrotechnical Commission, Technical Committee 11, Working Group 07, Venice, September 1979. [HBW Sect/Chairman]
38 An Attempt at Determining the Required Strength of Transmission Line Components.
Prepared April 1980. Presented as: An Overall Philosophy for Loadings and Strengths of Transmission Line Components, Canadian Electricity Association SM, Toronto, 17 March 1981.
39 Special Report to CIGRÉ, 1980. CIGRÉ, Paris.
40 Comments Regarding the Possibility of Vertical and Transverse Cascades.
Revision of manuscript submitted 21 February 1981 to CIGRÉ Study Committee 22, Working Group 06, 22 February 1982, Paper No. 22-81(WG 06)04. CIGRÉ, Paris.

41 An Overall Philosophy for Loadings and Strengths of Transmission Line Components.
Presentation: Canadian Electricity Association SM, Toronto, March 1981.

42 The Alicura 500 kV Transmission System.
Presentation: CIGRÉ, Paris, September 1984, Paper No. 22-02. [Co-author with R. Behncke]

43 A Practical Approach to Reliability (Based) Design.
Presentation: IEEE PES WM, New York, February 1985, Paper No. 85 WM 253-0.

44 Micro-computer Provides Optimized Tower Spotting.
Co-author with John Bates, *T + D*, April 1985.

45 *Recommendations for Overhead Lines – Document on Special Loadings (Failure Containment).*
International Electrotechnical Commission, Technical Committee 11, Working Group 07, issued 1982/83. [Written by HBW Secretary/Chairman]

46 Probability Based Line Design: A Draft Proposal for 230 kV and Above for CSA.
Presentation: First International Symposium, Probabilistic Methods Applied to Electric Power Systems (PMAPS), Toronto, July 1986. [Co-author with C. I. Orde]

47 Rational Limits and Expectations for Probability Based Design Criteria.
Presentation: Second International Symposium of Probabilistic Methods Applied to Electric Power Systems (PMAPS), Oakland, CA, 20–23 September 1988.

48 Uprating and Refurbishing of Transmission Lines.
Canadian Electricity Association, Vancouver, 1 April 1992. [Co-author with M. S. Nashid]

49 Offset Clipping and Some Useful Parabolic Relationships.
TLCADD Design Bulletin, July 1991.

50 Spans, the Long and the Short of It.
TLCADD Design Bulletin, January 1992.

51 Uprating of Transmission Lines.
TLCADD Design Bulletin, September 1992.
T + D, December 1992.

52 Structural Optimization of Transmission Line Conductors.
Engineering Structures **15(4)**: 247 (1993).

53 Guyed Structures for Transmission Lines.
Engineering Structures **15(4)**: 289 (1993).

54 An Alternative Approach to Transmission Line Design Based on Relative Reliability.
1994. [Co-author with R. H. Behncke]

55 High Intensity Wind and Relative Reliability Based Loads for Transmission Line Design.
22-205. CIGRÉ, Paris, 1994. [Co-author with R. H. Behncke and R. V. Milford]

56 Five Major Issues Facing Today's Transmission Line Engineer.
Presentation: T and D Conference, New Orleans, LA, March 1995.

57 Increasing Line Capacities.
T and D International, 3rd quarter, 1995.

58 Winds Wreak Havoc on Transmission Lines.
Transmission and Distribution World, June 1996.

59 Special Problems with In-cloud Icing and Transmission Lines.
Presentation: International Workshop on Atmospheric Icing of Structures (IWAIS), Chicoutimi, Quebec, June 1996.

60 Unique Suspension System Conquers Rugged Terrain.
Transmission and Distribution World, August 1997

61 Presentation: Ice and the Wire Systems of a Transmission Line.
American Society of Civil Engineers, Technical Council on Cold Regions Engineering, August 1999.

62 Galloping of Ice Covered Wires of a Transmission Line.
American Society of Civil Engineers, Technical Council on Cold Regions Engineering, August 1999.

63 Guidelines for Electrical Transmission Structural Loads.
Update, American Society of Civil Engineers, Manuals and Reports on Engineering Practices No. 74, 2000. [Co-author with M. Miller and C. J. Wong]

64 The Cross-Rope Suspension Structure.
[Co-author with R. H. Behncke]

65 Metal Towers for 650 kV Operation on Project EHV.
American Institute of Electrical Engineers, 1961. [Co-author]

66 Aluminum Towers.
Canadian Electricity Association, 1965.

67 Vibration of Metal Towers.
Institute of Electrical and Electronics Engineers, 1969.

68 Component Utilization and Related Problems.
Canadian Electricity Association, 1983.

69 Control of Sequence of Failure.
Canadian Electricity Association, 1985.

70 Spatial Influences of the Assessment of Loading Data.
Institute of Electrical and Electronics Engineers, 1997.

71 Application of the T/m Concept to a new 400 kV TL.
Institute of Electrical and Electronics Engineers, 1997.

72 Investigation of Transmission Line Failures.
1997.

73 Transverse Cascading Instigated by High Intensity Winds.
University of Western Ontario, 1997.

74 Limitations of the Ruling Span Method for Overhead Line Conductors at High Operating Temperatures.
Institute of Electrical and Electronics Engineers, 1999.

75 Damage to Overhead Lines caused by Conductor Galloping.
CIGRÉ, 1984. [Co-author]

Structural Engineering of Transmission Lines
ISBN 978-0-7277-5988-7

http://dx.doi.org/10.1680/setl.59887.325

Index

Page references in *italics* indicate a figure or table.

Printed and bound by CPI Group (UK) Ltd, Croydon, CR0 4YY
16/09/2026
14963399-0004